www.ingramcontent.com/pod-product-compliance
Lightning Source LLC
LaVergne TN
LVHW010623300926
849162LV00001B/41

Sustainability Matters

Environmental Management in the Anthropocene

Sustainability Matters

Collection of Best Papers from NUS Master of Science (Environmental Management) [MEM] Programme

---

Published

Vol. 1 Sustainability Matters: Environmental Management in Asia
Edited by: Lye Lin-Heng, George Ofori, Malone-Lee Lai Choo, Victor R Savage & Ting Yen-Peng
Published by World Scientific

Vol. 2 Sustainability Matters: Challenges and Opportunities in Environmental Management in Asia
Edited by: Lye Lin-Heng, Chou Loke-Ming, Audrey Chia, Asanga Gunawansa, Kua Harn-Wei
Published by Pearson

Vol. 3 Sustainability Matters: Asia's Green Challenges
Edited by: Lye Lin-Heng, Victor R Savage, Chou Loke-Ming, Yu Liya E & Kua Harn-Wei
Published by World Scientific

Vol. 4 Sustainability Matters: Asia's Energy Concerns, Green Policies and Environmental Advocacy
Edited by: Lye Lin-Heng, Victor R Savage, Chou Loke-Ming, Yu Liya E & Kua Harn-Wei
Published by World Scientific

Vol. 5 Sustainability Matters: Environmental and Climate Changes in the Asia-Pacific
Edited by: Lye Lin-Heng, Victor R Savage, Kua Harn-Wei, Chou Loke-Ming & Tan Puay-Yok
Published by World Scientific

Vol. 6 Sustainability Matters: Environmental Management in the Anthropocene
Edited by: Lye Lin-Heng, Harvey Neo, Sekhar Kondepudi, Yew Wen-Shan & Judy Sng Gek-Khim
Published by World Scientific

Sustainability Matters

Environmental Management in the Anthropocene

Editors

Lye Lin-Heng

Harvey Neo

Sekhar Kondepudi

Yew Wen-Shan

Judy Sng Gek-Khim

National University of Singapore

NUS National University of Singapore

World Scientific

Published by

World Scientific Publishing Co. Pte. Ltd.
5 Toh Tuck Link, Singapore 596224
USA office: 27 Warren Street, Suite 401-402, Hackensack, NJ 07601
UK office: 57 Shelton Street, Covent Garden, London WC2H 9HE

National Library Board, Singapore Cataloguing in Publication Data
Name(s): Lye, Lin Heng, editor. | Neo, Harvey, editor. | Kondepudi, Sekhar, editor. | Yew, Wen-Shan, editor. | Sng, Judy Gek-Khim, editor.
Title: Sustainability matters : environmental management in the Anthropocene / editors, Lye Lin-Heng, Harvey Neo, Sekhar Kondepudi, Yew Wen-Shan, Judy Sng Gek-Khim.
Description: Singapore : World Scientific Publishing Co Pte Ltd, 2017. | Includes bibliographical references.
Identifier(s): OCN 992793026 | ISBN 978-981-32-2988-4 (hardcover) | ISBN 978-981-32-3002-6 (paperback)
Subject(s): LCSH: Environmental management. | Sustainable development. | Environmental policy.
Classification: DDC 363.7--dc23

British Library Cataloguing-in-Publication Data
A catalogue record for this book is available from the British Library.

Desk Editor: Amanda Yun

PREFACE AND ACKNOWLEDGEMENTS

Introduction to the MEM Programme

The National University of Singapore (NUS)'s Master of Science (Environmental Management) "MEM" Programme was launched in July 2001. This is a multi-disciplinary and inter-disciplinary graduate program to provide education in environmental management for senior and mid-level managers in corporations, institutions, and government and non-government organisations. It aims to equip graduates with the necessary knowledge and skills to properly manage the environment and to deal with the challenges of an environmentally conscious society and international market. It enables graduates to assume responsible and influential roles in the public and private sectors, and to make environmentally sound decisions that support sustainable development and livelihoods.

The MEM Programme has the collaboration of **nine** Faculties and Schools in NUS, and is hosted by the School of Design and Environment (SDE). The other partners in this program are: the Faculties of Arts and Social Sciences, Engineering, Law, and Science; the NUS Business School, the Lee Kuan Yew School of Public Policy, the Saw Swee Hock School of Public Health, and the Yong Loo Lin School of Medicine.

The Programme is administered by a Director from the staff of the host Faculty/School, the SDE, and led by a Programme Management Committee comprising a representative from each Faculty/School at NUS. The Programme has formal links with the Yale School of Forestry and Environmental Studies (FES), USA, and Duke University's Nicholas School of the Environment and Earth Sciences, USA, providing for collaboration in teaching and research as well as academic staff and student exchanges. It has a distinguished Advisory Committee, chaired by Prof Tommy Koh, Ambassador-at-Large and former Dean of the Law School at NUS, comprising national and international leaders in the environment, from academia, industry, and government and non-government organisations.

The MEM programme can be undertaken either full-time (one academic year) or part-time (two academic years). It welcomes applicants from any discipline with a good honours degree from a reputable university, who meet the required standards for proficiency in the English language. Candidates are required to complete a 40 modular credit programme, of which the main components are a group of seven Core Modules, taught by staff from the different faculties/schools in NUS. These modules are:

- Business and the Environment
- Environmental Economics and Public Policy
- Environmental Law
- Environmental Management and Assessment
- Environmental Planning
- Environmental Science
- Environmental Technology

Students must also pursue substantive research, completing either a Study Report of 10,000 words plus one elective module, or a Dissertation of 20,000 words, under guided supervision. Students must also attend a series of seminars on current issues relating to the environment, across a wide spectrum of disciplines, conducted by experts in the relevant fields, to broaden their perspectives and widen their horizons. These are held every fortnight, and are also open to the public, as part of the University's efforts towards public education on the environment.

Through the years, the programme has been enriched by students from many disciplines and many nations—from Asia to Europe and North America, including the Middle East and Africa. Our graduates now hold positions in many different organisations, public and private, in various parts of the world. They bring the benefit of their multi-disciplinary experience in the MEM programme, to build a better and more sustainable world. The MEM Alumni Association plays an active role in the programme's activities, continuing to maintain close ties with alumni members as well as with the teaching staff.

This book is the sixth volume of some of the best research papers done by students in the programme for the academic years from 2014/2015 and 2015/16. They have been edited for brevity. Where two names appear as authors, the first name will be that of the student, with the supervisor's name thereafter, as co-author. The programme thanks these students and their supervisors for their valuable contributions to research in the many facets of environmental studies and management, and also acknowledges those students whose works do not appear in this volume, emphasising that every student in this programme has contributed to its success. Each paper has served to enhance our understanding of the environment and compelled us to think how we can manage it better.

Acknowledgements

Many people have contributed to the success of this programme. We wish to thank them most sincerely. They are:

1. President Prof Tan Chorh Chuan and Provost Prof Tan Eng Chye for continuing to support and fund the teaching of this program.
2. The Dean of the School of Design and Environment, Prof Lam Khee Poh and his team, for hosting the programme since its inception, and continuing to support it administratively and financially.
3. The Deans of the other eight Faculties and Schools in NUS who are partners and collaborators in this programme. They have supported us by mounting relevant courses in their faculties/schools and recognising the contributions of their staff in the teaching, supervision of research and the examination of our students.
4. The representatives of each Faculty, who serve on the Programme Management Committee (PMC). Their enthusiasm and passion for the environment and for this programme, have helped in immeasurable ways to ensure its sustainability and its constant improvement.
5. Our teaching staff from the different schools/faculties, adjunct teaching staff, as well as other staff members who served as supervisors, examiners and contribute in many different ways to the programme.
6. Members of our Advisory Committee, led by Prof Tommy Koh as Chair.
7. Shell Companies in Singapore for being the programme's stalwart supporter since its early years, providing bursaries for needy students and prizes for the best students. Shell also bears the costs of this publication and the earlier volumes in this series. We wish to especially thank Shell's Chairperson, Ms Goh Swee Chen for her support and

encouragement, and for being a member of the MEM's Advisory Committee. Thanks are also due to Shell's Mr Jason Leow, General Manager, Communications, and Ms Caroline Loke, Manager, Social Investment, for their continued support and encouragement.

8. The Tan Chay Bing Education Foundation and the National University of Singapore Society for providing scholarships and other financial assistance for our students.
9. Our alumni, who are spread across the globe in different capacities and disciplines, who are working to improve the environment, and whom we call upon to assist in various ways from time to time.
10. The many individuals we have called upon to speak to our students at fortnightly seminars and who willingly shared their expertise and time on Friday evenings from 6–8 pm.

As has earlier been mentioned, this is the sixth volume of the best student papers from the NUS MEM programme. There are 15 papers, on the following themes: Biodiversity and Conservation; Environmental Science; Environmental Governance and Management; Energy; and Urban Studies. We wish to congratulate the writers and their supervisors for their hard work and thank them for sharing their research findings and proposed solutions. These papers are another fine contribution to the vast field of sustainable environmental management, collected in these volumes entitled *Sustainability Matters*.

It should be noted that the MEM programme celebrated its 15th Anniversary last November (2016), together with the Bachelor in Environmental Studies (BES) programme's 5th Anniversary and the Law Faculty's Asia-Pacific Centre for Environmental Law (APCEL)'s 20th anniversary. The BES programme, like the MEM programme, is multi-disciplinary and involves the same nine faculties/schools at NUS. Apart from APCEL, NUS has also established other centres on the environment—these include the NUS Environment Research Institute (NERI), the Energy Studies Institute, the Centre for Sustainable Asian Cities (housed at the School of Design and Environment), the Tropical Marine Science Institute (TMSI) housed at the Faculty of Science, and the Lee Kong Chian Natural History Museum. Environmental Studies is also one of the areas of focus in the Yale-NUS programme.

It is clear that NUS is well on its way to be a centre of excellence in environmental studies for Asia and the world, and this latest volume of the best student papers from the MEM programme is a contribution towards this effort.

Assoc Prof Lye Lin-Heng
Chair, M.Sc. (Env. Mgt.) Programme Management Committee
School of Design and Environment,
&
Director, Asia-Pacific Centre for Environmental Law,
Faculty of Law
National University of Singapore

June 2017

MESSAGE FROM MS GOH SWEE-CHEN, CHAIRPERSON, SHELL COMPANIES IN SINGAPORE

This collection of the best student papers presents some of the most urgent and critical issues confronting our environment and the world. I congratulate the faculty and students for their excellent scholarship and dedication. It is exciting to know, from the range of writers presented here, that the NUS Master of Science (Environmental Management) programme continues to attract a diverse mix of students from across the globe.

Thought leadership from young talent is important, not least to an energy company like Shell, because it helps us understand a key demographic for our business—young people across the world—and how we can address the global energy transition together.

Royal Dutch Shell, our parent company, is embracing the energy transition. Although we believe oil and gas will remain in the energy mix, we need to accelerate our offering on renewables. This will allow us to become a key player in providing cleaner energy solutions.

Shell Singapore, specifically, is supportive of the government's national ambitions of reducing the country's carbon footprint and building a sustainable city. We want to help the government deliver economic growth and smart city solutions, particularly in personal transport, energy security through gas, and by growing a thriving and competitive petrochemical value chain. In particular, Shell Singapore aims to become a "living lab" of innovations, including digital and clean energy solutions, while providing thought leadership on the future of energy. At the same time, we are mindful of the increasing distrust towards Big Business, particularly among purpose-driven Millennials who frown on large corporations' monopoly of ideas, technologies and funding. We believe we are not that kind of company.

Instead, over the years, both Royal Dutch Shell and Shell Singapore have been actively reaching out to diverse communities to collaborate on environmental and energy initiatives, and put them into action where possible. Our support for the MEM programme is one such effort. We are grateful that the Singapore community supports us and are prepared to co-create with us. We will continue to seek broad partnerships to make the future together.

Goh Swee-Chen
Chairperson
Shell Companies in Singapore
&
Member, M.Sc. (Env. Mgt) [MEM] programme Advisory Committee

MESSAGE FROM PROF TOMMY KOH, FACULTY OF LAW, NATIONAL UNIVERSITY OF SINGAPORE

We are bombarded every day with bad news. One reason for this is that, from the point of view of the media, good news is not news. I would therefore like to use this message to share some good news relating to the environment.

The first good news relates to the oceans. We often forget that 70% of the earth's surface is covered by water. The ocean is the highway for shipping and maritime trade. The ocean provides us with food and energy. It is a carbon sink. We have, however, not been good stewards. As a result, the oceans are seriously threatened by pollution, over-fishing and climate change

Guided by Sustainable Development Goal No. 14, the UN convened an Ocean Conference in New York from 5 to 9 June 2017. The objective was to promote the sustainable use of the oceans, seas and marine resources.

The conference ended successfully by adopting, by consensus, a 14-point Call For Action by the 193 member states of the UN. Over 1,300 voluntary commitments were made, by Governments (44%) and Non-Governmental Organizations (19%), UN entities (9%) and the private sector (6%).

One problem which was highlighted at the conference was plastic waste. Each year, about 8 million metric tons of plastic waste is discharged into the oceans. They contaminate sediments at the bottom of the sea. They cause plastic waste to be embedded in the aquatic food chain. If we do not do something about this problem, the oceans may contain more plastic than fish by 2050. China, Thailand, Indonesia and the Philippine pledged to keep plastic out of the sea.

Another major problem highlighted at the conference was over-fishing. According to FAO, 90% of the world's fish stocks are over-fished. At the conference, nine of the world's biggest fishing companies from Asia, Europe and America signed the *Seafood for Ocean Stewardship* initiative, whose objective is to end unsustainable fishing practices.

The second good news relates to the African elephant. I am an elephant lover. I would like to do whatever I can to prevent the elephant from becoming extinct. Is there such a danger? The answer is yes. Due to the demand for ivory, 30,000 African elephants are being killed each year by poachers for their ivory. At this rate, the African elephant will become extinct in several generations.

The biggest market for ivory is China. The biggest hub for the trade in ivory is Hong Kong. The good news is that both China and Hong Kong are taking decisive action to combat the problem.

In 2015, in a historic joint statement, President Barack Obama and President Xi Jinping agreed to close down their domestic ivory markets. China is closing down its market by the end of this year. Hong Kong will end the trade in ivory by 2021.

The third good news is the opening of the Sustainability Academy. This wonderful achievement is due to the hard work of Esther An, from City Development Ltd (CDL), and Edwin Khew, from the Singapore Sustainable Energy Association. I am confident that the Academy will play a significant role in education, training and research.

My message to the students in the MEM programme is not to become disheartened by the bad news. We must remain positive and optimistic. We can make good news happen.

Tommy Koh
Ambassador-at-Large &
Professor, Faculty of Law
Chair, M.Sc. (Env. Mgt) [MEM] Programme Advisory Committee
National University of Singapore

MESSAGE FROM PROF LAM KHEE-POH, DEAN, SCHOOL OF DESIGN AND ENVIRONMENT, NUS

How time flies. It has been a year since my return to NUS to serve as Dean of the School of Design and Environment.

I am pleased that the Master of Science in Environmental Management programme is still thriving so well since its launch in 2001. As the host of the programme, SDE continues to work very closely with the Faculties of Arts & Social Sciences, Engineering, Law and Science, as well as the Schools including Business, Medicine, Public Health and Public Policy, to bring critical contemporary knowledge to the multi- and inter-disciplinary curricula of the programme, informed by cutting edge research that the faculty actively engaged in. The influence and impact of the collaborative effort is evident in the dissertations of the students that are included in this Sixth Volume of *Sustainability Matters*

Collectively, the work addressed the evolving challenges and opportunities confronting us in a comprehensive way, touching upon our planet (land, sea and air), our processes (science, technology, management, governance) and ultimately the inhabitants (human and non-human species). This volume is a great addition to the knowledge repository of the MEM programme.

However, time is of the essence in implementing many of the recommendations made. What is necessary now is "exponential transformation rather than incremental change", said Christiana Figueres the former United Nations Framework Convention on Climate Change executive secretary and architect of the Paris Agreement. Urgent and concerted efforts by all stakeholders in the entire value chain are required.

I offer my heartiest congratulations to the MEM graduating class of 2017, and especially to acknowledge the outstanding work of the students whose dissertations have been selected for inclusion in this volume. You will now join the NUS-MEM alumni family situated in different parts of the world, but united in one mission—to assume responsible and influential roles in the public and private sectors and to make sound decisions that support sustainable development in all countries.

Last but not least, I would like to thank the Deans of the collaborating Faculties and Schools for their continuous commitment and support, as well as the dedication of all staff who worked tirelessly to make the programme a success. I am also grateful to the Shell Group of Companies for providing a grant once again for the publication of this volume.

Lam Khee-Poh, PhD, FRIBA
Provost's Chair Professor of Architecture and Building
Dean
School of Design and Environment

PROGRAMME COMMITTEES

Programme Administration

Host Faculty of Programme: School of Design and Environment
Dean: Prof Lam Khee-Poh
Vice-Dean (Academic): Assoc Prof Joseph Ooi
Assistant Dean (Graduate Programmes): Assoc Prof Kua Harn-Wei
Programme Director: Assoc Prof Sekhar Kondepudi

MEM Programme Management Committee

Chair: **Assoc Prof Lye Lin-Heng, Irene**
Faculty of Law,
Director, Asia-Pacific Centre for Environmental Law (APCEL)

Deputy Chair: **Assoc Prof Yew Wen-Shan**
Yong Loo Lin School of Medicine, Dept of Biochemistry

Members: **Assoc Prof Audrey Chia**
NUS Business School

Assoc Prof Edward Layman Webb
Director, Bachelor in Environmental Studies (BES) program
Faculty of Science

Assoc Prof Harvey Neo
Faculty of Arts and Social Sciences

Assoc Prof Sekhar Kondepudi
School of Design and Environment, Dept of Building

Dr Judy Sng Gek-Khim
Saw Swee Hock School of Public Health

Assoc Prof Jesuthason Thampapillai
Lee Kuan Yew School of Public Policy

Assoc Prof Yu Liya
Faculty of Engineering, Dept of Civil and Environmental Engineering

MEM Advisory Committee

The MEM programme has an Advisory Committee comprising eminent members. The names of the chairman and members are shown below.

Chair:

Prof Tommy Koh
Ambassador at Large, Singapore;
Professor, Faculty of Law
&
Rector, Tembusu College
National University of Singapore

Members:

Sir Peter Crane
Former Dean, Yale School of Forestry and Environmental Studies
Yale University, USA

Mr Kenneth Er
Chief Executive Officer
National Parks Board

Mr Warren Fernandez
Editor
The Straits Times

Dr Geh Min
Former President, Nature Society Singapore,
Consultant, Eye Surgeon
Mt. Elizabeth Medical Centre
Mt. Elizabeth Hospital Ltd

Mr Edwin Khew
Chairman,
Sustainable Energy Association of Singapore

Prof Jack L. Knetsch
Professor Emeritus,
Economics and Resource Management,
Simon Fraser University, Canada

Ms Goh Swee-Chen
Chairman,
Shell Companies in Singapore

Mr Ng Lang
Chief Executive Officer
Urban Redevelopment Authority of Singapore

Mr Poon Hong-Yuen
Chief Executive
SPRING Singapore

Mr Andrew Tan
Chief Executive
Maritime & Port Authority of Singapore

ABOUT THE AUTHORS

ANG Chun-Wei Alan has been working in various operational and analytical capacities in PUB, Singapore's National Water Agency since 2012. His areas of responsibility include industrial and domestic water conservation projects, data analytics of water consumption patterns and habits, and public outreach programmes. Prior to serving in the public service, Alan was involved in developing and implementing green logistics initatives in a German Logistics MNC. His environmental interest lies in the areas of industrial ecology, sustainable development and behavioral economics. An avid volunteer, Alan has also been involved in a variety of community development and recycling projects with various religious, non-governmental and grassroot organizations. Alan holds a B.S. (Hons) in Logistics from the National University of Ireland, UK, and an M.S. (Environmental Management) from the National University of Singapore.

FAM Mei-Ling holds a B.Eng. in Environmental Engineering and M.Sc. in Environmental Management (Shell Medal and Prize, and Shell Best Dissertation Award winner 2014/2015) from the National University of Singapore. She is currently pursuing a Ph.D. at Nanyang Technological University, Singapore, and working as a research engineer at Lloyd's Register under the Singapore Economic Development Board Industrial Postgraduate Programme. Her Ph.D. research is in the area of risk assessment of decommissioning options of offshore facilities. Previously, she was an environmental engineer/project engineer at CH2M HILL Singapore, and was involved in environmental site assessments, decommissioning, demolition and decontamination projects.

KHO Yue-Min Veron was the recipient of the "Shell Best Dissertation Award, 2015–2016" and holds a B.Soc.Sci. (Hons) in Geography from the National University of Singapore. She also holds a Postgraduate Diploma in Education from the National Institute of Education, Singapore. She currently works as an Education Programmer where she designs National

Education programmes to engage youths. She believes that National Education can be looked at through the lens of sustainability and thus, she actively seeks opportunities and partnerships to infuse environmental literacy and the concept of environmental sustainability into the National Education programmes that she designs. After receiving the Ministry of Education Teaching Award in 2005, she pursued a teaching career in Geography at an 'A' Level Institution. During her teaching career, she helmed many environmental related programmes in the institution, (E.g. Environmental Representatives Leadership Development Programme, Earth Day, etc.) which ignited her passion to spread awareness of environmental sustainability. Today, she brings this passion with her regardless of the field of work that she engages in.

Carmelita LEOW holds a B.S.Sc. (Hons) in Geography and M.Sc. in Environmental Management from the National University of Singapore. She has 6 years of experience in the oil and gas sector as an analyst specializing in upstream intelligence. Her experience and research of energy policies has reinforced her certainty in the importance of environmental sustainability and the need for businesses and societies to start thinking geographically. She is currently working in ESRI (Environmental Systems Research Institute) Singapore in Business Development, spreading the message of the importance of geography, and helping clients leverage on their spatial advantage.

LIM Chin-Sing is a Research Associate at the Tropical Marine Science Institute (TMSI), National University of Singapore (NUS). He joined TMSI in 2006, after graduating from the Department of Mechanical Engineering, NUS with a B.Eng. Together with local and overseas research partners, he has worked mainly on marine biofouling organisms and assessment of antifouling technologies and surfaces. He maintains a keen interest in marine biosecurity issues and policies—especially how knowledge and understanding of our unique marine environment and its interaction with the regional society, can better inform decision-making in policies and stakeholders.

LOH Zhiyang is currently an internal auditor for the National Environment Agency of Singapore. He holds a B.A. (Hons) in Accountancy from Nanyang Technological University, Singapore. He started his career as a financial auditor in Ernst & Young LLP in 2008. He was active in advocating recycling when he was an undergraduate, and joined the university's environmental club, NTU Earthlink. With the support of the club and the facility management department of the university, he spearheaded the installation of recycling bins and coordinated the collection of recyclable waste resources for all 16 halls of residences in the varsity campus from 2007 to 2008. He continued to have a keen interest in environmental matters and decided to leave financial auditing in 2013 to further his studies at the National University of Singapore.

Deng MAO holds a B.Eng. in Environmental Science and Engineering from the National University of Singapore (NUS) and obtained his M.Sc. in Environmental Management (MEM) from NUS in 2016. He has worked as an environmental consultant since 2011 with particular focus on environmental assessment and monitoring, including all aspects of studies relating to environmental permits such as Environmental Impact Assessments, Environmental Baseline Studies, Quantitative Risk Assessments and Environmental Monitoring Plans in Singapore. Since 2014, he has worked as a founding member of a Singapore-based start-up technology company, Asia Pacific Air Quality Group Pte Ltd, specialising in air quality assessments, advanced monitoring systems, instrumentation, and technologies—serving the growing needs of regional industrialisation. He currently holds the position of Technical Area Manager, leading key project developments on system package and service contracts in Singapore and across the ASEAN countries.

Kaleem Anwar MIR is a Scientific Officer at the Global Change Impact Studies Centre, Ministry of Climate Change, Islamabad, Pakistan. He is also a qualified member of the UNFCCC expert review team and participated as energy expert in the annual reviews of GHG inventories of Annex I parties. He is involved in Energy-Environment Modelling and the preparation of national GHG inventories for Pakistan. He holds a Master's degree in Environmental Management (on ADB—Japan scholarship programme) from the National University of Singapore, and a Bachelor's degree in Chemical Engineering (with specialisation in Environmental Engineering) from the University of the Punjab, Lahore, Pakistan.

NG Chin-Soon Lionel holds a B.Sc. (Hons) in Life Sciences and an M.Sc. in Environmental Management from the National University of Singapore (NUS). From 2008 to 2016, he worked as a Research Assistant at the Tropical Marine Science Institute and the Department of Biological Sciences, NUS. He has a keen interest in coral reef ecology and marine conservation, and specializes in coral reef restoration. He led and co-authored numerous peer-reviewed, as well as popular science articles. He is currently pursuing a Ph.D. in Biological Sciences at NUS.

Prerna SHAH holds a Bachelors degree in Psychology from the University of Delhi, India, and is currently working as Youth Mentor with the POP (Protect our Planet) Movement in collaboration with BROAD Group, China, to introduce environmental education into the formal curricula of educational institutions in the Hunan province. She graduated from the National University of Singapore's M.Sc. (Environmental Management) programme in 2016. She has worked at the Institute of Water Policy (IWP), Lee Kuan Yew School of Public Policy, NUS, and the Centre for Liveable Cities (CLC), Ministry of National Development, Singapore, where she was engaged in social and environmental research projects primarily focused on risk, resilience and urban planning in Singapore. Deeply concerned by the environmental externalities afflicting the Himalayan region and driven by a desire to effect meaningful change, she left her executive job in India to pursue her passion in the field. Prerna was selected among a group of 81 students, researchers, and practitioners from 26 nations for the Inspire International Antarctic Expedition (IAE) 2015. Thoroughly inspired by the work of Sir Robert Swan, the IAE leader and the first person to walk to both poles, she raised the bulk of the financial resources for the expedition and aligned herself to a cause she believed in.

Nazifa Zabeen SIDDIQUA holds a Bachelor of Architecture degree from the Department of Architecture, Bangladesh University of Engineering & Technology, and graduated with an M.Sc. in Environmental Management from the National University of Singapore in 2016. She is currently working as a researcher at the Centre for Sustainable Asian Cities (CSAC), NUS, where she focuses on urban sustainability assessment of Singapore under various themes (e.g. Public Transportation, Air quality Waste management, Food and Economy) and benchmarking them against global cities like New York, London, Tokyo, Seoul and Hong Kong. Prior to joining CSAC, she worked as faculty in the Department of Architecture, at the American International University-Bangladesh for three years, where she grew an avid interest in understanding environmental sustainability issues which led her to join the M.Sc. programme. Her research interests include: neighbourhood sustainability assessment, climate change adaptation in dense urban settings, and environmental impact assessment.

TAN Choon-Puay holds a B.Eng. in Chemical Engineering from The University of Queensland, Australia, and an M.Sc. in Environmental Management from the National University of Singapore. She has been working in the water and wastewater treatment industry as a Resident Engineer in the field of Instrumentation, Control and Automation for the past twelve years. Among her employers are CH2M Hill and Black & Veatch, which are well-established process designer consultancy firms in water and wastewater treatment technology, head-quartered in United States of America, with international offices in many parts of the world. She was involved in many well-acclaimed Public Utilities Board water and network resource projects like the Deep Tunnel Sewerage System to the Changi Water Reclamation Plant, the upgrading of Chestnut Avenue Waterworks as a recirculation scheme for the Marina Reservoir, and the fairly recently constructed Lower Seletar Waterworks that houses facilities to produce chlorine (first of its kind in Singapore waterworks) which is a critical element in water treatment processes. Her involvement in the water and wastewater treatment arena will continue to hold high regard as Singapore is a country which places great importance in water security and strives to achieve water self-sufficiency as a small but well-developed first-world nation.

Sallie Chia-Wei YANG, M.Sc. (Environmental Management), National University of Singapore); LL.B. Hons (London), is the Director of Governance at Freeland and the Legal Specialist for the USAID Wildlife Asia program. She coordinated the "ASEAN Legal Support Task Force to Fight Transnational Organized Wildlife Crime" (Legal Task Force) under the Freeland-led ARREST on Program (Asia's Regional Response to Endangered Species Trafficking), a partnership between Freeland, ASEAN-WEN, UNODC, USAID, USFWS, and APCEL-NUS. The Legal Task Force includes ASEAN-wide capacity building and research initiatives that promote strategic legal remedies in prosecuting transnational wildlife trafficking in Southeast Asia. Under Legal Task Force, she managed the development and publication of the ASEAN Handbook on Legal Cooperation to Combat Wildlife Crime. She is a lawyer, qualified both in the UK as a barrister-at-law and in Singapore as an advocate and solicitor. Prior to joining Freeland, she has more than 18 years' experience as a litigation lawyer and as lead legal counsel for MNCs in infrastructure, environmental, and renewable energy sectors. She is currently based in

Bangkok. She is an appointed member of the IUCN World Commission on Environmental Law (WCEL) and an associate member of the Asia Pacific Centre for Environmental Law of the Faculty of Law, NUS.

ZHANG Yuzhe holds a Bachelor of Arts degree in Radio and Television Journalism from Nanjing Forestry University, China, and a Master of Science degree in Environmental Management from the National University of Singapore. She is interested in environmental issues, education and writing. She has had varied work experience, which has given her an open mind, including working as an assistant director, a trainee TV director, and a casual sales assistant. She was also a data clerk in Zhangjiakou Academy of Forestry (Hebei Province, China) for a year and a half. She hopes she can do something useful for the community in the environmental or education fields.

ABOUT THE EDITORS AND SUPERVISORS

Rajasekhar BALASUBRAMANIAN is an Associate Professor in the Department of Civil and Environmental Engineering, National University of Singapore (NUS). He has broad research experience and interests ranging across: Air Quality, Climate Change, Aerosol Science and Technology, Environmental and Health Risk Assessment, and Water Quality. Dr Bala has been actively involved in the development of the Environmental Engineering curriculum at NUS, at both undergraduate and graduate levels. He has taught a variety of courses in Environmental Science and Engineering. He has also been actively involved in promoting Environmental Education and Environmental Awareness, by giving public lectures and seminars, and contributing articles to magazines and newspapers.

Oscar CARRACEDO is an Architect, Urbanist and Professor at the National University of Singapore (NUS) where he directs the Inner-Hoods Lab. He holds an M.Sc. in Urban Design and Planning and an M. (Architecture) both with distinction, from the UPC_BarcelonaTECH. Later he earned a Postgraduate in Urban Design and Planning from the Massachusetts Institute of Technology (MIT). Professor Carracedo is the Programme Director and Chair of the Designing Resilience in Asia (DRIA); Technical Expert in Smart Cities and Urban Development of the Smart City Expo World Congress; Co-chair of the Great Asian Streets Symposium (GASS); Committee Member of the UN-Habitat Academia and Research Constituent Group and member of Habitat UNI. At the InnerHoods Lab he is leading research on Resilient and Liveable Neighbourhoods, Informal Urbanism and Integrated Planning. He has lectured in Europe, America and Asia, and has been invited as a UN-Habitat Expert in several high-level political meetings. Professor Carracedo is the author of numerous books, book chapters, and articles about urbanism and architecture. He is now member of the Board of Editors of the *International Journal of Architecture and Sustainable Urbanism* (IJASU) in Cairo. Professor

Carracedo's personal career has been awarded on several occasions; in 2002, he won the Regional Planning Catalan Society Award (SCOT), and in 2014 he won the Barcelona Award to the Best World Store. He is the curator of "50 × 55. 50 Years, 50 Neighbourhoods", exhibited at the National Library of Singapore.

Audrey CHIA is Associate Professor at NUS Business School, with a joint appointment at the Saw Swee Hock School of Public Health. Using leadership and change as theoretical foundations, Audrey studies how social and health problems can be addressed by social entrepreneurship and innovative philanthropy. Audrey has served as invited expert on social entrepreneurship, philanthropy, innovative financing and inclusive growth at meetings of the Asian Development Bank, OECD, United Nations and the European Union. Her work has been presented at international conferences such as the Academy of Management and the World Health Summit; and published in international journals like *Journal of the American Medical Association* (JAMA).

Namrata CHINDARKAR is an Assistant Professor at the Lee Kuan Yew School of Public Policy, National University of Singapore (NUS). She received her Ph.D. from the School of Public Policy, University of Maryland, USA, concentrating in international development policy. Her fields of interest include international development and policy areas such as poverty and inequality, water and energy, policy impact evaluation, public health, and gender and development. She holds an MA in Development Studies from the University of Manchester, UK, and an M.Phil. in Social Sciences from the Tata Institute of Social Sciences, Mumbai, India.

CHOU Loke-Ming is currently an Adjunct Research Professor at the Tropical Marine Science Institute, National University of Singapore. His research interests include coral reef ecology and restoration, and integrated coastal management. He has provided environmental management consultancy services to international agencies such as United Nations Environment Programme, as well as many national agencies. Presently a Fellow of the Singapore National Academy of Sciences and an Honorary Fellow of the Singapore Institute of Biology, he has over 100 publications in international journals and over 50 book chapters on the marine environment.

Sekhar KONDEPUDI is Associate Professor in the Department of Building, School of Design and Environment, National University of Singapore (NUS). He holds a Ph.D. in HVAC and Energy Systems from

Texas A & M University, USA, and over 20 years of global business and product experience in energy, buildings, green solutions, utilities, software and high tech. Prof Kondepudi has held leadership roles in Fortune 100 companies (Cisco, Motorola, Sprint) and start-ups, and has experience in business development, strategy, product design, development and management. His research and teaching interests are sustainable design, intelligent facilities, smart buildings energy systems.

Cecilia LIM is an Associate Professor in the Philosophy Department at the National University of Singapore (NUS). She did her B.A. and M.A. in Philosophy at NUS, and Ph.D. in Philosophy at the University of Pittsburgh, USA. Since coming back to Singapore in 1994, she has worked at NUS. Her main teaching areas include History of Philosophy (Founders of Modern Philosophy, Hume and Kant, Scepticism in the 17th–18th Century, Aristotle's and Descartes' Psychology), Environmental Ethics, Comparative Environmental Ethics, Topics in Environmental Philosophy, Graduate Research Seminar, Introduction to Philosophy (for Special Term), and tutorials for 1st year modules. Her main interests are in early modern philosophy and environmental ethics. She is a member of the editorial boards for *Environmental Ethics* and *The British Journal for the History of Philosophy*.

LYE Lin-Heng LL.B. (Sing), LL.M. (King's College, London, UK), L.L.M. (Harvard, USA) is an Associate Professor at the Faculty of Law, National University of Singapore (NUS) and Director of its Asia-Pacific Centre for Environmental Law (APCEL). Her areas of teaching and research are Environmental Law and Property Law. She chairs the Program Management Committee for the NUS M.Sc. (Environmental Management) [MEM] Programme. She is a member of the Program Management Committee for the Bachelor in Environmental Studies (BES) program. She is former Vice-Dean of the Law School at NUS, former Vice-Chair of the IUCN Academy of Environmental Law and a current member of the Academy's Board of Governors, representing Southeast Asia. She is Visiting Associate Professor at the Yale School of Forestry & Environmental Studies. She was Honorary Legal Advisor to the Nature Society Singapore for many years and is now a member of the Board of Directors for WWF Singapore. She is a member of the Ministry of Environment and Water Resources Committee on Clean Drinking Water Standards for Singapore. She is also a member of the Strata Titles Board, and a former Board Member of the Housing and Development Board (HDB), Singapore.

Harvey NEO is an Associate Professor of Geography at the National University of Singapore (NUS), where he teaches courses on development, environmental politics and nature/society issues. Broadly trained as a socio-economic geographer, his research interests include the political economy of the livestock industry, meat consumption, animal geographies and the development of social enterprises. Winner of multiple teaching awards, Dr Neo is also one of the editors of the leading geography journal, *Geoforum*. His most recent co-authored book, *The Geographies of Meat*, was published by Taylor and Francis in 2017.

POW Choon-Piew is an Associate Professor in the Department of Geography at the National University of Singapore (NUS). He received his Ph.D. from the University of California Los Angeles (UCLA), USA. His research centres on examining the critical geographies of the urban built environment, and interrogating the inter-relationship between urban forms and the social-spatial reproduction of urban life and urbanism in Asian cities. His ongoing research examines the planning practices and ideologies revolving around the idea of urban sustainability and the construction of 'eco-cities' in Asia. He currently serves on the Research Advisory Panel of the Housing Development Board (HDB) and is an international board member of the Urban Studies Foundation.

Rick REIDINGER is an Adjunct Senior Lecturer at the School of Design and Environment at the National University of Singapore (NUS). He teaches the core modules of "Business and the Environment" and "Environmental Management and Assessment" in the Master of Science (Environmental Management) (MEM) programme. He has an M.Sc. from the London School of Economics and Political Science, UK, and more than 20 years of experience in the environmental industry, primarily based in Asia. He is currently the CEO of Eco Special Waste Management, Singapore's largest hazardous waste management company, with 270 staff and annual revenues of US$45 million. Prior to joining Eco, Rick was Regional Director of AECOM's environmental consulting business in Asia, worked at SembCorp, Singapore's waste management business, and spent 10 years with Environmental Resources Management (ERM), Singapore, the global environmental consulting company.

Victor R SAVAGE retired from the National University of Singapore (NUS), Geography Department, in June 2016 and is currently adviser at the NUS Office of Alumni Relations and visiting senior fellow at S. Rajaratnam School of International Studies, Nanyang Technological

University (NTU), Singapore. He has previously served as the Director of NUS Office of Alumni Relations, Deputy Chair of the M.Sc. (Environmental Management) [MEM] Programme Management Committee, and Chair of the FASS Environmental Research Cluster. Dr Savage's research interest is mainly on Singapore and the Southeast Asian region—historical and cultural landscapes, sustainable development, environmental education, sustainable urban development, and cross-cultural issues and cultural identity. Among his books are: *Western Impressions of Nature and Landscape in Southeast Asia* (1984); *Environmental Stakes: Myanmar and Agenda 21* (jointly edited with Lily Kong, 1997); *The Naga Awakens: Growth and Change in Southeast Asia* (co-edited with Lily Kong and Warwick Neville, 1998); *Mad About Green* (jointly edited with Penelope Phoon-Cohen; 2001); *Toponymics: A Study of Singapore Street Names* (co-authored with Brenda Yeoh, 2004), *The Naga Challenged: Southeast Asia in the Winds of Change* (jointly edited with May Tan-Mullins, 2005) and *Sustainability Matters: Environmental Management in Asia* (co-edited with Lye Lin Heng, George Ofori, Malone-Lee Lai Choo, and Ting Yen-Ping, 2010).

Judy SNG Gek-Khim is an occupational medicine specialist and Senior Lecturer in Saw Swee Hock School of Public Health, at the National University of Singapore (NUS). She is Visiting Consultant for the Occupational Health Clinic in the National University Hospital. She was involved in the early conceptualisation and piloting of what is now known as Total Workplace Safety and Health, a comprehensive and integrated approach to managing workplace health risks. Her interests include work-environment health interactions and emerging occupational health challenges.

Sanjay SWARUP is an Associate Professor at the Department of Biological Sciences at the National University of Singapore (NUS). He is also the Deputy Director of the NUS Environmental Research Institute (NERI) and the Director of the Graduate Programme at the Singapore Centre for Environmental Life Sciences Engineering (SCELSE) at Nanyang Technological University, Singapore. He received his Ph.D. in Plant Pathology from the University of Florida, USA, and Ph.D.in Genetics and Plant Breeding from the Indian Agricultural Research Institute, India. His research interests lie in the understanding of biological role of small biomolecules, such as the metabolic networks in environmental bacteria communities. He has published in various journals such as *Plant Physiology*, *Biogeosciences* and *Microbiology*. He has also contributed book chapters in several books on molecular biology.

TAN Koh-Siang is a Senior Research Fellow and the Head of the Marine Biology and Ecology Laboratory at the Tropical Marine Science Institute (TMSI), National University of Singapore (NUS). He obtained his Ph.D. (Zoology) at NUS in 1996 and was postdoctoral scholar at the Swire Institute of Marine Science, University of Hong Kong before joining TMSI in 1997. Deeply passionate and inquisitive about local and regional marine biodiversity and environment, he has led the successful completion of the five-year Comprehensive Marine Biodiversity Survey (CMBS) in 2016, which have described some 24 marine invertebrate species new to science and documented more than 260 new records for Singapore. His current research projects include deep-sea biodiversity assessment of the Singapore claim area in the Clarion-Clipperton Zone, eastern Pacific Ocean, biology of native and introduced mussels, and invertebrate interactions on artificial seawall surfaces; while still actively sharing his wealth of knowledge with undergraduates, post-graduates and internship students alike.

TAN Puay-Yok is an Associate Professor in the Department of Architecture in the School of Design and Environment of the National University of Singapore (NUS). He currently serves as the Programme Director for Master of Landscape Architecture and leader of the research and teaching group for landscape studies. He obtained his Ph.D. from Cornell University and his academic training was in horticulture science and plant physiology. His research, teaching and professional activities focus on the science, policies, and practices of urban greening and ecology of the built environment. He combines his background in the sciences, experience in urban governance from the public sector, and interactions with practitioners to apply knowledge for urban greening to improve environmental quality and societal well-being. He is active in international collaborations, grant reviews and serves as editorial board member in a number of international journals. He also advises on landscape design and planning projects in the region as means of transferring knowledge from the academia to practice.

TONG Yen-Wah is an Associate Professor in the Department of Chemical & Biomolecular Engineering, Faculty of Engineering at the National University of Singapore (NUS). He obtained his Ph.D. at the University of Toronto, Canada. His main research interest is in the field of polymer chemistry, biomedical and tissue engineering. The major focus of his Biomimetic Materials and Systems Laboratory is thus on applying biomimetic principles for various applications, including membranes, scaffolds and processes. Recently, his group has expanded research into bioenergy using biomass wastes and municipal solid wastes, converting these into renewable fuels and chemicals.

Wen-Shan YEW is an Associate Professor with the Department of Biochemistry, Yong Loo Lin School of Medicine, National University of Singapore (NUS). He is also a core Principal Investigator of the NUS Synthetic Biology for Clinical and Technological Innovation (SynCTI). He trained as a mechanistic enzymologist with Professor John A. Gerlt from the University of Illinois at Urbana-Champaign, USA. As a continuation and elaboration of his interests in enzymology, his current and future research is centred on expanding the multi-faceted interface between enzymology, biological chemistry and synthetic biology, in an area he terms "synthetic enzymology". In particular, his research is focused on the use of structural and mechanistic enzymological knowledge for protein engineering, biocatalysis, therapeutic development (against metabolic and infectious diseases, cancer and ageing) and bioremediation (towards urban sustainability). His research is closely aligned with sustainable industrial needs and solutions. Amongst his scientific and pedagogical publications, he has authored a chapter in the textbook *Microbial Biotechnology* for the NUS Life Sciences module on Environmental Biochemistry.

CONTENTS

INTRODUCTION: ENVIRONMENTAL MANAGEMENT AND MITIGATION IN THE ANTHROPOCENE

LYE LIN-HENG, HARVEY NEO, SEKHAR KONDEPUDI,
YEW WEN-SHAN and JUDY SNG GEK-KHIM

As early as 2000, geologists Paul Crutzen and Eugene Stoermer mooted the concept of "Anthropocene". The Anthropocene was then suggested as an epoch that is substantially different from the Holocene epoch, in that it is marked by extensive and irreversible impacts made by the human race to the environment they live in. This was a provocative suggestion as previous events and characteristics that differentiate different periods in the geological timescale were all rooted in changes in the environment itself and made no reference to human activities. Put simply, the acknowledgment of the Anthropocene is the acceptance of human's immense control and destruction of Earth as well as the realisation that we have reached a point of no-return in human–nature relations.

In 2016, during the 35th International Geological Congress, held in South Africa, the majority of the delegates voted to formalise the Anthropocene epoch to follow the Holocene. What does this formalisation of Man's impact on our living environment mean to environmental management? To put it differently, does this signal—and if so, in what ways—a dramatically different approach to the way humans relate to, use and rehabilitate compromised landscapes?

To be sure, even before this paradigmatic shift in society–nature relations, as represented by the Anthropocene, scholars have drawn our attention to the destruction humans have wrought onto the environment. In popular literature, as early as 1962, books such as Rachel Carson's *Silent Spring* have cautioned us on the widespread damage human actions have had on earth. In the academia, since the early 1980s, a series of programmatic texts have educated successive generations of students on the environmental impacts of human development. These include Andrew Goudie's *The Human Impact on the Natural Environment*, first published in 1981, and *The Earth as Transformed by Human Action*, edited by Billie Lee Turner II *et al.*, in 1993.

Tracing the impacts of human actions on the environment is imperative; yet, equally important, is how we might face, resolve or mitigate such impacts. This has always been a key driving question in the field of environmental management. Under the shadow of the Anthropocene, more than ever, we (as part of humanity) need to come to terms with our own flawed power, our responsibility to future generations and living environments, as well as to tap into our collective wisdom and imagination to face a future that will be constantly changing and uncertain. The Anthropocene is hence the beginning of our collective introspection and not an excuse for us to trumpet how the environment is under our "control".

This is the sixth consecutive volume showcasing the best student theses from the Masters in Environmental Management programme (MEM) at the National University of Singapore. This collection is testament to the range of environmental challenges faced by diverse places around the world, and how through human ingenuity and the appropriate application of science and technology, one might yet manage the environment for the betterment of posterity.

Technology and Human Possibilism: A Double-edged Sword?

The seeds of the Anthropocene were planted long before the publication of the works mentioned earlier. That human beings had the ability and appetite to transform the environment dramatically—sometimes irreversibly—was most forcefully argued by Vidal de la Blache (1845–1948) who denounced the prevailing environmental determinist belief of his time. Instead, he asserted that human actions are neither shaped nor dictated by the environment (Cloke *et al.*, 1991). Yet, for de la Blache, such human actions are not unfettered and our increasing use of technology to transform the physical world does not mean that we are completely free to determine our own directions. Rather, there is (or should be) an "ongoing 'dialogue' between natural environments and the human communities they support" (ibid), with the hope that the most appropriate environments are produced as a result of human actions. Notwithstanding its vagueness, Vidal de la Blache did highlight two crucial points in development and environmental management: first, that increasing "power" needs concomitant and attuned responsibility; and second, the ambiguous role played by technology.

The latter point is particularly instructive. History has shown time and again how the use of science and technology to solve egregious environmental problems or to purportedly advance human welfare often comes with unintended and unanticipated consequences. Rachel Carson's ground-breaking book, mentioned earlier, demonstrates how in the drive for agricultural productivity through the use of pesticides, the resultant environmental consequences have not been duly considered. To give another more contemporary example, while solar energy has been promoted in recent years as an exemplary source of clean energy to wean society off fossil fuels, criticisms of the harvesting of solar energy through more efficient solar panels have emerged. Specifically, there have been concerns over the use of cadmium telluride, which produces panels significantly thinner and easier to set up and which can convert solar into power much more efficiently than existing technologies, in solar energy technologies. The problem with tellurium is that it is notoriously difficult to come by, and erstwhile have been harvested as a by-product of copper mining. The discovery of a massive tellurium deposit under the seabed in the Atlantic Oceans means that the reality of efficient and likely cheaper solar panels has to be weighed against the potential environmental destruction that might arise due to its extraction. Such cautionary dilemmas are all too common and replayed in many instances of technological innovations and advances. These include fracking as well as wind energy. The latter, especially when sited along the coastline, often results in the degradation of the marine environment and is a threat to bird life.

These are questions that extend beyond science, and implicates structures of governance, questions about ethics, as well as the influence "big industries" might have on the regulatory authorities. Nonetheless, regulation and constant monitoring, now more so than ever before, are needed to contextualise science and technologies for better environmental management.

Reining in the Exuberance of Technology and Capital

The pursuit for new technologies in the way we manage and extract resources from the environment is an unceasing process. It is also almost always an expensive process that has to be under-written by considerable capital. This is one of the reasons why regulatory measures are often slow, and sometimes almost reluctant to go beyond a light touch in overseeing such developments. Other reasons include the struggle of regulatory bodies to keep up with unfolding new technologies, and the challenges of monitoring, regulation and enforcement.

The example of food animal industry is instructive here. As Neo and Emel (2017) have recently argued, the food animal industry is a global industry buttressed by strong capital. The reason why many leading global meat industries are able to lobby for greater space in the ways they manage their animal, environment, as well as their use of animal feed and growth hormones can be attributed to the comprehensive manner in which they influence all facets of the meat supply chain. In their book, they show how scientific knowledge and technical practices surrounding the meat industry are (re)produced to make commodified food animals the norm. Such a commodification process is done with little regard to the myriad of socio-economic and environmental impacts of intensified meat production. Some of these impacts include the plight of small-scale farmers; the contested transformations of rural communities; the increased health risks and environmental pollution, as well as the ethical and just treatment of livestock (for example, the way they are reared and the modification to their genetic make-up). The modification of food animals reflects the Anthropocene moment, in that food animals have been irreversibly transformed by modern animal sciences, backed by food conglomerates.

The food animal industry also cautions us to the extent to which these powerful conglomerates might side-step or even ignore precautionary regulations that impede their pursuit for profit. We see similar battles being fought in the realm of energy extraction, and the numerous battles over nature conservation. It suffices to say that contemporary environmental management is a complex architecture of values, knowledge, science/technology, capital power and regulation. The Anthropocene, in highlighting the irreversible changes to the environment, wrought by an ascendant and ambitious humanity, valorises the critical roles played by environmental managers in all their varied guises. As our collection of essays will show, it is not about bringing society–environment relationships to some past mythical or romanticised state. Rather, it is to face the myriad of environment challenges faced by a diverse world with considered solutions underpinned by the appropriate technology and strong science.

Sustaining Natural Heritage, Innovating Urban Environments and Managing Risks

Our essays open with a focus on the importance of natural heritage—a perennial concern in the field of environmental management. The two essays represented in this theme illustrate both the importance and difficulties in protecting and ensuring the thriving of natural resources, in the face of a rapidly urbanising world. Ng Chin Soon Lionel and Chou Loke Ming (Chapter 1) provide us with a primer on the state of coral reef population in Singapore, highlighting the urgent need for their comprehensive conservation, while in Chapter 2, Yaw Chia Wei and Lye Lin Heng clearly lie out the limits and potential of CITES in managing the wildlife trade in China and Vietnam.

Although diverse in its empirical focus and subject matter, many of the essays in this volume share an ethical concern to make the environment more resilient. Such resiliencies are undergirded by careful science, which is the theme of our next section, represented by three essays. In Chapter 3, Fam Meiling and Sanjay Swarup explore the possibility of employing phytoremediation techniques in various places in Singapore, arguing that while much work needs to be done to harness the natural abilities of plants to "clean up" dirty environments, it is an option that is certainly worth considering. On the other hand, Veron Kho and Tan Puay Yok (Chapter 5) investigate the less known, negative impacts of some plant species when they emit biogenic volatile organic compounds (BVOCs) like isoprene and monoterpenes into the atmosphere that could have harmful impacts on humans. While their research, which was based in Singapore, shows no evidence of higher than usual emittance, theirs was a timely reminder to consider any potential side effects of seemingly positive environmental management strategies. Lim Chin Sing and Tan Koh Siang (Chapter 4) explore the potential impact of non-indigenous species on the marine ecosystem as a result of vessel movements in Singapore.

Several essays make up the third section of the book which highlights the fact that effective environmental management requires clear regulatory and reporting frameworks. Indeed, managing and regulating the environment have to go hand in hand. In their chapter (Chapter 9), Deng Mao and Richard Reidinger critically examine the strength of environment impact assessment policies in China and Singapore so as to determine their effectiveness in different contexts. Similarly, Loh Zhiyang and Audrey Chia (Chapter 6), while cognizant of the merits of sustainability reporting by private firms, nonetheless highlight the practical challenges encountered by Singapore-based companies in producing meaningful reports. Turning to air emissions, Kaleem Anwar Mir and Rajasekhar Balasubramanian in Chapter 7 demonstrate the importance of establishing a consistent and comprehensive framework to enforce greenhouse gas emissions in Pakistan. Digging deeper into what values drive effective environmental management,

Zhang Yuzhe and Cecilia Lim (Chapter 8) provide us with a useful overview of the basics of environmental philosophy and how it is applied in real life, drawing on Tibetan and Japanese examples. The section ends with a chapter by Ang Chun Wei and Victor Savage (Chapter 10). Their chapter goes to the core of environmental management by asking how big a population can a resource base support. Using the case study of Singapore and its use of energy, water and other resources, they caution against the view that technology or affluence can resolve all resource problems that result from a larger population.

The two chapters in Part 4 deal with energy issues in contrasting manners. In Chapter 12, Carmelita Leow and Namrata Chindarkar take a realist view of the development of fracking by acknowledging its inevitable expansion. In so doing, they recognise the urgent need to advance a set of robust regulatory framework for fracking. This is done through a comparison of emergent American and European regulations that govern fracking. On the other hand, Tan Choon Puay and Tong Yen Wah (Chapter 11) turn their attention on the most common types of renewable energies to analyse their strengths and weaknesses when used in a hypothetical industrial plant. In so doing, they offer us clear insights on the limitations and possibilities of such forms of energy.

We began this introductory chapter by considering the extent to which the Anthropocene might reshape our knowledge and practice of environmental management. Amongst other things, the Anthropocene is the outcome of rapid urbanisation and the explosion of a global population that is concomitant to such a process. In Part 5, we consider the urban question in three different chapters. Nazifa Zabeen Siddiqua and Oscar Carracedo (Chapter 13) show why smart and sustainable urbanism can and should be actualised at the neighbourhood level through a careful contextualisation of established principles and lived experiences of residents in Dhaka, Bangladesh. Prerna Shah and Pow Choon Piew's chapter highlight the need to understand the socio-politics of the construction of risks in the urbanisation of the Himalayan highlands (Chapter 14). Only when one understands the social basis of risk perception can better policies be formulated to build a more resilient future for the marginalised communities. To round off the volume, Supriyanto and Tan Puay Kok (Chapter 15) show us why is it important to maintain and expand on protected forested areas in purportedly urban areas in Jakarta, Indonesia; highlighting one of the key dilemmas facing the expansion of urban spaces.

Conclusion

As the brief introduction to the essays has shown, the environmental challenges we face are numerous, yet it should be clear that each is as important as the other. With wisdom, knowledge and a particular sensitivity to the workings of politics and policy, we might collectively make a better world for everyone—the age of Anthropocene notwithstanding.

References

Cloke, P., Crang, P., and Goodwin, M. (1991). *Introducing Human Geographies* [1st edition]. London: Routledge.

Neo, H., and Emel, J. (2017). *Geographies of Meat: Politics, Economy and Culture.* London: Routledge (Critical Food Studies Series).

Part 1

Biodiversity & Conservation

1

CORAL REEF RESTORATION IN SINGAPORE — PAST, PRESENT AND FUTURE

NG CHIN-SOON LIONEL and CHOU LOKE-MING

The decline of coral reefs has necessitated the implementation of management strategies such as reef restoration to maintain and enhance the resiliency of these ecosystems. Singapore's reefs have been chronically impacted by high sedimentation and unstable substrate arising from extensive coastal development work, and it is critical that effective strategies be applied to assist with reef recovery. These strategies include: 1) substrate modification to enhance coral settlement and survivorship, 2) deployment of artificial reefs to create stable platforms for the colonisation of reef biota, 3) establishment of nurseries for propagating coral material, and 4) transplantation to help increase coral cover in degraded areas. These efforts are beneficial towards augmenting reef diversity and restoring ecosystem services. To improve coastal management and reduce further loss of reefs, other strategies to enhance current restoration protocols could include implementing long-term monitoring regimes, incorporating novel techniques, mitigating climate change impacts, engaging and educating stakeholders, and re-assessing the legal protection of corals. The cooperation and commitment of all stakeholders (policy makers, scientists, civil society) are critical to ensuring the future of Singapore's reefs.

1.1. Coral Reefs: Importance and Impacts

Coral reefs are unparalleled providers of shelter, food, and nursery grounds for an immense number of marine organisms (Nagelkerken *et al.*, 2000; Hoeksema, 2007; Wilkinson 2008). Despite occupying less than 0.1% of the world's oceans, they support over 800 species of scleractinian corals and 4000 species of reef fish (Burke *et al.*, 2011), and supply essential services such as coastal protection, nutrient cycling, maintenance of genetic resources, and the general regulation of ecosystem functions (Moberg and Rönnbäck, 2003; Ferrario *et al.*, 2014). Additionally, reefs sustain coastal livelihoods and industries such as seafood, curio, recreation, and pharmaceuticals (Moberg and Falke, 1999). These ecosystem goods and services are worth more than US$375 billion annually (Costanza *et al.*, 1997). Coral reefs are thus critically important, especially to 40% of the global population who reside within 100 km of coasts, and even more so for some 275 million people living 30 km from the reefs (Burke *et al.*, 2011).

However, an estimated 19% of the world's reefs have already been lost (Wilkinson, 2008), and up to 75% are imperilled by local and global threats (Burke *et al.*, 2011). For example, unsustainable fishing practices have led to the loss of key species and reef structure, resulting in algae overproliferation, which smothers existing corals and inhibits the establishment of new reefs (Done, 1992; Jackson *et al.*, 2001). Land reclamation and dredging directly bury reefs and increase sedimentation and turbidity levels, impairing physiological processes such as photosynthesis (Erftemeijer *et al.*, 2012). Additionally, stressors such as ocean acidification weakens reefs by reducing calcium carbonate accretion rates (Hoegh-Guldberg *et al.*, 2007; Hoegh-Guldberg, 2011), while higher than normal sea surface temperatures have resulted in three episodes of mass coral bleaching and mortality within the last two decades alone (Wilkinson, 1998; Tun *et al.*, 2010; Great Barrier Reef Marine Park Authority, 2016).

Management strategies such as coral reef restoration and rehabilitation are thus urgently required to hasten the recovery of damaged reefs, and enhance overall resiliency. These actions attempt to bring the degraded habitats and diminished ecosystem functions back to their original condition (Edwards and Gomez, 2010), While conservation should be prioritised over restoration, the latter is necessary as natural recovery processes might be too slow in the light of rapid environmental degradation, and hence ineffective (Edwards and Gomez, 2010; Chou, 2011). Despite the infancy of the field of reef restoration, a variety of techniques have been tested, including substrate stabilisation, deployment of artificial reef structures, establishment of coral nurseries, and the transplantation of coral material to target localities (e.g. Clark and Edwards, 1995; Rinkevich, 2005; Raymundo *et al.*, 2007; Edwards, 2010; Chavanich *et al.*, 2014).

1.2. Status of Coral Reefs in Singapore and the Need for Restoration

Over 80% of Singapore's territorial seas are managed as port waters, with the remainder utilised by sectors ranging from the military, to petrochemical industries, aquaculture, and recreation (Chou, 2008). Extensive reclamation carried out since the 1960s has increased the total land area from 581.5 km^2 in the 1960s to 719.2 km^2 in 2016 (Tun, 2012; Singapore Department of Statistics, 2017). Seawalls currently line almost two-thirds of the country's coastlines (Lai *et al.*, 2015). Based on projections by the Singapore government, land area is expected to increase to 766 km^2 by the year 2030 (Ministry of National Development, 2013).

Singapore's population growth and economic progress have come at a heavy price. Substantial loss and degradation of Singapore's coral reefs have been documented (Chou, 2006). The total reef area in 2012, at 13.25 km^2, was a stark contrast to an estimated 39.85 km^2 in 1953 (Tun, 2012). High sedimentation rates of 44.64 mg/cm^2/day increased the smothering and mortality of corals (Low and Chou, 1994; Erftemeijer *et al.*, 2012), and decreased underwater visibility (Chou, 1996). The drastic reduction in light penetration has restricted reef establishment to shallower depths (Chou and Tun, 2012). The reef substrate has also destabilised, and loose rubble pieces are frequently moved by strong currents (Tun, 2012), compromising the survival of coral larvae that inadvertently settle on them (Fox, 2004). Further declines in reef area are expected due to the proposed land reclamation to support future needs, including plans to create a mega terminal at Tuas on the western coast (Lam, 2016).

It is in the interest of Parties to the Convention on Biological Diversity to be proactive in reef rehabilitation. Specifically, Aichi Biodiversity Target 15 in the Strategic Plan for Biodiversity 2011–2020 stated that: "By 2020, ecosystem resilience and the contribution of biodiversity to carbon stocks [should] have been enhanced, through conservation and restoration, including [the] restoration of at least 15% of degraded ecosystems, thereby contributing to climate change mitigation and adaptation and to combating desertification" (Convention on Biological Diversity COP-10, 2010). This is a very clear indication that enhancing the resiliency of Singapore's marine environment by conserving its remaining natural habitats and carrying out active habitat restoration is relevant to addressing the global agenda of international and inter-governmental bodies. It is thus essential that the most effective restoration techniques be applied to assist the recovery of Singapore's remaining reefs.

This study aims to: 1) assess the reef restoration strategies used since the 1990s and evaluate their contribution to coral reef conservation in Singapore; 2) synthesise the overall findings to identify potential management strategies and provide recommendations for future reef restoration work.

2.1. Reef Restoration: Past and Present

Should reef habitats be damaged by human activities, restitution procedures are necessary to compensate for the loss (Edwards and Gomez, 2010), but this should be viewed as a last resort and clearly distinguished from efforts at reef restoration (Chou *et al.*, 2009). In Singapore, such mitigation measures typically entailed the relocation of entire coral colonies away from sites slated for land reclamation to secure locations where they can be expected to establish and continue to grow. At least five major coral relocation exercises have been conducted (Chew, 2014). In the early 1990s, the Marine Conservation Group of the Nature Society (Singapore) (or NSS) mobilised over 400 volunteers to shift coral colonies and reef invertebrates from Buran Darat (now part of Sentosa Cove) and Pulau Ayer Chawan (now part of Jurong Island) to Sentosa Island (The Straits Times, 1993; Chou and Tun, 1997). However, the efforts were largely unsuccessful. Less than 11% of the transplants survived due to the poor choice of recipient site (a shallow area easily overgrown with algae and which was constantly subjected to surge and heavy sedimentation) as well as improper securing methods (transplanted colonies were only wedged between boulders without the use of adhesives, and thus easily dislodged by wave action) (Chou and Tun, 1997). This episode led to improvements in coral relocation techniques and subsequent mitigation projects. This included the relocation of corals from Labrador Park (for the installation of new power and water cables; Kesava, 2007a), Sultan Shoal (for the construction of a mega container terminal; Chew, 2014), and Pulau Semakau (for the disposal of incinerated waste; Tan, 2015). Transplant survival rates for coral relocation projects are currently an estimated 70–80% (Chew, 2014).

Singapore's approach to reef restoration and rehabilitation evolved with the measures at damage mitigation. In a bid to incorporate rigorous scientific approaches to guide management, the reef restoration practices of today are distilled from the findings of individual studies conducted since the 1990s. As differences in local environments and scales of disturbance affect the extent of the reef impacted, it is important to customise restoration strategies to achieve the best outcomes. For Singapore's reefs, this meant targeting the problems of chronic sedimentation and substrate instability. The plethora of reef restoration projects attempted have since varied in location, scale and design, and strategies can be broadly categorised as the following: substrate modification, artificial reefs, coral nurseries, and transplantation.

2.2. Substrate Modification

An unstable reef substrate leads to current-induced abrasion, which is detrimental to the post-settlement survivorship of corals (Fox, 2004). Coral larvae availability is not a limiting factor in Singapore (Guest *et al.*, 2005; Tay *et al.*, 2012), and researchers explored ways to modify substrate quality via biological and physical enhancements. Cement blocks covered with mature layers of marine biofilm, as well as those impregnated or encrusted with crustose coralline algae, registered higher rates of coral recruitment and survivorship (Seow *et al.*, 2008; Tun *et al.*, 2008). Crustose coralline algae and marine biofilm are known to enhance the settlement and metamorphosis of a number of marine invertebrates including corals (Morse *et al.*, 1996; Webster *et al.*, 2004).

Separate experiments also revealed that coral recruitment was highest on terracotta tiles placed on the reef flat and crest compared to the slope; recruits also preferred settling on tiles angled 60° and 90° (fully vertical) to the horizontal, compared to 30° and 0° (fully horizontal) (Tanzil JTI, unpublished data). This knowledge led to the development of custom-designed artificial reef structures and coral nurseries that encouraged coral settlement and growth.

2.3. Artificial Reefs

Artificial reefs (ARs) are structures introduced to areas that are degraded or less productive to enhance marine resources (e.g. fish) (White *et al.*, 1990; Charbonnel *et al.*, 2002). A wide variety of materials, designs and scales have been used to create additional habitat for biotic establishment, with some going as far as sinking cars and train carriages (Baine, 2001). In the mid-1980s, precast hollow concrete modules and pyramidal modules assembled from disused rubber tyres were deployed off Pulau Hantu to improve local fish stocks (Tan *et al.*, 2010). Sixty-seven fish species from 26 families were recorded over seven years, with the concrete modules supporting higher fish densities and diversity (55 species) than the tyre modules (48 species) (Low, 1999). Sessile organisms also recruited on the artificial reefs, including sponges, hydroids, algae, ascidians, octocorals, and bryozoans (Low, 1999). However, the reduced light availability at the depths that these ARs were deployed ($\sim$10 m) limited hard coral growth ($<$10% cover) to the genera *Dendrophyllia* and *Tubastraea*, both of which form small colonies that contribute minimally to reef-building.

The use of ARs specifically as a tool for reef restoration in Singapore started in the early 2000s. The objective was to provide coral larvae with elevated and stable settlement surfaces which reduced smothering by sediment or scouring by rubble. The installation of bulky concrete and tyre AR structures required the use of barges, and was deemed expensive and potentially destructive if operated in shallow areas where the structures are recommended to be sited. Dome-shaped fibreglass modules ('Reef Enhancement Units', or REUs), light enough to be transported and stabilized at exact locations by SCUBA divers, were thus fabricated (Loh *et al.*, 2006). The perforated and steep sloping surfaces of each REU reduced sediment accumulation and facilitated water movement through the structure. The REUs were anchored to bare patches on reefs fringing various southern offshore islands at depths of 2–7 m. Within six months, they were readily colonised by coralline algae. The survival and growth rates of corals that recruited on the REUs were higher compared to those that had settled on adjacent rubble areas, illustrating the suitability of stabilised fibreglass substrates in supporting marine life.

The biological communities on these REUs were surveyed again a decade after they had been deployed (Ng *et al.*, 2017). While the sessile communities that colonised the REUs varied across sites, the overall community was more diverse in 2014 (including hard corals, soft corals, seafans, sponges, sea squirts, anemones, hydroids, zoanthids, bryozoans and algae) than in 2004 (mainly coralline and algal turfs). Colonies from ten of the 30 coral genera documented were sexually mature, indicating that they participated in annual coral mass spawning events and contributed genetic material to seed the surrounding reefs. Mobile fauna also interacted with the REUs for food and habitat. This included wrasses and damselfishes that used the cavity of the REUs for shelter, while brittlestars and commensal crustaceans sought refuge among the branches of the corals which had recruited on the structures. Featherstars perched on the REUs and extended their arms to feed in the strong current. Herbivorous sea urchins and gastropods grazed the algae growing on the REUs. Reef fish and nudibranchs also laid eggs on the interior walls of the REUs. In total, at least 119 taxa were recorded interacting with the REUs. The findings strongly demonstrated that the REUs complemented habitat rehabilitation efforts in Singapore by helping to improve reef life and restore ecosystem services (Ng *et al.*, 2017).

2.4. Coral Nurseries

Coral nurseries allow coral propagules to grow in a sheltered environment which improves their survival chances before they are transplanted to target localities (Epstein *et al.*, 2003; Rinkevich 2005). Nurseries have important roles as repositories of coral material from reefs threatened by degradation or destruction (Epstein *et al.*, 2001; Schopmeyer *et al.*, 2012). To date, both *in situ* (sea-based) and *ex situ* (land-based) versions have been tested, and have had substantial roles in shaping Singapore's reef restoration strategies.

Two *in situ* coral nurseries established on the reefs fringing St John's and Lazarus Islands in 2004 supported the rearing of over 2900 small fragments (known as 'nubbins'; ~1 cm) from 13 hard coral species (Bongiorni *et al.*, 2011). Coral mortality rates were initially high due to sediment smothering and heavy fouling, but the nubbins that survived had manifold increases in colony size after just a year of rearing. Subsequently, another *in situ* nursery was established at Pulau Semakau in 2007 to rear 'corals of opportunity' (i.e. naturally occurring coral fragments, or loose rubble pieces bearing juvenile corals, that would likely perish if left on their own in an environment of high sedimentation and low substrate consolidation; Edwards and Gomez, 2007). The approach reduced the need to directly harvest from healthy colonies. More than 600 fragments and recruits from 36 genera were collected and secured on modular frames that were elevated to minimise sediment accumulation (Chou *et al.*, 2009). Trained volunteer divers assisted with maintenance and monitoring work, and appreciable survivorship and growth of the COPs were recorded over the course of the project (Ng and Chou, 2014). The results collectively showed that *in situ* nursery-rearing was viable in environments chronically impacted by high sedimentation, and that COPs are a viable source material to supplement reef restoration in areas with unstable substrate.

The St John's Island National Marine Laboratory (SJINML), which supplies flow-through, sand-filtered seawater, has functioned as an effective *ex situ* coral nursery by providing a relatively controlled environment with minimal stressors such as sedimentation. While *ex situ* rearing is expensive (Shafir *et al.*, 2010), the Laboratory has facilitated the propagation of corals through asexual and sexual techniques. Data on the growth and developmental patterns of more than 20 species of corals have been obtained

(Ng *et al.*, 2012; Toh *et al.*, 2012a; Toh *et al.*, 2013), and this will help to shape future coral rearing programmes.

2.5. Coral Transplantation

Coral transplantation is a commonly employed tool in reef restoration that involves securing whole colonies, fragments or juveniles on an area with adhesives or other binding materials such as marine epoxy and wires (Gomez *et al.*, 2010). It hastens recovery processes of damaged reefs by directly enhancing coral cover, biodiversity and habitat complexity (Lindahl, 2003; Shaish *et al.*, 2010a; Gomez *et al.*, 2014). Notably, the recent two decades have seen the advent and application of the coral 'gardening' concept (Rinkevich, 1995; Epstein *et al.*, 2001), in which corals are first reared in nurseries to a suitable size before they are transplanted to degraded areas.

Although coral transplantation has been practised as a compensatory mitigation measure, its use as a reef restoration tool in Singapore is relatively new. Targeted projects only began since the mid-2000s, with the transplantation of nursery-reared corals onto natural reefs and artificial structures. At Kusu Island, researchers observed that six months after transplantation, the survival rates of sexually-propagated corals that were fed brine shrimp were significantly higher (56–63%) than those which were unfed (38%) (Toh *et al.*, 2014). Feeding brine shrimp to juvenile corals at higher densities was also cheaper in the long run, demonstrating that nutritional enhancement was biologically and economically feasible for increasing transplantation success. Unfortunately, the site where these corals had been transplanted to has since been destroyed by a ship grounding incident.

Nine genera of coral nubbins and fragments have also been transplanted on dead coral boulders at Pulau Semakau and P. Hantu in 2007, as well as on granite seawalls north of mainland Singapore (Ng *et al.*, 2015; Ng *et al.*, 2016). Corals with massive and submassive growth forms generally fared better in terms of growth and survivorship than branching coral transplants which were damaged easily and had high mortality rates. This suggested that the former group was more suitable than the latter for creating an initial reef framework for degraded areas in Singapore, before other corals can be introduced. The surviving transplants also established themselves in their respective environments and provided food and shelter to reef fish and gastropods.

While transplantation has only been carried out on limited spatial and temporal scales in Singapore, there are strong indications that if carried out properly, active transplantation of nursery-reared corals is a viable approach for rehabilitating Singapore's coral habitats. A number of lessons have been learnt: e.g. that it is important to use sufficient and appropriate adhesives to reduce the chances of transplants dislodging from the substrate, and that a regular monitoring regime should be in place so that reinforcements or reattachments can be carried out when necessary. There is a need for more species to be tested so that fundamental biological information such as growth trends can be obtained to facilitate planning and prioritisation of subsequent restoration efforts.

3.1. Looking Forward

The restoration strategies and techniques attempted in Singapore had varying degrees of success, and it is apparent that no single most effective restoration strategy exists for all situations. ARs are most effectively deployed in areas with unconsolidated substrate, but such sites can still be affected by impacts such as strong currents or anchor damage that destabilise the reef framework. Transplantation could help to kick-start reef development in denuded areas, but coral species are not all equally effective. A rehabilitation plan for sites affected by one or more stressors would likely require a combination of different strategies to achieve the most optimal outcomes.

However, as further loss of Singapore's reefs is predicted over the next two decades, there is an urgent need to explore other aspects to augment current restoration protocols. The overall strategy for reef restoration in the future would benefit from the incorporation of long term monitoring regimes, the use of new tools and techniques, mitigation of the effects of climate change, stakeholder engagement, and a re-assessment of the legal protection status of Singapore's coral reefs. The successful restoration of degraded areas and conservation of Singapore's remaining reefs requires not only scientific expertise, but also the cooperation and commitment of other stakeholder entities, including policy makers and civil society.

3.2. Implementation of Long Term Monitoring

Jaap (2000) noted that "if a project is worthy of executing a restoration, it is worthy of monitoring the progress of recovery", emphasising the importance of follow-up actions on reef restoration projects during and after their completion. The slow development and growth of corals mean that successes (e.g. ecosystem service provision) or failures (e.g. mass transplant mortality) are typically not apparent over the usual duration of these projects. For example, sexually propagated corals have been documented to require at least three to five years of growth before they attain sexual maturity and begin producing gametes (Baria *et al.*, 2012; Toh, 2014). Basic biological information such as scleractinian life histories and population dynamics on impacted reefs can only be gleaned if the corals are tracked (Vermeij, 2006).

The dislodgement of transplanted nubbins at P. Hantu and discovery of the destroyed transplant site at Kusu would have remained unnoticed if not for continual monitoring. As regular monitoring can help to capture acute or chronic disturbance events, determine if additional maintenance is required at a restoration site, and provide information on whether a particular technique is feasible elsewhere (Spieler *et al.*, 2001), it is critical for reef restoration initiatives to be assessed over longer timescales (e.g. at least five to ten years) so that their relevance and efficacy may be adequately evaluated. While long-term commitments of funding and manpower may be difficult to ensure, the value of rigorous monitoring programmes even beyond the completion of projects, as well as the implementation of adaptive management protocols where necessary, cannot be overemphasised. The wealth of information gathered can then be deposited with a central body which will coordinate future restoration projects so that valuable data will not be lost over time, thus enhancing the management of Singapore's coral reefs.

3.3. Incorporation of Novel Techniques into Restoration Design

Over the years, reef restoration has evolved from a descriptive discipline to one that is driven by ecological concepts (Rinkevich, 2005). As resources are finite and restoration work is expensive, there is a need to explore the use of other techniques to enhance outcomes (Aronson and Alexander, 2013). There exists a plethora of potential new tools, ranging from modelling to phylogenetic analyses, and cost-benefit studies.

Modelling can optimise resource usage by providing information on the arrangement and density of corals to be transplanted (Sleeman *et al.*, 2005), choice of species to be used (Muko and Iwasa, 2011), and reefs to be selected for conservation (Tay *et al.*, 2015). For example, Tay *et al.* (2012)

used a hydrodynamic model to recommend the best source (the islands of Sudong, Pawai, and Senang) and sink (Kusu and the Sisters') reefs in Singapore for conservation. Cost-benefit analyses also represent a useful tool for monetising strategies so that reef restoration can be carried out more effectively at lower cost (Edwards *et al.*, 2010b). Toh *et al.* (2014) showed that while labour and equipment contributed significantly to the cost of producing sexually propagated corals for transplantation, greater economies of scale could be achieved by introducing feeds to augment coral growth during the nursery phase. Additionally, factors such as the evolutionary distinctiveness or vulnerability of a species to global extinction can be considered when prioritising corals for restoration (Curnick *et al.*, 2015; Huang and Roy, 2015), to supplement broad-based proxies such as site-specific species rarity.

3.4. Mitigating the Effects of Climate Change

Reef restoration strategies, initially designed to counter anthropogenic impacts, must be re-assessed to surmount the challenges posed by climate change. While coral reefs and restoration efforts in Singapore have not been affected by frequent natural catastrophes such as typhoons (Shaish *et al.*, 2010b), they are still subject to other effects of climate change. In 1998 and 2010, sea surface temperatures in Singapore rose 2°C above the average monthly readings, causing widespread bleaching and up to 20% mortality of corals in both incidents (Guest *et al.*, 2012; Toh *et al.*, 2012b). The 2010 episode also affected ongoing restoration projects, with both nursery-reared and transplanted corals exhibiting varying degrees of bleaching and recovery (Ng *et al.*, 2013).

Instead of being viewed as setbacks, valuable lessons can be learnt to improve the outcomes of future restoration efforts (Rinkevich, 2015). Concerted efforts by scientists and policy makers can help in mitigating the effects of climate change. When rising sea surface temperatures are detected, a monitoring regime will provide useful information on corals' susceptibilities to bleaching, while adaptive measures such as increasing water flow in mariculture tanks or shading to reduce insolation can be introduced. Nursery-rearing efforts can be focused on species or genotypes which, after transplantation, demonstrated tolerance or recovery from increased temperatures. Research should also gear towards adapting restoration protocols in response to other climate change-related impacts such as increasing occurrences of algal blooms, sea level rise, or ocean acidification.

3.5. Stakeholder Engagement and Education

It is recognised that stakeholder involvement can benefit reef restoration programmes (Fisk and Edwards, 2010). However, while information on reef restoration may be readily available and accessible, interpretations can vary among the stakeholders. As reef restoration is often synonymous with coral transplantation or the use of artificial reefs (Gomez *et al.*, 2010), an over-simplification and misinterpretation of concepts can result, leading to the proliferation of quick-fix solutions. In 2007, volunteers installed large concrete slabs with PVC pipes on the intertidal flat at Labrador Beach "to create an underwater garden and open-sea aquarium" (Tjan and Tan, 2007). The plan was for the structures to mimic a natural reef and provide surfaces for the colonisation of marine life. However, members of the public observed that the structures had disintegrated, and that pieces of plastic and concrete, pushed about by wave action, ended up damaging the adjacent seagrass beds and other marine life (Kesava, 2007b). The failure of the project was attributed to the lack of rigorous research into the suitability of materials and local conditions, as well as the lack of consultation with marine biologists on the feasibility of the approach (The Hantu Bloggers, 2011; The Lazy Lizard's Tales, 2011). In a separate incident, nature enthusiasts criticised the deployment of the REUs as they felt that it was more beneficial to allow coral reefs to naturally recover rather than introduce artificial structures into the marine environment (Lau and Leong, 2005). Chou (2005) responded then that such interventions were necessary given the declining state of reef health in Singapore. Surveys a decade after the REUs were deployed (this study) showed that these ARs were a viable reef restoration option for Singapore's marine environment.

While an awareness of marine conservation issues is evidently greater in recent years, these episodes highlighted the need to properly educate stakeholders on the scientific basis and appropriate techniques for reef restoration. An example of how this might be done is the public engagement programme that was held in conjunction with the relocation of corals from Sultan Shoal (Feng, 2014). Subject experts from TMSI and DHI Water & Environment (an environmental consultant) conducted training workshops and field activities to expose members of the public to the theoretical and practical aspects of reef restoration in Singapore. Social media outlets manned by non-governmental organisations or nature enthusiasts can also be useful avenues for raising awareness on matters relating to restoration, such as WildSingapore (see http://wildshores.blogspot.sg/2013/11/volunteer-wanted-for-coral-relocation.html#.VQb0V8ai3W4) and The Hantu Bloggers (see http://www.pulauhantu.sg/coral-relocation-at-sultan-shoal). A pool of trained volunteers who are cognizant of proper restoration protocols may then serve as indispensible helpers for the labour-intensive work.

The importance of educating professionals in proper reef restoration techniques must also not be neglected. Globally-funded training programmes such as the European Commission project "Developing ubiquitous restoration practices for Indo-Pacific coral reefs" in 2004 (Chou, 2011) and the IOC/WESTPAC Workshop on Coral Reef Restoration Techniques in the Western Pacific Region in 2012 (Chavanich *et al.*, 2014) have played key roles in capacity building for Southeast Asian marine scientists in the recent decade.

3.6. Legal Protection of Corals

In Singapore, the laws that pertain to reef conservation and restoration are at best, broad definitions. Under the State Lands Encroachment Act, any person who "cuts, digs or takes... coral... shall be guilty of an offence and shall be liable on conviction to a fine not exceeding $2,000" (Cap 315, 1985 Rev. Ed. Sing., ss. 14,"). It is unclear if the loss of coral from development works or ship grounding incidents then constitutes as cutting, digging or taking. However, it is apparent that the vagueness of the law and the leniency of the penalty understate the ecological consequences of coral reef loss. Additionally, while Singapore is Party to the Convention on Biological Diversity which stipulates a requirement for environmental impacts assessments (EIAs) to be carried out for "projects that are likely to have significant adverse effects on biological diversity" and "allow for public participation in such procedures" (Convention on Biological Diversity Article 14), there is no local instrument mandating that EIAs be conducted before proceeding with development (Lye, 2008). The nearest approximations for EIAs are stated in the Environmental Protection and Management Act, and are more relevant for pollution control in industrial situations (Lye, 2008). It can thus be said that a national policy on safeguarding coral reef habitats does not exist (Tun, 2012). The diverse ecosystem services that reefs provide warrant a re-evaluation of the legal protection status of Singapore's reefs. New environmental protection laws, stricter enforcement and harsher penalties may be required.

There are however promising signs, as various stakeholders have called for the improved management of Singapore's reefs. For example, the

Singapore Blue Plan 2009, a proposal submitted to policy makers by the civil society, recommends key steps to conserve Singapore's marine environment. Some recommendations appear to have been taken into consideration, such as the designation of the Sisters' Islands as a Marine Park. Additionally, EIAs have been increasingly carried out prior to coastal development work, such as the case for Sultan Shoal and Pulau Semakau, following which, corals were relocated from these sites.

4. Conclusions

Despite the varying degrees of success in restoration strategies, efforts over the past two decades have generally helped to improve reef life in Singapore's heavily sedimented marine environment. Existing and new restoration approaches must be integrated in a comprehensive coastal management framework to more effectively conserve Singapore's remaining reef diversity. The lessons learnt are relevant and applicable to other coastal locales impacted by intense development.

5. References

Aronson, J., and Alexander, S. (2013). Ecosystem restoration is now a global priority: time to roll up our sleeves. *Restoration Ecology,* **21**(3), 293–296.

Baine, M. (2001). Artificial reefs: a review of their design, application, management and performance. *Ocean and Coastal Management,* **44**(3), 241–259.

Baria, M.V.B., dela Cruz, D. W., Villanueva, R. D., and Guest, J. R. (2012). Spawning of three-year-old *Acropora millepora* corals reared from larvae in northwestern Philippines. *Bulletin of Marine Science,* **88**(1), 61–62.

Bongiorni, L., Giovanelli, D., Rinkevich, B., Pusceddu, A., Chou, L.M., and Danovaro, R. (2011). First step in the restoration of a highly degraded coral reef (Singapore) by *in situ* coral intensive farming. *Aquaculture,* **322–323**, 191–200.

Burke, L.M., Reytar, K., Spalding, M., and Perry, A. (2011). Reefs at Risk Revisited. USA World Resources Institute, Washington, DC.

Charbonnel, E., Serre, C., Ruitton, S., Harmelin, J.G., and Jensen, A. (2002). Effects of increased habitat complexity on fish assemblages associated with large artificial reef units (French Mediterranean coast). *ICES Journal of Marine Science,* **59**(suppl), S208–S213.

Chavanich, S., Gomez, E., Chou, L.M., Goh, B., Tan, L.T., Tun, K., Toh, T.C., Cabaitan, P., Guest, J., Ng, C.S.L., Omori, M., Thongtham, N., Chankong, A., Viyakarn, V., and Zhu, W. (2014). Coral restoration techniques in the Western Pacific region. UNESCO-IOC/WESTPAC, Bangkok, Thailand. 8pp.

Chew, H.M. (2014). Corals go spawning, but El Nino may turn them white. *MyPaper*, 26 May 2014.

Chou, L.M. (1996). Response of Singapore reefs to land reclamation. *Galaxea, Journal of Coral Reef Studies,* **13**, 85–92.

Chou, L.M. (2005). Coral reefs need help to restore themselves. *The Straits Times*, 12 September 2005.

Chou, L.M. (2006). Marine habitats in one of the world's busiest harbours. In: Wolanski, E. (ed.) *The Environment in Asia Pacific Harbours.* Springer, Netherlands, pp. 377–391.

Chou, L.M. (2008). Nature and sustainability of the marine environment. In: Wong, T.C., Yuen, B., and Goldblum, C. (eds.) *Spatial Planning for a Sustainable Singapore.* Springer, Netherlands, pp. 169–182.

Chou, L.M. (2011) Southeast Asia's coral reef biodiversity — can restoration reverse the loss? *Innovation,* **10**(1), 36–39.

Chou, L.M., and Tun, K.P.P. (1997). Coral transplantation as a reef conservation tool — the Singapore experience. *PACON 97 Proceedings. PACON International*, Hawaii. 12pp.

Chou, L.M., and Tun, K. (2012). Status and challenges of coral reef monitoring in East Asia: Singapore. *8th ICRI East Asia Regional Workshop*, 3–5th September 2012, Jeju, Republic of Korea.

Chou, L.M., Yeemin, T., Abdul Rahim, B.G.Y., Vo, S.T., Alino, P. and Suharsono (2009). Coral reef restoration in the South China Sea. *Galaxea, Journal of Coral Reef Studies*, **11**(2), 67–74.

Clark, S., and Edwards, A.J. (1995). Coral transplantation as an aid to reef rehabilitation: evaluation of a case study in the Maldive Islands. *Coral Reefs*, **14**(4), 201–213.

Convention on Biological Diversity. (1992). (Article 14).

Convention on Biological Diversity. (2010). Conference of the Parties 10 Decision X/2 <https://www.cbd.int/decision/cop/?id=12268> (Retrieved 22 March 2015).

Costanza, R., d'Arge, R., de Groot, R., Farber, S., Grasso, M., Hannon, B., Limburg, K., Naeem, S., O'Neill, R.V., Paruelo, J., Raskin, R.G., Sutton, P., and van den Belt, M. (1997). The value of the world's ecosystem services and natural capital. *Nature,* **387**, 253–260.

Curnick, D.J., Head, C.E.I., Huang, D., Crabbe, M.J.C., Gollock, M., Hoeksema, B.W., Johnson, K.G., Jones, R., Koldewey, H.J., Obura, D.O., Rosen, B.R., Smith, D.J., Taylor, M.L., Turner, J.R., Wren, S., and Redding, D.W. (2015). Setting evolutionary-based conservation priorities for a phylogenetically data-poor taxonomic group (Scleractinia). *Animal Conservation* DOI: 10.1111/acv.12185

Done, T.J. (1992). Phase shifts in coral reef communities and their ecological significance. In: Jaccarini, V. and Martens. E. (eds.) *The Ecology of Mangrove and Related Ecosystems.* Springer, Netherlands, pp. 121–132.

Edwards, A. (2010). *Reef Rehabilitation Manual.* The Coral Reef Targeted Research and Capacity Building for Management Program. St Lucia, Australia. ii + 166pp.

Edwards, A.J., and Gomez, E.D. (2007). *Reef restoration concepts and guidelines: making sensible management choices in the face of uncertainty.* Coral Reef Targeted Research and Capacity Building for Management Program. St Lucia, Australia. iv + 38pp.

Edwards, A., and Gomez, E. (2010). Objectives and scope of this manual. In: Edwards, A. (ed.) *Reef Rehabilitation Manual.* The Coral Reef Targeted Research and Capacity Building for Management Program. St Lucia, Australia, pp. 1–6.

Edwards. A., Guest, J., Rinkevich, B., Omori, M., Iwao, K., Levy, G., and Shaish, L. (2010b). Evaluating costs of restoration. In: Edwards, A. (ed.) *Reef Rehabilitation Manual.* The Coral Reef Targeted Research and Capacity Building for Management Program. St Lucia, Australia, pp. 113–128.

Epstein, N., Bak, R.P.M., and Rinkevich, B. (2001). Strategies for gardening denuded coral reef areas: the applicability of using different types of coral material for reef restoration. *Restoration Ecology,* **9**(4), 432–442.

Epstein, N., Bak, R.P.M., and Rinkevich, B. (2003). Applying forest restoration principles to coral reef rehabilitation. *Aquatic Conservation: Marine and Freshwater Ecosystems,* **13**, 387–395.

Erftemeijer, P.L., Riegl, B., Hoeksema, B.W., and Todd, P.A. (2012). Environmental impacts of dredging and other sediment disturbances on corals: a review. *Marine Pollution Bulletin,* **64**(9), 1737–1765.

Feng, Z. (2014). Local coral colonies getting a new home. *The Straits Times*, 27 April 2014.

Ferrario, F., Beck, M.W., Storlazzi, C.D., Micheli, F., Shepard, C.C., and Airoldi, L. (2014). The effectiveness of coral reefs for coastal hazard risk reduction and adaptation. *Nature Communications*, **5**. doi:10.1038/ncomms4794

Fisk, D., and Edwards, A. (2010). In: Edwards, A. (ed.) *Reef Rehabilitation Manual.* The Coral Reef Targeted Research and Capacity Building for Management Program. St Lucia, Australia, pp. 29–48.

Fox. H.E. (2004). Coral recruitment in blasted and unblasted sites in Indonesia: assessing rehabilitation potential. *Marine Ecology Progress Series,* **269**, 131–139.

Gomez, E.D., Cabaitan, P.C., Yap, H.T., and Dizon, R.M. (2014). Can coral cover be restored in the absence of natural recruitment and reef recovery? *Restoration Ecology,* **22**(2), 142–150.

Gomez, E., Dizon, R., Edwards, A. (2010). Methods of coral transplantation. In: Edwards, A. (ed.) *Reef Rehabilitation Manual.* The Coral Reef Targeted Research and Capacity Building for Management Program. St Lucia, Australia, pp. 99–112.

Great Barrier Reef Marine Park Authority. (2016). Interim Report: 2016 Coral bleaching event on the Great Barrier Reef, GBRMPA, Townsville.

Guest, J.R., Baird, A.H., Goh, B.P.L., and Chou, L.M. (2005). Reproductive seasonality in an equatorial assemblage of scleractinian corals. *Coral Reefs,* **24**(1), 112–116.

Guest, J.R., Baird, A.H., Maynard, J.A., Muttaqin, E., Edwards, A.J., Campbell, S. J., Yewdall, K., Affendi, Y.A., and Chou, L.M. (2012). Contrasting patterns of coral bleaching susceptibility in 2010 suggest an adaptive response to thermal stress. *PLoS One*, **7**(3), e33353.

Hoegh-Guldberg, O. (2011). Coral reef ecosystems and anthropogenic climate change. *Regional Environmental Change*, **11**(1), 215–227.

Hoegh-Guldberg, O., Mumby, P.J., Hooten, A.J., Steneck, R.S., Greenfield, P., Gomez, E., Harvell, C.D., Sale, P.F., Edwards, A.J., Caldeira, K., Knowlton, N., Eakin, C.M., Iglesias-Prieto, R., Muthiga, N., Bradbury, R.H., Dubi, A., and Hatziolos, M.E. (2007). Coral reefs under rapid climate change and ocean acidification. *Science*, **318**(5857), 1737–1742.

Hoeksema, B.W. (2007). Delineation of the Indo-Malayan centre of maximum marine biodiversity: the Coral Triangle. In: Renema, W. (ed.) *Biogeography, time, and place: distributions, barriers, and islands.* Springer, Netherlands, pp. 117–178.

Huang, D., and Roy, K. (2015). The future of evolutionary diversity in reef corals. *Philosophical Transactions of the Royal Society B: Biological Sciences*, **370**(1662), 20140010.

Jaap, W.C. (2000). Coral reef restoration. *Ecological Engineering*, **15**(3), 345–364.

Jackson, J.B., Kirby, M.X., Berger, W.H., Bjorndal, K.A., Botsford, L.W., Bourque, B.J., Bradbury, R.H., Cooke, R., Erlandson, J., Estes, J.A., Hughes, T.P., Kidwell, S., Lange, C.B., Lenihan, H.S., Pandolfi, J.M., Peterson, C.H., Steneck, R.S., Tegner, M.J. and Warner, R.R. (2001). Historical overfishing and the recent collapse of coastal ecosystems. *Science*, **293**(5530), 629–637.

Kesava, S. (2007a). Undersea garden takes root. *The Straits Times* 29 October 2007.

Kesava, S. (2007b). Labrador Beach: Nature lovers fear coral project will cause damage. *The Straits Times*, 26 November 2007.

Lai, S., Loke, L.H., Hilton, M.J., Bouma, T.J., and Todd, P.A. (2015). The effects of urbanisation on coastal habitats and the potential for ecological engineering: A Singapore case study. *Ocean and Coastal Management*, **103**, 78–85.

Lam, J.S.L. (2016). Strategy of a Transhipment Hub: The Case of Port of Singapore. In: Lee, P.T.-W. and Cullinane, K. *Dynamic Shipping and Port Development in the Globalized Economy.* Palgrave Macmillan, UK, pp. 12–38.

Lau, P.K.S., and Leong, K.P. (2005). Better to leave natural coral reefs untouched <http://www.wildsingapore.com/news/20050910/050907-1.htm>. *Straits Times Forum Online*, 7 September 2005 (Retrieved 23 March 2015).

Lindahl, U. (2003). Coral reef rehabilitation through transplantation of staghorn corals: effects of artificial stabilization and mechanical damages. *Coral Reefs*, **22**(3), 217–223.

Loh, T.-L., Tanzil, J.T.I., and Chou, L.M. (2006). Preliminary study of community development and scleractinian recruitment on fibreglass artificial reef units in the sedimented waters of Singapore. *Aquatic Conservation: Marine and Freshwater Ecosystems,* **16**, 61–76.

Low, J.K.Y. (1999). Artificial reefs as a marine resource enhancement tool. MSc thesis, Department of Biological Sciences, National University of Singapore. 91pp.

Low, J.K.Y., and Chou, L.M. (1994). Sedimentation rates in Singapore waters. In: Sudara, S., Wilkinson, C.R., and Chou, L.M. *Proceedings of the Third ASEAN–Australian Symposium on Living Coastal Resources, Vol. 2 Research Papers.* Chulalongkorn University, Bangkok, Thailand, pp. 697–701.

Lye, L.H. (2008). A fine city in a garden — environmental law and governance in Singapore. *Singapore Journal of Legal Studies*, 68–117.

Ministry of National Development. (2013). Our Land Use Plan <http://www.mnd.gov.sg/landuseplan/> (Retrieved 22 March 2015).

Moberg, F., and Folke, C. (1999). Ecological goods and services of coral reef ecosystems. *Ecological Economics*, **29**(2), 215–233.

Moberg, F., and Rönnbäck, P. (2003). Ecosystem services of the tropical seascape: interactions, substitutions and restoration. *Ocean and Coastal Management*, **46**(1), 27–46.

Morse, A.N.C., Iwao, K., Baba, M., Shimoike, K., Hayashibara, T., and Omori, M. (1996) An ancient chemosensory mechanism brings new life to coral reefs. *The Biological Bulletin,* **191**, 149–154.

Muko, S., and Iwasa, Y. (2011). Long-term effect of coral transplantation: Restoration goals and the choice of species. *Journal of Theoretical Biology*, **280**(1), 127–138.

Nagelkerken, I., Van der Velde, G., Gorissen, M.W., Meijer, G.J., Van't Hof, T., and Den Hartog, C. (2000). Importance of mangroves, seagrass beds and the shallow coral reef as a nursery for important coral reef fishes, using a visual census technique. *Estuarine, Coastal and Shelf Science*, **51**(1), 31–44.

Ng, C.S.L., and Chou, L.M. (2014). Rearing juvenile 'corals of opportunity' in *in situ* nurseries–A reef rehabilitation approach for sediment-impacted environments. *Marine Biology Research*, **10**(8), 833–838.

Ng, C.S.L., Lim, S.C., Ong, J.Y., Teo, L.M.S, Chou, L.M., Chua, K.E., and Tan, K.S. (2015). Enhancing the biodiversity of coastal defence structures: transplantation of nursery-reared reef biota onto intertidal seawalls. *Ecological Engineering,* **82**, 480–486.

Ng, C.S.L., Ng, S.Z., and Chou, L.M. (2012). Does an *ex situ* coral nursery facilitate reef restoration in Singapore's waters? In: Tan, K.S. (ed.), *Tropical Marine Science Institute St John's Island 10th Anniversary Commemorative Volume: Contributions to Marine Science*, pp. 95–100.

Ng, C.S.L., Toh, T.C., and Chou, L.M. (2013). Current status of coral reef restoration in Singapore. *The Asian Conference on Sustainability, Energy and the Environment Official Conference Proceedings 2013*, pp. 546–558.

Ng, C.S.L., Toh, T.C., and Chou, L.M. (2016). Coral restoration in Singapore's sediment-challenged sea. *Regional Studies in Marine Science*, **8**, 422–429.

Ng, C.S.L., Toh, T.C., and Chou, L.M. (2017). Artificial reefs as a reef restoration strategy in sediment-impacted environments: Insights from long-term monitoring. *Aquatic Conservation: Marine and Freshwater Ecosystems* DOI: 10.1002/aqc.2755.

Raymundo, L.J., Maypa, A.P., Gomez, E.D., and Cadiz, P. (2007). Can dynamite-blasted reefs recover? A novel, low-tech approach to stimulating natural recovery in fish and coral populations. *Marine Pollution Bulletin,* **54**(7), 1009–1019.

Rinkevich, B. (1995). Restoration strategies for coral reefs damaged by recreational activities: the use of sexual and asexual recruits. *Restoration Ecology,* **3**(4), 241–251.

Rinkevich, B. (2005). Conservation of coral reefs through active restoration measures: recent approaches and last decade progress. *Environmental Science and Technology,* **39**(12), 4333–4342.

Rinkevich, B. (2015). Climate change and active reef restoration — ways of constructing the "reefs of tomorrow". *Journal of Marine Science and Engineering,* **3**(1), 111–127.

Schopmeyer, S.A., Lirman, D., Bartels, E., Byrne, J., Gilliam, D.S., Hunt, J., Johnson, M.E., Larson, E.A., Maxwell, K., Nedimyer, K., and Walter, C. (2012). In situ coral nurseries serve as genetic repositories for coral reef restoration after an extreme cold-water event. *Restoration Ecology,* **20**(6), 696–703.

Seow, L.A., Ng, C.S.L., Neo, K.Y.D., Chou, L.M., and Tun, P.P.K. (2008). Reef restoration in Singapore's high sediment environment. *Developing ubiquitous practices for restoration of Indo-Pacific reefs (REEFRES)* (student presentations).

Shafir, S., Edwards, A., Rinkevich, B., Bongiorni, L., Levy, G., and Shaish, L. (2010). Constructing and managing nurseries for asexual rearing of corals. *Reef Rehabilitation Manual.* The Coral Reef Targeted Research and Capacity Building for Management Program. St Lucia, Australia, pp. 49–72.

Shaish, L., Levy, G., Katzir, G., and Rinkevich, B. (2010a). Employing a highly fragmented, weedy coral species in reef restoration. *Ecological Engineering,* **36**(10), 1424–1432.

Shaish, L., Levy, G., Katzir, G., and Rinkevich, B. (2010b). Coral reef restoration (Bolinao, Philippines) in the face of frequent natural catastrophes. *Restoration Ecology,* **18**(3), 285–299.

Singapore Department of Statistics (2017). <http://www.singstat.gov.sg/statistics/latest-data#14> (Retrieved 24 February 2017).

Sleeman, J.C., Boggs, G.S., Radford, B.C., and Kendrick, G.A. (2005). Using agent-based models to aid reef restoration: Enhancing coral cover and topographic complexity through the spatial arrangement of coral transplants. *Restoration Ecology,* **13**(4), 685–694.

Spieler, R.E., Gilliam, D.S., and Sherman, R.L. (2001). Artificial substrate and coral reef restoration: what do we need to know to know what we need. *Bulletin of Marine Science,* **69**(2), 1013–1030.

State Lands Encroachments Act (Cap 315, 1985 Rev. Ed. Sing.)

Tan, A. (2015). More than 700 coral colonies relocated successfully from Semakau Landfill to marine park. *The Straits Times,* 20 January 2015.

Tan, H.T.W., Chou, L.M., Yeo, D.C.J., and Ng, P.K.L. (2010). *The Natural Heritage of Singapore 3rd edition.* Prentice Hall-Pearson Education South Asia Pte Ltd, Singapore. 323pp.

Tay, Y.C., Noreen, A.M.E., Chou, L.M., and Todd, P.A. (2015). Genetic connectivity of the broadcast spawning reef coral *Platygyra sinensis* on impacted reefs, and the description of new microsatellite markers. *Coral Reefs*, **34**, 301–311.

Tay, Y.C., Todd, P.A., Rosshaug, P.S., and Chou, L.M. (2012). Simulating the transport of broadcast coral larvae among the Southern Islands of Singapore. *Aquatic Biology*, **15**, 283–297.

The Hantu Bloggers. (2011). 27 Jul (Wed): Discussion on 'The Seacil Artificial Reef' hosted by the Nature Society (Singapore) <http://www.pulauhantu.sg/27-jul-wed-discussion-on-the-seacil-artificial-reef-hosted-by-the-nature-society-singapore/> (Retrieved 23 March 2015).

The Lazy Lizard's Tales. (2011). Saga of the Seacil. <http://lazy-lizard-tales.blogspot.sg/2011/07/saga-of-seacil.html> (Retrieved 23 March 2015).

The Straits Times. (1993). 400 volunteers in project to save corals. *The Straits Times,* 31 May 1993.

Tjan, K.W., and Tan, L.L. (2007). Artificial reef revitalises marine life <http://www.innovationmagazine.com/innovation/volumes/v7n1/feature3.shtml>. *Innovation,* **7**(1). (Retrieved 22 March 2015).

Toh, K.B., Chou, L.M., Tay, Y.C., and Phang, V.X.H. (2012b). The impacts of climatic extremes on coastal and marine biodiversity in Singapore and management challenges. *The Asian Conference on Sustainability, Energy and the Environment Official Conference Proceedings 2012*, pp. 423–430.

Toh, T.C. (2014). The use of sexually propagated scleractinian corals for reef restoration. PhD thesis, Department of Biological Sciences, National University of Singapore. 178pp.

Toh, T.C., Guest, J., and Chou, L.M. (2012a). Coral larval rearing in Singapore: observations on spawning timing, larval development and settlement of two common scleractinian coral species. In: Tan, K.S. (ed.), *Tropical Marine Science Institute St John's Island 10th Anniversary Commemorative Volume: Contributions to Marine Science*, pp. 81–87.

Toh, T.C., Ng, C.S.L., Peh, J.W.K., Toh, K.B., and Chou, L.M. (2014). Augmenting the Post-Transplantation Growth and Survivorship of Juvenile Scleractinian Corals via Nutritional Enhancement. *PloS one*, **9**(6), e98529.

Toh, T.C., Peh, J.W.K., and Chou, L.M. (2013). Early onset of zooplanktivory in equatorial reef coral recruits. *Marine Biodiversity*, **43**(3), 177–178.

Tun, K.P.P. (2012). Optimisation of reef survey methods and application of reef metrics and biocriteria for the monitoring of sediment-impacted reefs. PhD thesis, Department of Biological Sciences, National University of Singapore. 208pp.

Tun, K., Chou, L.M., Low, J., Yeemin, T., Phongsuwan, N., Setiasih, N., Wilson, J., Amri, A.Y., Adzis, K.A.A., Lane, D., van Bochove, J.-W., Kluskens, B., Long, N.V., Vo, S.T., and Gomez, E. (2010). A regional review on the 2010 coral bleaching event in Southeast Asia. In: *Status of Coral Reefs in East Asian Seas Region: 2010.* Ministry of the Environment, Tokyo, Japan, pp. 9–27.

Tun, K., Seow, A., and Chou, L.M. (2008). The influence of overlying sediments and underlying substrate type on the settlement and survivorship of *Pocillopora damicornis* planulae and recruits. *Developing ubiquitous practices for restoration of Indo-Pacific reefs (REEFRES)* (student presentations).

Vermeij, M.J.A. (2006). Early life-history dynamics of Caribbean coral species on artificial substratum: the importance of competition, growth and variation in life-history strategy. *Coral Reefs*, **25**(1), 59–71.

Webster, N.S., Smith, L.D., Heyward, A.J., Watts, J.E., Webb, R.I., Blackall, L.L., and Negri, A.P. (2004). Metamorphosis of a scleractinian coral in response to microbial biofilms. *Applied and Environmental Microbiology*, **70**(2), 1213–1221.

White, A., Chou, L.M., De Silva, M.W.R.N., and Guarrin, F.Y. (1990). Artificial reefs for marine habitat enhancement in Southeast Asia. *ICLARM Education Series 11.* International Center for Living Aquatic Resources Management, Philippines. 45pp.

Wilkinson, C. (1998). *Status of coral reefs of the world: 1998.* Australian Institute of Marine Science, Townsville, Australia. 181pp.

Wilkinson, C. (2008). *Status of Coral Reefs of the World: 2008.* Global Coral Reef Monitoring Network and Reef and Rainforest Research Center, Townsville, Australia. 296pp.

2

WILDLIFE TRADE IN CHINA AND VIETNAM: A COMPARATIVE STUDY OF THE IMPLEMENTATION OF THE CONVENTION ON INTERNATIONAL TRADE IN ENDANGERED SPECIES OF WILD FAUNA AND FLORA(CITES)

SALLIE CHIA-WEI, YANG[1] and LYE LIN-HENG[2]

The Convention on International Trade in Endangered Species of Wild Fauna and Flora ("CITES") is an international legal agreement to ensure that international trade in specimens of wild animals and plants does not threaten their survival. CITES is regarded as one of the most important international conservation instruments. It serves to both facilitate legal, sustainable and traceable trade and to intercept illegal wildlife trade. In this sense, CITES is both a conservation and a trade convention.

Fundamental to the implementation of CITES goals and effectiveness of the wildlife enforcement networks, is the efficacy of the national laws of each member state. This paper seeks to examine and discuss the implementation and enforcement efforts of CITES in China and Vietnam[3] by examining their national legal and regulatory frameworks; their national agencies; and enforcement challenges and issues. Recommendations are made as to possible ways forward for these two countries to improve the implementation and enforcement of their CITES obligations.

Introduction to CITES

The text of the Convention on International Trade in Endangered Species of Wild Flora and Fauna (CITES) was agreed in Washington D.C, USA, on 3 March 1973. It entered into force on 1 July 1975. Since then, some 180 member states are party to CITES. Its aim is to ensure that international trade in specimens of wild animals and plants does not threaten their survival. CITES is regarded as one of the most important international conservation instruments. It serves to both facilitate legal, sustainable

[1]LLB (Hons) London, MSc (Env Mgt) NUS, Director of Governance, Freeland Foundation, Bangkok, Thailand.

[2]LLB (Hons) NUS, LLM (London), LLM (Harvard), Director, Asia-Pacific Centre for Environmental Law (APCEL), Law Faculty, National University of Singapore.

[3]The cutoff date for research was January 2015.

and traceable trade, and to intercept illegal wildlife trade as well as punish offenders. In this sense, CITES is both a conservation and a trade convention.

CITES regulates international trade in endangered species listed in three Appendices by providing for import, export and re-export licensing, permit and certification requirements. CITES regulates international trade in approximately 35,000 species of animals and plants.[4]

Each Party to CITES must designate one or more Management Authorities for its administration of the licensing requirements, and one or more Scientific Authorities for advice on the effect of trade on the status of the species.[5]

CITES provides a framework for each Party, which has to adopt its own domestic legislation to ensure that CITES is implemented at the national level. In 1992, CITES introduced its National Legislation Project to help Parties enact national laws for its implementation.[6]

In 1992, CITES introduced its National Legislation Project, where Parties are assessed based on three categories. Category One is where a Party's legislation is believed to generally meet all four requirements for the effective implementation of CITES;[7] Category Two is where its legislation is believed to generally meet one to three of the four requirements for the effective implementation of CITES; and Category Three is where its legislation is believed generally to not meet any of the four requirements for the effective implementation of CITES.

The methods adopted by CITES to encourage compliance include technical assistance to facilitate capacity building, political missions, warnings, compliance action plans, and as a last resort, recommendations to suspend trade in CITES species if the party concerned

---

[4]The three Appendices are: Appendix I — includes species threatened with extinction. Trade in wild-taken specimens is generally prohibited species and is only permitted in exceptional circumstances; Appendix II — includes species that are not necessarily threatened with extinction but for which trade must be controlled to avoid their becoming so. International commercial trade in Appendix II species is allowed but regulated; and Appendix III — includes species for which a country is asking the Parties to help with its protection. These species that are subject to national protection and where the cooperation of the other Parties is needed to control the trade.

[5]Article IX, CITES.

[6]CoP12 Doc.28, "National laws for implementation of the Convention".

[7]These are: (1) the appointment of national CITES authorities, (2) prohibition of trade in violation of CITES, (3) penalisation of illegal trade, (4) authorization to confiscate specimens illegally traded or possessed.

persists in non-compliance and shows no intention to achieve compliance.[8] The recommendations are made either by the CITES Secretariat or by the Standing Committee. In practice, such decisions are made by consensus.[9]

CITES has been in force for almost forty years, and despite obvious efforts and actions, illegal wildlife trade has been estimated to be worth between US$8 to 10 billion a year,[10] making it a very lucrative criminal activity.

In East Asia and the Pacific alone, the estimated value of the illegal trade in wildlife is US$2.5 billion a year[11] (excluding illegal timber and off-shore fishing).

A conservative estimate for the illegal trade in selected mammals (see Fig. 2.1) is US$400 million, with the ivory trade accounting for more than half of this amount.[12]

In WWF's Wildlife Scorecard,[13] China ranked number two in having the highest level of illegal trade in elephant ivory, rhino horn and tiger parts among twenty-three countries from Asia and Africa.[14] Vietnam ranked number one in WWF's Wildlife scorecard.

In this paper, we shall examine the nature of illegal wildlife trade in China and Vietnam and the efforts and challenges faced by these two countries in the implementation and enforcement of the CITES regime.

---

[8]Trade sanctions are the suspension of commercial or all trade in either particular CITES species or country specific bans (that is, denying recognition to all CITES export permits issued by the targeted country as valid documentation for entry. Such trade sanctions may be recommended by the CITES Secretariat (pursuant to Article XIII of CITES) or the Standing Committee (pursuant to its functions as delegated by the Conference of the Partoes (CoP). See Article XIV.1(a) CITES, codified in the 2007 Guidelines on Compliance with the Convention: CITES Resolution Conf. 14.3, CITES Compliance Procedures (2007), Annex: Guide to CITES Compliance Procedures ('CITES Compliance Procedures').

[9]Reeve, R. (2006). "Wildlife Trade, Sanctions and Compliance: Lessons from the Cites Regime." *International Affairs* (Royal Institute of International Affairs 1944–), **82**(5), 881–97.

[10]Global Financial Integrity, Jeremy Haken, 8 February 2011, "Transnational Crime in the Developing World", http://www.gfintegrity.org/report/briefing-paper-transnational-crime/.

[11]UNDOC (2013). "UNODC Transnational Organized Crime Threat Assessment Chapter 7: The illegal wildlife trade in East Asia and Pacific.", p86.

[12]Ibid.

[13]WWF, K. N. f. (2012). "Wildlife Crime Scorecard."

[14]Ibid, Table 10 p29.

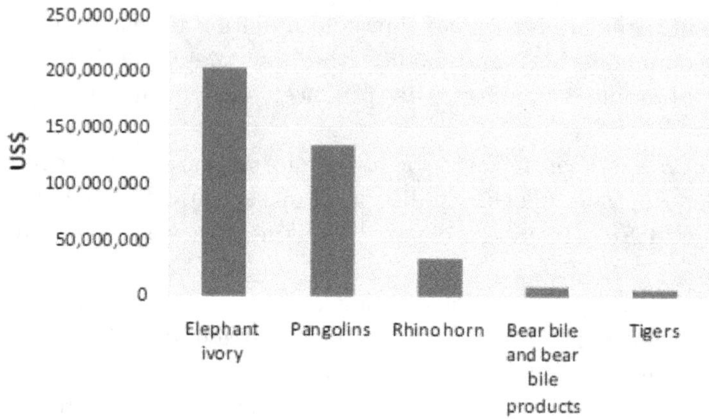

Fig. 2.1. Value of illegal wildlife trade in East Asia and the Pacific[15] (UNODC, 2013).

China: CITES Implementation and Enforcement Efforts

Introduction

China acceded to CITES in 1981 and celebrated its 30th year with CITES in 2011.[16] It is a Category One Party under the National Legislation Project (NLP).[17]

Throughout China's history, wildlife has been hunted for their skin, for food and for medicinal use. The most notorious example is the trade in tiger parts: in particular, tiger skin, bones, teeth, claws and organs.

In recent years, the illegal wildlife trade in China has been driven mainly by demand for food (delicacies), traditional medicine (TCM), crafts and souvenirs, garments, decorations and exotic pets.[18]

The Policy and Legal Framework

In 1981, China, eager to be a part of the international community, joined CITES. China has since enacted a series of statutes, local laws and regulations for the protection of its natural environment.

---

[15] Ibid

[16] http://www.traffic.org/home/2011/4/8/china-celebrates-30-years-of-cites-membership.html.

[17] See n. 6.

[18] Zhang, L. and F. Yin (2014). "Wildlife consumption and conservation awareness in China: a long way to go." *Biodiversity & Conservation*, 1–11.

The precursor to the Wildlife Protection Law 1988[19] is the Circular of the State Council on Stringent Protection for Rare and Precious Wildlife (1983), issued by the State Council. The Circular banned hunting and trade in endangered species, and imposed punishments for illegal activities. The export of rare wildlife and wildlife products without a special permit from the Office for the Management of Import and Export of Endangered Species was also prohibited.

In 1988, the National People's Congress (NPC), enacted the Wild Animals Protection Law.[20] As with its previous position, the State is declared the owner of all wild animals;[21] the Forestry Administration and the Fishery Department of the State Council were the regulatory authorities. The law contains five chapters and 42 articles, the scope of which covers wildlife ownership, scope of protection, management mechanisms, and penalties for violations. The law also provides for wildlife domestication and breeding.[22] Since then, the Wildlife Protection Law has been revised in 2004[23] (hereinafter referred to as the "Wildlife Protection Law").

The Wildlife Protection Law and its Key Provisions

Currently, the main wildlife laws implementing CITES are the Wildlife Protection Law and the Regulations of the People's Republic of China on Administration of Import and Export of Endangered Wild Animals and Plants.[24]

---

[19]Law of The People's Republic of China (PRC) on the Protection of Wildlife (Adopted at the Fourth Meeting of the Standing Committee of the Seventh National People's Congress and promulgated by Order No. 9 of the President of the PRC on November 8, 1988, and effective as of March 1, 1989.

[20]Ibid.

[21]Ante, note 31, Article 3.

[22]Articles 17 and 22, Wildlife Protection Laws.

[23]Law of the PRC on The Protection of Wildlife (Adopted at the Fourth Meeting of the Standing Committee of the Seventh National People's Congress and promulgated by Order No. 9 of the President of the PRC on November 8, 1988, and effective as of March 1, 1989; Revised at the 11th session of the standing committee of the 10th National People's Congress of the PRC on August 28, 2004).

[24]Regulations of the PRC on Administration of Import and Export of Endangered Wild Animals and Plants (Adopted at the 131st Executive Meeting of the State Council on April 12, 2006, promulgated by Decree No. 465 of the State Council of the PRC on April 29, 2006, and effective as of September 1, 2006).

In terms of wildlife trade, the Wildlife Protection Law protects species which are *"rare or near extinction which are beneficial or of important economic or scientific value"*(Article 2).Such species are classified under two classes — species listed in the first class are protected by the national government, and species in the second class, by provincial government protection (Article 9).

The Wildlife Protection Law prohibits the hunting and killing of protected wildlife, as well as their sale and purchase. However, it was only in 1993, under pressure from international communities and in the bid to win the Olympic games hosting rights, that the Chinese government placed the rhinoceros, ostrich, African elephant, zebra, and other non-Chinese species under China's protection.[25]

While the enabling provisions are contained in the Wildlife Protection Laws, the criminalisation of the offences falls under the Criminal Law of the People's Republic of China (1997 revision) (hereinafter referred to as the "Criminal Law").

Penalties

The penalties for criminal violation of the Wildlife Protection Law can be found in China's Criminal Law.[26]

Based on the seriousness of the crime, penalties can range from not more than a three-year jail term to life imprisonment.[27]

In the most recent meeting of the Standing Committee of the National People's Congress (NPC) in April 2014, perhaps under the pressure of the impending 65th SC meeting in July 2014, Articles 312 and 341 of the Criminal Law[28] were adopted to deal with the increasing illegal

[25]Forestry Ministry, "Linyebu guanyu hezhun bufen binwei yesheng dongwu wei guojia zhongdian baohu yesheng dongwu de tongzhi" (Notice on approving the inclusion of certain endangered species in the state-protected lists), Forestry Ministry Notice no. 48, 14 April 1993, <http://vip.chinalawinfo.com/Newlaw2002/SLC/SLC.asp?Db=chl&Gid=9694>.

[26]Criminal Law of the People's Republic of China (1997 revision) Section 6: Crimes of undermining protection of environmental resources. Issuing authority National People's Congress. Date issued 03-14-1997. Note however that other sources of laws, legislative decisions and court ruling do play a part too.

[27]Articles 151, 312 and 341 of the Criminal Laws.

[28]Article 312 provides that *"Whoever conceals, transfers, purchases, or acts as an agent to sell something he clearly knows as booty which have been gained through committing a crime is to be sentenced to not more than three years of fixed-term imprisonment,*

sale of protected wildlife for consumption within China[29] — i.e. that consumers of illegally hunted wildlife who knowingly or should have known that what is purchased or is to be purchased for eating is a rare or endangered wild animal or its product, shall be liable to prosecution and punishment as stipulated under Article 312 or Article 341 of the Criminal Law as the case may be. This is a big step forward as China has a large population of wildlife meat consumers and so far, law enforcers have only been focusing on punishing the suppliers but not the consumers.

Enforcement and Implementation

The Forestry Administration is responsible for the investigation and administration of compliance with the Protection of Wildlife Law.[30] It works together with the industry and commerce, public security, and customs sectors in law enforcement.

The Forestry Police (public security) is responsible for forest wildlife crime while the Anti–smuggling Department of the Customs sector is primarily responsible for the criminal enforcement of laws limiting/prohibiting wildlife smuggling.

In December 2011, a National Inter-agency CITES Enforcement Collaboration Group (NICECG) was established to form a cooperative law enforcement platform for the relevant agencies. Similar groups are also set up at the provincial level.

---

criminal detention, or control and may in addition or exclusively be sentenced to a fine."; Article 341 provides that *"Those who illegally hunt and kill rare and endangered wild animals which are under the state key production plan or illegally purchase, transport or sell those rare and endangered wild animals and their manufactured products are to be sentenced to not more than five years of fixed-term imprisonment or criminal detention, and may in addition be sentenced to a fine. In serious cases, those law offenders are to be sentenced to not less than five years and not more than 10 years of fixed-term imprisonment, and may in addition be sentenced to a fine. In especially serious cases, those law offenders are to be sentenced to more than 10 years of fixed-term imprisonment, and in addition be sentenced to a fine and confiscation of their properties. . . ."*

[29] Interpretation of the Standing Committee of the National People's Congress on Articles 341 and 312 of the Criminal Law of the People's Republic of China Issuing authority Standing Committee of the National People's Congress. Date issued: 04-24-2014. Level of Authority: Legal Interpretation (Adopted at the 8th Session of the Standing Committee of the Twelfth National People's Congress on April 24, 2014).

[30] Article 7.

At regional levels, China is actively engaging with the ASEAN Wildlife Enforcement Network[31] (ASEAN-WEN) by attending annual working meetings and taking part in training practices.

Besides a series of memoranda of understanding with several countries, China has been taking the lead in organising enforcement operations, the most notable being Operation Cobra, which is a one-month transcontinental multinational joint enforcement operation involving twenty-two countries from Asia, Africa and America, carried out between 6 January 2013 and 5 February 2013. The operation resulted in seizures of assorted wildlife specimens and the arrest of more than one hundred criminals in Asia and Africa.[32]

At the 65th CITES Standing Committee meeting, China reported that between 2012 to October 2013, China investigated and dealt with 11,836 cases of destruction of wildlife resources and 14,279 criminals.[33]

From media reports,[34] China has demonstrated that it is willing to take strong actions against wildlife offenders. For example, in 2013, eight Chinese citizens were convicted and sentenced to three to fifteen years' imprisonment in east China's Anhui province for smuggling a total of 3.2 tons of ivory between 2010 and 2012. In Zhejiang Province, ten individuals were sentenced to prison for terms ranging between six and a half to fifteen years. In early 2013, three Chinese citizens in Fujian Province were sentenced to periods of seven to fifteen years' imprisonment for smuggling 7.7 tons of ivory from Africa. On 8 November, 2013, the Supreme

[31]ASEAN-WEN was formed in 2005 pursuant to the ASEAN Regional Action Plan on Trade in Wild Fauna and Flora (2005–2010), and developed and adopted by the Special Meeting of the ASEAN Experts Group on CITES (AEG-CITES) in Jakarta. The establishment of ASEAN-WEN is to address common issues of law enforcement in respect of networking, inter-agency co-operation, strengthening of national legislation and increasing the availability of scientific information to guide wildlife trade management by CITES authorities. It is an integrated network among law enforcement agencies and involves CITES authorities, customs, police, prosecutors, specialised governmental wildlife-law enforcement organisations and other relevant national law enforcement agencies.

[32]http://cites.org/eng/news/pr/2013/20130218_operation-cobra.php; http://www.asean-wen.org/index.php/news-announcements/359-cites-secretariat-praises-asean-wen-and-partner-enforcement-networks.

[33]This was achieved with the collaboration of the various agencies in NICECG. It should be noted that the statistics are incomplete — see Report on the Implementation of Resolution Conf, 12.5 (Rev. CoP16) on Conservation of and trade in tigers and other Appendix-I Asian big cat species — SC 65 Doc 38 Annex 3-p.3; https://cites.org/sites/default/files/eng/com/sc/65/EFS-SC65-38-A03.pdf.

[34]http://www.cites.org/eng/news/sundry/2013/20131128_china_ivory_prosecutions.php.

Court of southern China's Guangdong Province upheld the judgment of the "court of first instance" in an ivory smuggling case, as a result of which two ivory smugglers was jailed for twelve and fourteen years for smuggling 1.04 tons of ivory.

Criticisms and challenges

With these initiatives and successful enforcement operations, why is illegal wildlife trade still proliferating in China?

One of the main criticisms of the Wildlife Protection Law is the fact that it is built upon the principle that the protection of wildlife is primarily for purposes of human use.[35]

With this mindset, the implementation of wildlife laws becomes complicated, particularly in the context of wildlife domestication and breeding. Article 17 of the Wildlife Protection Law actually encourages wildlife domestication and breeding to meet market demands and at the same time ease the pressure on wildlife in the wild.

There is a need to review and update the list of protected wildlife species. There is also a need to expand the protection list and the elevate species from Second Class to First Class. There is a need for more distinct definitions of wildlife species — between those in the wild and those in captivity; and between endangered species and not-yet-endangered species.

While the National Wildlife Protection Department under the State Forestry Bureau is responsible for the management of wildlife, the provincial agencies are the implementing agencies at local levels. Provincial wildlife protection departments are dependent on local officials who determine their budget. Very often, local growth takes priority over wildlife protection. For example, in Shanghai, wildlife protection is not part of the budget proposal.[36] In XingJiang's Aerjin Mountain Reserve, encompassing some 45,000km, there was only a husband and wife team patrolling an area of seven to eight square km per day.[37]

This issue also highlights a fundamental issue with wildlife enforcement: Lack of funding. The Chinese government needs to allocate higher budgets to improve and sustain wildlife protection and enforcement efforts.

[35]Li, P. J. (2007). "Enforcing Wildlife Protection in China: The Legislative and Political Solutions." *China Information*, **21**(1), 71–107.
[36]Ibid.
[37]Ibid.

One way to counter this is for the State government to impose wildlife protection key performance indicators ("KPIs"), and not merely economic KPIs on local officials.

Corruption is another institutional issue that China faces. According to Transparency International 2013, China ranked 80/177 and scored 40 out of 100.[38] This means that China has a serious problem with corruption. This makes it easier for wildlife traffickers to remain undetected or unpunished.

Vietnam: CITES Implementation and Enforcement Efforts

Introduction

Vietnam celebrated its 20th year with CITES in May 2014. It has achieved Category One status under CITES. This mean that Vietnam's legal framework has fully complied with and addressed the four requirements under the National Legislation Project for successful implementation of CITES.[39]

Yet Vietnam is both a major consumer country and an important trade conduit to China. In 2000, the estimated revenue generated by the illegal wildlife trade in Vietnam totalled US$67 million — more than 12 times the value of the legal wildlife trade in that country. On average, 3,500 to 4,000 tons of illegal wildlife (foodstuffs and forest products) are trafficked in and out of Vietnam each year.[40]

There are several factors leading to the increase in illegal wildlife trade:

1. High domestic and international demand for wildlife meat and products;
2. Lack of data on volume and distribution of endangered species;
3. Lack of resource and funding;
4. Corruption.

Like China, wildlife consumption in Vietnam has been part of its customary practice. The species include snakes, palm civets, monitor lizards, porcupines, leopards, pangolins, monkeys, forest pigs, hard-shell turtles, soft-shell turtles, civets, boas, and birds. Of these, the most common

[38]Transparency International, http://www.transparency.org/country#idx99.
[39]See n. 8.
[40]Nguyen, V. S. (2008). "Wildlife Trading in Vietnam Situation, Causes, and Solutions." *The Journal of Environment & Development* **17**(2), 145–165.

and largest in volume are snakes, civets, forest pigs, and birds.[41] Most illegal wildlife trade go through the Vietnam-China border.[42]

Vietnam's Wildlife Policy and Legal Framework

Since its accession to CITES in April 1994, Vietnam has been developing its policy on the management and protection of wildlife.

Vietnam currently meets its obligations under CITES through three legislative instruments relating to international wildlife trade and one for domestic wildlife exploitation and trade: Decree No.82/2006/NĐ-CP (relating to CITES requirements for international trade generally), Decree No.59/2005/NĐ-CP (relating to export of endangered aquatic species), Circular No.02/2006/TT-BTS (containing instructions on implementing No.59/2005/NĐ-CP), and Decree No. 32/2006/ND-CP (domestic exploitation and trade).

The main legal document to regulate domestic wildlife exploitation and trade is Decree 32/2006/NĐ-CP (2006).[43] It is illegal to hunt, transport, keep, advertise, sell, purchase and consume rare and endangered species or their parts and derivatives.

International wildlife trade and businesses in Vietnam are regulated under Decree No.82/2006/NĐ-CP. This Decree issued by the Ministry of Agriculture and Development is the main instrument to implement CITES. It covers all species in Appendices I, II and III of CITES and applies to all the specimens of endangered species of precious and rare wild fauna and flora in Vietnam.[44]

The penalties for violation of these laws are set out in Article 190 of The Penal Code No. 15/1999/QH10 (hereinafter referred to as the "Penal Code").

[41] Ibid, p. 154.

[42] Supra note 75, p. 161.

[43] Its predecessor was Decree No.18/HĐBT (1992), and was based on Forest Protection and Development Law (1991) It was the first legal document to provide definitions for endangered and common wildlife. It was subsequently amended by Decree 48/2002/NDP (2002) to update the list of protected species. The new Decree 32/2006/NĐ-CP became necessary and was also timely in replacing its predecessors as the latter decrees were outdated and limited in their effectiveness. It also removed duplication of national law — for example fishery species such as sea turtles and fish species were removed to avoid overlap with the legislation under the Ministry of Fishery; http://vbqppl.moj.gov.vn/vbpq/en/Lists/Vn%20bn%20php%20lut/View_Detail.aspx?ItemID=5787.

[44] Article 25(3), Decree No.82/2006/NĐ-CP.

When the amendments to the Penal Code came into force in 2009,[45] it became a criminal offence to breach regulations on the protection of animals in the List of Endangered, Precious, and Rare Species Prioritized for Protection.[46] Accordingly, the acts of illegal hunting, killing, transporting, smuggling, raising and keeping of endangered, rare and precious wild animals prioritised for protection and the act of illegal transporting and smuggling of their products/ body parts are prohibited.

Depending on the severity and circumstances of the offences, the offenders are subject to a fine from VND 50,000,000 (approximately US$2,400) to VND 500,000,000 (approximately US$24,000),[47] non-custodial reform for up to three years or a prison term up to seven years.

Enforcement and Implementation

Many agencies and branches of government are involved in Vietnam's wildlife management and law enforcement structure. The CITES Management Authority, under Vietnam Administration of Forestry in the Ministry of Agriculture and Rural Development, is responsible for CITES management. The Ministry of Public Security's Environmental Police are charged with investigation, while the Ministry of Finance's Department of Customs plays a key role monitoring the import and export of goods.

The Vietnam Wildlife Enforcement Network (Vietnam-WEN) was established in 2010 to improve the monitoring and enforcement capacity of the central government on wildlife law enforcement activities. The members of Vietnam-WEN include representatives from the CITES Management Authority, Forest Protection Department, Environmental Police, General Department of Customs, Border Security Department, Biodiversity Conservation Agency and the Supreme People's Procuracy. The Vietnam-WEN has identified its priorities as monitoring the illegal trade on Asian big cats, elephants, pangolins, and rhino specimens.

---

[45] Law Amending and Supplementing a Number of Articles of the Penal Code (No. 37/2009/QH12) Pursuant to the 1992 Constitution of the Socialist Republic of Vietnam, which was amended and supplemented under Resolution No. 51/2001/QH10; The National Assembly promulgates the Law Amending and Supplementing a Number of Articles of Penal Code No. 15/1999/QH10.

[46] Article 190, Penal Code.

[47] The 2015, GDP per capita for Vietnam was US$2,110.9 according to World Bank. http://data.worldbank.org/indicator/NY.GDP.PCAP.CD?locations=VN&name_desc=true.

It is difficult to obtain comprehensive prosecution data. By its own admission at the CITES 65th Standing Committee meeting (2014), Vietnam acknowledged that achieving successful prosecutions with effective deterrent sentences was an area that has had limited success to date and needed improvement. Vietnam was only able to provide seizure numbers in its report to the Standing Committee, CITES.

The conclusions from the findings of Education for Nature (ENV)[48] were:

1. The 93 major cases reviewed did not include any major criminal figures that were known to the authorities to be involved in illegal smuggling and trade of protected wildlife. The offenders were low-level criminal operators such as hunters, drivers, and middlemen.
2. The current punishment regime will not act as a deterrent as two-thirds of the prosecuted defendants did not serve jail time.

A possible explanation for the inadequate punishment imposed is the inherent flaw in the implementation of the relevant regulations. The current regime was developed based on violations relating to timber and plants, which differs from offences relating to wild animals. For example, the level of violation is calculated by weight and price. However, this becomes illogical when assessing the punishment for trafficking in animals, as the severity for smaller species should not be less than for larger species. The criteria need to be amended to assess punishment based on category of species.[49]

Challenges

As we can see from the above discussion, Vietnam is facing a number of criticism and challenges, namely inconsistent policies and legislation, inadequate penalties and institutional challenges such as corruption.

One of the criticisms levelled at the legislation of Vietnam regarding wildlife protection is the myriad of legal documents, issued at different

[48]In 2014, Education for Nature (ENV) had published a summary report entitled "An evaluation of the outcomes of prosecutions for serious wildlife crimes in Vietnam", which reviewed the criminal prosecutions involving rare and endangered wildlife species that are currently listed in Group 1B of Decree 32/2006 and are fully protected under Vietnamese law. The review was carried out to evaluate the outcome of these criminal cases and assess whether the punishment that was applied for serious criminal offenses involving wildlife may be effective in deterring criminal behaviour and meeting the full protection needs of Vietnam's most critically endangered species as intended under the law.

[49]CITES, V. (2008). "Report on Vietnam's Wildlife Trade Policy.", p58.

levels, by different ministries on a piece-meal basis, creating overlaps and ambiguity in the process.

Further, the terms use in the different legal documents should be consistent. For example, the term "rare" and "endangered" species have been used differently in different legal documents: e.g. "rare forest plants and animals",[50] "rare wild animals and plants",[51] "rare and endangered forest animals and plants",[52] "rare endangered wild animals and plants"[53] and "endangered species which face extinction".[54]

Current penalties for wildlife crime are too lenient. As discussed earlier, the penalties range from a fine from fifty million VND (US$480) to five hundred million VND (US$24,000), non-custodial reform for up to three years or a prison term up to seven years. This is highly inadequate, given the fact that wildlife trafficking has been increasingly run by syndicated criminals who also traffic drugs and humans, using the same network and resources. Further, the penalty imposed depends on the seriousness of the offence. There is lack of consistency and understanding between the prosecution and judiciary in terms of what amounts to a serious offence.

Adding to the above challenges is an institutional issue; Vietnam has been rated as having high levels of corruption (serious corruption problem) — scoring 40 points in the public sector and ranked at 116 out of 177 countries surveyed[55] (Transparency International 2013), an issue that is known to contribute to the proliferation of wildlife crime.[56]

With the growing affluence of the middle class working population, the demand and consumption of wildlife have increased. There is an urgent need to create awareness and educate them on illegal wildlife trade and the consequences.

[50]Decree No.18-HĐBT (1992) (has since been superseded).

[51]No.48/2002/NĐ-CP (has since been superseded).

[52]Decree No.32/2006/NĐ-CP.

[53]The National Action Plan on Biodiversity towards 2010 and orientation towards 2020 for the implementation of CITES and Cartagena Protocol on Biological Safety.

[54]Ibid.

[55]Transparency International 2013, http://www.transparency.org/country#VNM.

[56]Horne, D. 2013, 'Policy Responses to Transnational Wildlife Crime in the Asia-Pacific Region, Part 1: Global and Regional Policy Context and a Potential Framework for Optimal National Policy', Transnational Environmental Crime Project Working Paper 1/2013 (Canberra: Department of International Relations Australian National University).

Implementation of CITES in China and Vietnam: The Issues

In many ways, China and Vietnam share the same profile and challenges when it comes to illegal wildlife trade.

In terms of achievement, both are Category One countries under CITES but both are facing pressure from the proliferation of illegal wildlife trade perpetrated by syndicated criminals.

There are inconsistencies in the laws, which need to be revised and upgraded. For both China and Vietnam, the many layers of hierarchy of structure and laws issued by different ministries and agencies have created a situation where there is confusion because of overlapping jurisdictions and inconsistencies in the laws and terms used.

In the case of Vietnam, policies and legislation are issued in piecemeal, leading to overlaps and incompatibility between the different ministries and levels of authorities. This makes monitoring and evaluation difficult. It would appear that there is currently no evaluation or review process to assess the implementation, effect and impacts of these policies.[57]

An example of inconsistency in China's policy and laws can be found in the captive breeding of tigers. This is permitted by law but trade in tiger parts is not. If conservation is truly the goal of tiger breeding facilities, it can be done in designated government-run breeding centres with a definitive policy on rehabilitation and conservation. The success story of bringing the Giant Panda back from the brink of extinction is a good example. It is noted that panda parts are not sought after in traditional Chinese medicine (TCM), unlike tiger parts. Allowing tiger breeding (where the clear motivation is for economic gain) while making trade in tiger parts illegal only drives the market underground and does not eliminate it. China should acknowledge these inconsistencies and repeal Articles 17 and 22 of the Wildlife Protection Laws and ban wildlife breeding altogether, at the very least for CITES Appendices I and II species. If there is a genuine need to breed for conservation purposes, it can be done by establishing special units such as those set up for the Giant Panda.

As stated earlier, one aspect of wildlife protection that has been neglected in the wildlife laws of both China and Vietnam, is that of the welfare of wildlife. Whether we are dealing with captive tigers, or wildlife being transported whether legally or illegally, many of these animals suffer

[57] CITES Report on Vietnam's Wildlife Trade 2008.

or die as a result of stress and poor conditions. Penalties for causing harm and suffering to animals should be treated as part of the overall legal policy for the protection of wildlife.

In terms of wildlife consumption habits, both China and Vietnam have a long history of eating wildlife, especially along the China-Vietnam border areas. The two countries have a long tradition of using similar wildlife as traditional medicine. In fact, in Vietnam, traditional Chinese medicine is referred to as "North medicine". Both countries use bear bile, tiger parts and rhino horns for their purported medicinal properties. Rhino horn in particular has the reputation of a miracle cure-all, and a cure for cancer, which has driven the illegal trade in rhino horns to a new level.[58]

From a social point of view, both countries have emerging newly rich (typically young middleclass, high income, well-educated) consumers who are eager to show off their new found status in the form of consuming, keeping or using expensive and rare wildlife (in the case of exotic pets) and wildlife products.

On the issue of wildlife breeding, both China and Vietnam are proponents of wildlife domestication and breeding. In their view, the domestication and breeding of wildlife will boost the countries' income and at the same time alleviate the pressures on wildlife. Wildlife breeders/farmers are given tax incentives to set up wildlife breeding farms in Vietnam. In China, it is provided under the law that wildlife breeders should be rewarded, and awards for achievements in wildlife breeding are given.[59] Many species, protected and unprotected (snakes, turtles, porcupines, palm civets, crocodiles) are farmed for commercial use in China and Vietnam. If we are to tackle the issue of wildlife domestication and breeding, we shall have to deal with the appetites of these two countries.

The involvement of syndicated criminals in wildlife trafficking has been recognised to be on the rise. For high valued wildlife products, such as ivory, rhino horns and tiger parts, syndicated networks are increasingly involved[60] as trading in high valued wildlife products is lucrative and does not attract as high a penalty as trafficking in drugs and humans. Many

[58]'Cure for Cancer' rumours killed off Vietnam's rhinos, The Guardian, 25 November 2011; https://www.theguardian.com/environment/2011/nov/25/cure-cancer-rhino-horn-vietnam; see also "Vietnam's Illegal Trade in Rhino Horn" 9 February 2014; http://www.bbc.com/news/magazine-26081168.

[59]Article 17, Wildlife Protection Law.

[60]Zimmerman, M. E. (2003). "The Black Market for Wildlife: Combating Transnational Organized Crime in the Illegal Wildlife Trade." *Journal of Transnational Law*, **36**, 1657.

countries have yet to enact effective laws with penalties that reflect the severity of wildlife crimes. Thus, trafficking in wildlife is an activity that is low-risk with high returns. Because the trade routes are similar, as the trafficking of wildlife involves similar intricate networks of smugglers and routes, trafficking in wildlife becomes a convenient and natural add-on to the illegal activities of criminal syndicates. Syndicated criminals have a large network that facilitate wildlife trafficking. As many of the countries involved are developing countries, bribery, intimidation and forged documents are commonly used and seldom detected.[61] The advent and convenience of Internet trading make such transactions easier, borderless and anonymous. According to the International Fund for Animal Welfare (IFAW),[62] its online monitoring found nearly 18,000 ivory products listed on Chinese websites.[63]

This is a problem shared globally in the efforts to counter wildlife trafficking as well as other crimes perpetrated by organised criminals, such as human trafficking, drug trafficking and the smuggling of counterfeit goods.

Recommendations: Going Forward

Demand Reduction

For both China and Vietnam, the recent profile of consumers of wildlife is the young, well-educated, high-income, urban-dwelling population.[64] These persons are motivated by different reasons and respond to peer pressure. Therefore, the approach has to be different — such as using key opinion leaders in public service announcements to influence their lifestyle choices.

Then of course there is the next generation: their children. It is important to incorporate conservation and environmental education into the school curriculum. It has also been suggested that education programmes on wildlife conservation should be developed for university students.[65]

---

[61]Ibid at p. 1679.

[62]IFAW, Ending Wildlife Crime online.

[63]http://www.ifaw.org/united-states/our-work/wildlife-trade/ending-wildlife-crime-online; For a detailed report of an investigation conducted by IFAW in 2008 please see IFAW publication, "Killing with key strokes: An Investigation of the Illegal Wildlife Trade on the World Wide Web."

[64]Gwin, P. (2012). Rhino Wars. National Geographic, March 2012. National Geographic Society, Washington, DC., pp. 106–125.

[65]Zhang, L., *et al.* (2008). "Wildlife trade, consumption and conservation awareness in southwest China." *Biodiversity & Conservation*, **17**(6), 1493–1516.

Improve Efficacy of CITES Laws

Given the value of the illegal wildlife trade, the fines and penalties need to be revised upwards in order to have a deterrent effect on the criminals.[66]

Given the high black market value of the illegal trade, the laws need to take into consideration:

1. Not having any limitations/caps on the fines imposed;
2. Calculating the penalties based on the black market value of the illegal wildlife on a species specific and per specimen basis;
3. Empowering the judiciary to impose additional damages to be paid by the offenders to compensate for the costs [67] incurred in the repatriation, rehabilitation, loss of wildlife and other conservation and compensatory matters.

While there are laws both in Vietnam and China dealing with possession and consumption of illegally wildlife, much of the laws focus on the perpetrators of illegal wildlife trade; the consumers have not been targeted by the law enforcers. If consumers are also prosecuted, and severe sentences are imposed by the courts, this will send a strong message to the public of the gravity of wildlife crime, resulting in a fall in demand for wildlife and their products.

Going forward, it is essential for both Vietnam and China to build capacity in their judiciary in wildlife and conservation laws. While China has established many "green" or environmental tribunals, it would appear that the focus is on pollution and not nature conservation.[68]

Use of Other Laws and Inter-agency Co-operation

Wildlife crime also requires that one looks at other laws and agencies. This is especially true in the case of transnational wildlife crime. This

---

[66]The GDP per capita (current US $) is 2,110.9 for Vietnam and 8,069.2 for China according to World Banks' 2015 figures: http://data.worldbank.org/indicator/NY.GDP.PCAP.CD?locations=VN-CN& name_desc=false. http://data.worldbank.org/indicator/NY.GDP.PCAP.PP.CD? order=wbapi_data_value_2013+wbapi_data_value+wbapi_data_value-last&sort=desc

[67]Please also see 5.8 of this Chapter for discussion on restitutionary damages and creative sentencing.

[68]Chang Chun, "Growing pains for China's new environmental courts", 17 June 2015; https://www.chinadialogue.net/article/show/single/en/7972-Growing-pains-for-China-s-new-environmental-courts; Yuqian Peng, "Understanding the emergence of China's environmental courts", April 14, 2016, http://environment.yale.edu/yer/article/understanding-the-emergence-of-chinas-environmental-courts#gsc.tab=0.

is particularly helpful where the penalties under wildlife laws may be insufficient or where another agency may have more resources and capacity to investigate, arrest and prosecute the offenders. Laws such as customs laws, anti-money laundering laws, and anti-corruption laws are particularly useful as they are often linked to organised crimes and the penalties are much higher. For example, violation of corruption laws in both China and Vietnam could lead to life imprisonment and capital punishment.

With wildlife trafficking increasingly being transnational in nature, and given the high value and profit margins of such illegal trade, the necessity for robust policies and laws on anti-money laundering becomes more pertinent. Fortunately, both China and Vietnam,[69] are parties to the Financial Action Task Force on Money Laundering (FATF),[70] which has developed a series of recommendations that are recognised as "the international standard" for combating money laundering, amongst other things. According to the evaluation report by FATF,[71] China has made significant progress since it became a full member in 2007. Vietnam has in fact done so well in establishing a legal and regulatory framework to meet its commitments that it is no longer on FATF's global AML/CFT compliance process.[72]

A good case example on how anti-money laundering agencies can help combat wildlife crime is the Chiamat case in Thailand.[73] In May 2014, following the arrest of Kampanart Chaimat by the Thai Royal Police, Thailand's Anti-Money Laundering Office (AMLO) seized over 1 billion Thai baht (US$36.5 million) from a wildlife trafficking syndicate, marking the first ever multi-million-dollar seizure from wildlife

---

[69]Vietnam's participation through membership with the Asia-Pacific Group on Money Laundering (APG): http://www.fatf-gafi.org/pages/asiapacificgrouponmoney launderingapg.html.

[70]The Financial Action Task Force (FATF) is an inter-governmental body established in 1989 by the Ministers of its Member jurisdictions. The objectives of the FATF are to set standards and promote effective implementation of legal, regulatory and operational measures for combating money laundering, terrorist financing and other related threats to the integrity of the international financial system. The FATF is therefore a "policy-making body" which works to generate the necessary political will to bring about national legislative and regulatory reforms in these areas: For more information, see http://www.fatf-gafi.org/pages/aboutus/.

[71]FATF Follow-up Report to the Mutual Evaluation Report of China: http://www.fatf-gafi.org/countries/a-c/china/documents/follow-upreporttothemutualevaluationreportof china.html.

[72]http://www.fatf-gafi.org/countries/u-z/vietnam/documents/fatf-compliance-feb-2014. html.

[73]http://www.freeland.org/#!wildlife-trafficking-gang-floored-by-maj/czys.

criminals in Asia following close co-operation and enforcement efforts by both agencies. Kampanart and his sister were known wildlife traffickers but eluded conviction until the arrest and seizure by the AMLO. This kind of inter-agency cooperation sets the standard on how effective inter-agency co-operation can be and how non-CITES laws can be applied.

Consultation with Stakeholders

Very often, key stakeholders, such as rural villagers with rights to harvest and use wildlife and who depend on wildlife hunting for a living, are not consulted nor considered in the national policy. Authorities need to be sensitive to the needs of these communities and ensure that wildlife conservation policies involve local communities as partners in safeguarding these valuable resources.[74]

Forensic Capability and Capacity

One of the problems of criminal syndicates getting away with illegal trade is the lack of traceability of the illegal item. How does one distinguish whether an item is illegal or legal? This is made even more difficult when a country decides to sell confiscated items. Often this has resulted in an increased demand for the product, thereby allowing into the market illegal items that are mixed with legitimate items or passed off as such[75] (for example, ivory in China). There is a need for control and monitoring mechanisms. Capacities for DNA testing, micro-chipping and other forensic methods need to be built as criminals, especially from syndicates, are masters at evolving innovative methods of avoiding detection. China has started using sniffer dogs in the tracing of illegal wildlife smuggling at custom points. It is also micro chipping its captive tigers in the hope of ensuring that captive tiger parts do not leak into the market. However, the tracing of movement of illegal items within the country is still lacking. This is particularly true in the case of illegal tiger parts.

---

[74]See IUCN Guidelines for applying Protected Area Management Categories https://cmsdata.iucn.org/downloads/iucn_assignment_1.pdf.

[75]Solomon Hsiang and Nitin Sekar "Does Legalization Reduce Black Market Activity? Evidence from a Global Ivory Experiment and Elephant Poaching Data", NBER Working Paper No. 22314 June 2016 http://www.nber.org/papers/w22314.pdf.

Recognizing Wildlife Trafficking as a Transnational Organized Crime: Regional and International Collaboration

It is important to recognise that wildlife trafficking has evolved into a lucrative trade for organised criminals. As such, the role that ASEAN plays in combatting transnational wildlife crime is crucial. The ASEAN Wildlife Enforcement Network (ASEAN-WEN) was formed in 2005[76] to address common issues of law enforcement, particularly networking, inter-agency co-operation, strengthening of national legislation and increasing the availability of scientific information to guide wildlife trade management by CITES authorities. It is an integrated network among law enforcement agencies and involves CITES authorities, customs, police, prosecutors, specialised governmental wildlife-law enforcement organisations, and other relevant national law enforcement agencies. This ten-member network seeks to facilitate cross-border collaboration to fight the region's illegal wildlife trade.

As a concerned party, China has been invited to ASEAN-WEN's annual meetings to share information and ideas. It is important that China continues to participate, if not take the lead in such meetings and initiatives in order to tackle transnational organised wildlife trafficking. The success of Operation Cobra is evidence of such needs.

Realising the need for coordination amongst different agencies internationally, the International Consortium on Combatting Wildlife Crime (ICCWC) was formed in 2010 as a collaboration between five inter-governmental organisations working to bring coordinated support to national wildlife law enforcement agencies and to sub-regional and regional networks that act in defence of natural resources. The five inter-governmental organisations are the CITES Secretariat, INTERPOL, the United Nations Office on Drugs and Crime (UNODC), the World Bank and the World Customs Organization. Together with its partners, ICCWC published the ICCWC Wildlife and Forest Crime Analytic Toolkit to strengthen national enforcement capacities to fight wildlife crime.[77]

---

[76]Pursuant to the ASEAN Regional Action Plan on Trade in Wild Fauna and Flora (2005–2010)[76], developed and adopted by the Special Meeting of the ASEAN Experts Group on CITES (AEG-CITES) in Jakarta.

[77]http://www.unodc.org/documents/Wildlife/Toolkit_e.pdf.

One aspect that can definitely help Vietnam and China in their implementation and enforcement of their CITES obligations is collaboration with non-government organisations (NGOs). While there are NGOs that focus on highlighting the inadequacies of governments and putting pressure on them to make changes and improve, there are other NGOs that collaborate with governments to address demand issues, build capacity, train enforcement units, and provide technical support. In many instances, the NGOs also procure the funds for these activities.

There are organisations that provide specialised training for border petrol/rangers and investigation teams. For example, Freeland and UNODC jointly conducted the PATROL program[78] to train border enforcement officers on tackling trans-border crime.

These are but a few examples of how the role of NGOs can and have evolved. Increasingly, NGOs are recognised to be valuable partners and collaborators in a government's fight against wildlife crime. China and Vietnam should continue this collaboration to help them effectively implement and enforce their respective CITES obligations.

Establishing bilateral and multi-lateral agreements with countries involved in specific issues is another way of encouraging international or regional co-operation. For example, to deal with the issue of the poaching of rhinos for their horns, both China (on 27 March 2013) and Vietnam (on 10 December 2012) have separately signed a memorandum of understanding (MOU) with South Africa. While both agreements are designed to promote cooperation between South Africa, China and Vietnam on broader issues of conservation, they are particularly aimed at curbing rhino poaching through cooperation in law enforcement, compliance with international conventions and other relevant legislation on the basis of equality and mutual benefit, reaffirming the goals of CITES.

Vietnam has also signed MOUs with Lao PDR[79] and Indonesia[80] to improve cooperation in forest protection, the enforcement and control of wildlife, and timber trade.

While Vietnam and China have signed trade agreements, they have yet to sign a definitive wildlife-related cooperation agreement. Given the

[78]https://www.unodc.org/southeastasiaandpacific/en/patrol.html.

[79]Signed on 12 July 2012 to improve the cooperation between the two countries in forest protection, enforcement and control of wildlife and timber trade.

[80]Signed on 27 December 2012 to strengthen the cooperation between enforcement agencies of the two countries in terms of information sharing, capacity building, law enforcement and awareness raising.

proximity of these two countries and commonalities when dealing with illegal wildlife trade, it is crucial that concrete steps be taken to collaborate and harmonise their laws and enforcement efforts.

Tackling Corruption

As mentioned earlier, in many instances, part of the problem contributing to the proliferation of illegal wildlife trade and the lack of enforcement efforts is corruption. China and Vietnam have problems with public sector corruption and will need to ensure that this is effectively addressed.

For China, its penalties for corruption are severe, attracting the death penalty in the most serious case. It is encouraging to see that the Chinese government, under the leadership of President Xi Jinping, has maintained the momentum in investigating and prosecuting corrupt officials over the past two years. The successful prosecution and conviction of a high level official, Bo Xilai, to life imprisonment for bribe-taking[81] showed the determination of this anti-corruption campaign in China and was the start of a series of investigations and prosecutions on several high profile corrupt officials.[82] This shows the importance of political will.

On 9 December 2013, the Vietnamese government, together with the World Bank and other partners, launched the Vietnam Anti–Corruption Initiative Program 2014 under the theme "Transparency, Integrity and Accountability". Since then, Vietnamese courts have held at least 278 trials and prosecuted corporate corruption, with several convictions (of former officials) leading to death penalties.[83]

While this is to be applauded, the motivation behind these high profile cases was to improve China and Vietnam's corporate credibility in their economic development. What is needed is to elevate the role of corruption in transnational organised crimes involving wildlife trafficking so that priority will be accorded to investigating and punishing corrupt officials involved in illegal wildlife trade.

Funding and Resources

All the above measures need funding.

[81] http://edition.cnn.com/2013/09/21/world/asia/china-bo-xilai-verdict/.

[82] http://www.scmp.com/news/china/article/1678586/chinas-anti-graft-agency-makes-crackdown-political-factions-and.

[83] The Economist, 27 June 2014: Graft-busting, Wrist–slapping. Available at: http://www.economist.com/blogs/banyan/2014/06/graft-busting-vietnam.

Many of the issues are due to inadequacies in the laws, as well as the country in question's lack of proper enforcement capacity. There is a need to enhance penalties, to prosecute not just the trafficker but also the consumer, to reverse the onus of proof and to build capacity in the enforcement agencies. There is also a need to empower the courts to award additional damages and/or restitution against the perpetrators and to direct the proceeds of crime to wildlife-related funds.[84]

A possible solution is to establish a conservation fund whereby proceeds from illegal wildlife trade and fines can be ordered to be paid into (this fund). The fund shall be used to provide better facilities for seized wildlife, reward informants, and strengthen the capacity of enforcement officers towards combating wildlife crime.[85]

Conclusion

It is clear from the discussions in this paper that notwithstanding that China and Vietnam are accorded CITES Category One status in terms of legislative implementation, both countries still have many inadequacies when implementing the CITES regime in practice.

This is a criticism one can level on the CITES regime, as it gives the Category One countries a false sense of compliance. However, it should be noted that the main premise of the CITES regime is not one of criminal justice and law enforcement but rather one of providing an international framework for *"defining and implementing national wildlife trade policies"*.[86] If we revisit the four requirements for implementation of CITES, this becomes clear:

(1) Appointment of national CITES authorities — this requirement only ensures that under the national laws, the CITES authorities have the

[84]See Lacey Act (16 U.S.C. §§3371–3378); Mandatory Victims Restitution Act, 18 U.S.C. §3663A; Victim and Witness Protection Act, 18 U.S.C. §3663; United States v. Bengis, 43 ELR 20133?No. 1-03-cf-00308, (S.D.N.Y., 06/14/2013) (Kaplan, J.); Section 84.1, Wildlife Act [RSBC 1996] Chapter 488.

[85]In the Philippines, section 29 of the Republic Act No. 9147 (2001) provides for the establishment of a Wildlife Management Fund for the financing of rehabilitation or restoration of habitats affected by acts committed in violation of this Act. This is funded by fines imposed and damages awarded (amongst other contribution).

[86]In CITES CoP12 Doc 28, Twelfth meeting of the Conference of the Parties Santiago (Chile), 3–15 November 2002, Interpretation and Implementation of the Convention, General Compliance issues National Laws for Implementation of the Convention, it was clarified *that "CITES-implementing legislation should not be seen as a burdensome...*

necessary powers to carry out their responsibilities, such as granting permits and certificates, and have power to establish export quotas;
(2) Prohibition of trade in violation of CITES — this requirement only prohibits trade of endangered species without a valid permit;
(3) Penalisation of illegal trade — the premise of this requirement is the penalisation of trade without a valid permit;
(4) Authorisation to confiscate specimens illegally traded or possessed — this is an ancillary requirement, which does not have an impact on preventing illegal trade in endangered species.

The main deterrent mechanism for CITES is thus trade sanctions, which have limited effect, as it requires consensus at party level and is used only as a last resort Since 1985, China was only embargoed once in 1993[87] (which resulted in China's tiger trade ban policy in 1993) and Vietnam in 2002[88] for non-compliance. Further, such sanctions may instead drive the wildlife trade underground and fuel the black market.

Although the obvious recommendation is for CITES to review and make the implementation requirements more stringent, one must also recognise that with just the current four requirements, 93 out of 180 CITES parties (that is more than 50%) are still in Categories 2 or 3 in their legislative progress for implementing CITES.[89] It would not be realistic at this point to raise the bar of compliance.

Besides having adequate laws, it is clear that as shown in Operation Cobra I and Cobra II, that enforcing wildlife laws requires multi-agency and multi-regional co-operation to achieve results. This is particularly pertinent as organised criminals get involved in illegal wildlife trade. The formation of the International Consortium on Combating Wildlife Crime is an indication of such recognition by CITES.

---

but rather as the necessary framework for defining and implementing national wildlife trade policies for the conservation of and non-detrimental trade in all CITES-listed species. Legislation sets forth what citizens and enterprises are allowed to do in relation to the international trade in such species, that is, what behaviour is legal or illegal in the context of CITES."

[87] 30th Standing Committee meeting, Decision No. 6. See CITES Notification to the Parties No. 1993/774, Decisions of the Standing Committee on Trade in Rhinoceros Horn and Tiger Specimens (15 October 1993).

[88] The ban was lifted at the 46th Standing Committee meeting upon Vietnam's national legislation for implementing CTIES entering into force on 7 February 2002.

[89] CITES SCC65 Doc 22, Sixty-fifth meeting of the Standing Committee Geneva (Switzerland), 7–11 July 2014, Interpretation and implementation of the Convention Compliance and enforcement, National Laws for Implementation of the Convention.

On the political front, what is encouraging is the recent commitments made by high-level government ministers to combat transnational wildlife crime.

On 12 and 13 February 2014, representatives from 46 countries (including China and Vietnam) and 11 international organisations attended the London Conference on the Illegal Wildlife Trade, hosted by the Government of the United Kingdom (UK) and the British Royal Family in London to inject further high-level political commitment into efforts to tackle wildlife crime. The representatives signed a declaration with a series of action plans to combat wildlife crime and to elevate wildlife crime to a "serious crime" pursuant to the United Nations Convention against Transnational Organized Crime (UNTOC).[90] This requires a minimum prison sentence of four years. Not many laws implementing CITES have this penalty.

On 11 November 2014, at the 22nd APEC[91] Economic Leaders' Meeting, the leaders declared their commitment to combat illegal wildlife trade under paragraph 39 of the declaration:

> *"39. We commit to continue our efforts in combating wildlife trafficking. We will take steps to combat wildlife trafficking by enhancing international cooperation through Wildlife Enforcement Networks (WENs)and other existing mechanisms, reducing the supply of and demand for illegally traded wildlife, increasing public awareness and education related to wildlife trafficking and its impacts, and treating wildlife trafficking crimes seriously."*[92]

On 13 November 2014, at the East Asia Summit ("EAS"),[93] EAS leaders signed a declaration to combat wildlife crime — declaring their support and recognition of wildlife crime as a serious transnational crime,

[90] A copy of the London Declaration can be found here: http://cites.org/sites/default/files/eng/news/sundry/2014/london-wildlife-conference-declaration-140213.pdf.

[91] APEC is the world's largest regional economic group. It consists of 21 diverse economies which straddle the Pacific and work together to boost sustainable economic growth and prosperity by advancing free and open trade and investment, deepening regional economic integration, fostering a business environment that delivers greater economic opportunities to more people, and tackling common development challenges.

[92] http://www.apec.org/Meeting-Papers/Leaders-Declarations/2014/2014_aelm.aspx.

[93] The East Asia Summit (EAS) is a regional leaders' forum for strategic dialogue and cooperation on key challenges facing the East Asian region. The EAS is a significant regional grouping with an important role to play in advancing closer regional integration and cooperation at a time of particular dynamism in East Asia. Membership of the EAS comprises the ten ASEAN countries (Brunei, Cambodia, Indonesia, Laos, Malaysia, Myanmar, the Philippines, Singapore, Thailand, Vietnam), Australia, China, India, Japan, New Zealand, the Republic of Korea, the United States and Russia.

and requesting for the ASEAN ministers meeting on transnational crime to consider and recognise environmental crime as a serious transnational crime.[94]

The EAS leaders also supported the harmonisation of environmental laws to combat transnational crime and linking wildlife crime with the UN Convention Against Transnational Organized Crime and the UN Convention Against Corruption.

Both China and Vietnam are parties to these declarations. By signing these declarations, the leaders of China and Vietnam are setting the right example and hopefully also setting in motion the commitments made in the declarations.

In conclusion, while well-drafted laws and the capacity to implement and enforce the CITES regime is important, political will is a vital factor in the success of any implementation process. We hope with the recent public show of commitment by government leaders, the promises made will be transformed into actions and results, not just for China and Vietnam but all the other States involved.

References

Cites

CITES (2002). "12th CoP: National laws for implementation of the Convention (CoP12 Doc.28)."

CITES (2014). "Agenda — 65th meeting of the CITES Standing Committee."

CITES (2014). "Report of working group Rhinoceros."

CITES-China (2014). "Report on Big Cats (65th Standing Committee)."

CITES-Vietnam (2014). "Vietnam's Compliance with Resolution Conf. 9.14 (Rev.CoP15) on the Conservation of and Trade in African and Asian Rhinoceroses."

Hargreaves, R. (2014). "Independent Pro Bono Legal Analysis for Item 38 on the Agenda for the 65th Meeting of the Cites Standing Committee Concerning Asian Big Cats."

Horton, J. B. (2007). Controlling the wildlife trade: CITES, regime effectiveness, and the global market for wildlife products. Ann Arbor, The Johns Hopkins University. **3240731,** 348–348.

ICCWC (2014). "Strategic Mission 2014–2016."

ICCWC (2014). "ICCWC Annual Report 2013."

[94]A copy of the declaration can be found on this weblink: http://www.asean.org/images/pdf/2014_upload/EAS%20decleration%20on%20combating%20wldlife%20trafficking.pdf.

Reeve, R. (2006). "Wildlife Trade, Sanctions and Compliance: Lessons from the CITES Regime." *International Affairs* (Royal Institute of International Affairs 1944–) **82**(5), 881–897.

Regional, C. (2014). "Regional Report - Asia (65th Standing Committee)."

Sajeva, M., Augugliaro, C., Smith, M. J., and Oddo, E. (2013). "Regulating Internet Trade in CITES Species." *Conservation Biology*, **27**(2), 429–430.

Sand, Peter H. (2013). "Enforcing CITES: The Rise and Fall of Trade Sanctions." *Review of European, Comparative & International Environmental Law.* **22**(3), 251–263.

Wijnstekers, W. (2011). *The Evolution of CITES*, 9th Edition. Hungary: CIC — International Council for Game and Wildlife Conservation.

Yeater, M., and Vasquez, J. (2001). "Demystifying the Relationship Between CITES and the WTO." *Review of European Community & International Environmental Law*, **10**(3), 271–276.

China

Abbott, B., and Kooten, v. G. C. (2011). "Can Domestication of Wildlife Lead to Conservation? The Economics of Tiger Farming in China." *Ecological Economics*, **70**(4), 721–728.

Chen, G. (2009). *Politics of China's Environmental Protection: problems and progress.* Singapore: World Scientific.

Chow, A. T., Cheung, S., and Yip, P. K. (2014). "Wildlife markets in South China." *Human–Wildlife Interactions*, **8**(1), 108–112.

CITES (2013). "China increases prosecutions in response to illegal trade in elephant ivory."

CITES-China (2014). "Report on Big Cats (65th Standing Committee)."

Grumbine, E. R., and Xu, J. (2011). "Creating a 'Conservation with Chinese Characteristics." *Biological Conservation*, **144**(5), 1347–1355.

EIA (2013). "Hidden in Plain Sight China's Clandestine Tiger Trade."

Ellis, R. (2005). *Tiger Bone & Rhino Horn: the destruction of wildlife for traditional Chinese medicine.* Washington: Island Press.

Haibin, W., and Kunming, G. (1999). "Wildlife trade in Yunnan Province, China, at the border with Vietnam." *The Rufford Foundation*, **18**(1), 21.

Hong-xia, Q. (2010). "A Study on the Legal System of Wildlife Protection." *Journal of Hebei Polytechnic University (Social Science Edition)*, **3**, 012.

IFAW, A. D. f. (2013). "Working together with INTERPOL to ensue fewer wild animals are killed needlessly."

Jiang, Z., Zhou, Z., Meng, Z., Meng, X., Li, L., Ping, X., Zeng, Y., and Mallon, D.P. (2013). "Domestic and CITES regulations controlling the international snake trade in China." *Oryx*, **47**(4), 532.

Li, L., and Z. Jiang (2014). "International Trade of CITES Listed Bird Species in China." *PloS one*, **9**(2).

Li, P. J. (2007). "Enforcing Wildlife Protection in China: The Legislative and Political Solutions." *China Information*, **21**(1), 71–107.

Ng, C.-Y., and Wu, Z. (2005). A comparative study of the wildlife trade in Southern China and the bushmeat trade in Africa. (Thesis), The University of Hong Kong, Pokfulam, Hong Kong.

NICECG, W. Z. f. "China Approach to Combat Wildlife trafficking." China National inter-agency CITES Enforcement Coordination Group.

Nijman, V., and Shepherd, C. R. (2012). "China's role in trade in ivory and elephant parts from Lao PDR." *Oryx*, **46**(02), 172–173.

Okore, G. (2014). Knocking Poachers — China steps up seizures in the fight against illegal wildlife trade. Beijing Review. Beijing, Benchmark Information Limited.

Regional, C. (2014). "Regional Report — Asia (65th Standing Committee)."

Sharma, C. (2005). "Chinese Endangered Species at the Brink of Extinction: A Critical Look at the Current Law and Policy in China." *Animal Law*, **11**, 215.

Still, J. (2003). "Use of animal products in traditional Chinese medicine: environmental impact and health hazards." *Complementary Therapies in Medicine*, **11**(2), 118–122.

Swan, K., and Conrad, K. (2014). "The Conflict between Chinese Cultural and Environmental Values in Wildlife Consumption," *Routledge Handbook of Environment and Society in Asia*. London: Routledge.

UNDOC (2013). Transnational Organized Crime in East Asia and the Pacific. A Threat Assessment, Chapter 7.

Wang, H. (2013). Jamming Wildlife Trafficking — China steps up efforts to combat wildlife smuggling. *Beijing Review*. Beijing, Benchmark Information Limited.

Watters, L., and Xi, W. (2002). "Protection of Wildlife and Endangered Species in China." *Georgetown International Environmental Law Review*, **14**, 489.

Whitfort, A. (2012). "Evaluating China's draft animal protection law." *The Sydney Law Review*, **34**(2), 347–370.

WHO (2005). National policy on traditional medicine and regulation of herbal medicines — A Global Survey."

Wu, X.-P., and Hu, M.-L. (2004). "On Wild Animals Protection by Criminal Laws in China [J]." *Journal of Central South Forestry University*, **6**, 002.

Yang, D., Dai, X., Deng, Y., Lu, W., and Jiang, Z. (2007). "Changes in attitudes toward wildlife and wildlife meats in Hunan Province, central China, before and after the severe acute respiratory syndrome outbreak." *Integrative Zoology*, **2**(1), 19–25.

Yi-Ming, L., Zenxiang, G., Xinhai, L., Sung, W., and Niemelä, J. (2000). "Illegal wildlife trade in the Himalayan region of China." *Biodiversity & Conservation*, **9**(7), 901–918.

Yiming, L., and Dianmo, L. (1997). "Status and strategies for control of live wildlife trade across the Sino-Vietnamese border." *Conserving China'Biodiversity: Reports of the Biodiversity Working Group (BWG), China Council for International Cooperation on Environment and Development (CCICED)(1992 ± 1996)*, 128–140.

Yiming, L., and Dianmo, L. (1998). "The dynamics of trade in live wildlife across the Guangxi border between China and Vietnam during 1993–1996 and its control strategies." *Biodiversity & Conservation*, **7**(7), 895–914.

Yu, W., and Czarnezki, J. J. (2013). "Challenges in China's natural resources conservation & biodiversity legislation." *Environmental Law*, **43**(1), 125.

Zhang, L., He, G., Mol, A. P., and Zhu, X. (2013). "Power politics in the revision of China's Environmental Protection Law." *Environmental Politics*, **22**(6), 1029–1035.

Zhang, L., Hua, N., and Sun, S. (2008). "Wildlife trade, consumption and conservation awareness in southwest China." *Biodiversity & Conservation*, **17**(6), 1493–1516.

Zhang, L., and Yin, F. (2014). "Wildlife consumption and conservation awareness in China: a long way to go." *Biodiversity and Conservation*, **23**(9), 2371–2381.

Zheng, Y., Zhang, H., Niu, Z., & Gong, P. (2012). "Protection efficacy of national wetland reserves in China." *Chinese Science Bulletin*, **57**(10), 1116–1134.

Vietnam

Biggs, D., Courchamp, F., Martin, R., and Possingham, H. P. (2013). "Legal trade of Africa's rhino horns." *Science*. **339**(6123), 1038–1039.

"MOU between Government of The Socialist Republic of Vietnam and the Government of The Republic of South Africa on Cooperation in Biodiversity Conservation and Protection." (2012).

Badenoch, N. (2002). "Transboundary Environmental Governance — Principles and Practice in Mainland Southeast Asia." Washington D.C.: World Resources Institute.

Brooks, E. G., Roberton, S. I., and Bell, D. J. (2010). "The conservation impact of commercial wildlife farming of porcupines in Vietnam." *Biological Conservation*, **143**(11), 2808–2814.

Bryant, T. L., and Akers K. (1999). "Environmental controls in Vietnam." *Environmental Law*, **29**(1), 133.

Capistrano, D. (2008). "Decentralization and Forest Governance in Asia and the Pacific: Trends, Lessons and Continuing Challenges." *Lessons from Forest Decentralization: Money, Justice and the Quest for Good Governance in Asia-Pacific*, 211.

Chandran, R., Krishnan, P., and Nguyen, K. (2011). "Wildlife Enforcement Monitoring System (WEMS): A solution to support compliance of multilateral environmental agreements." *Government Information Quarterly*, **28**(2) 231–238.

CITES (2014). "Agenda — 65th meeting of the CITES Standing Committee."

CITES (2014). "Report of working group Rhinoceroes."

CITES, Vietnam (2008). "Report on Vietnam's Wildlife trade policy."

CITES, Vietnam (2014). "Vietnam's Compliance with Resolution Conf. 9.14 (Rev.CoP15) on the Conservation of and Trade in African and Asian Rhinoceroses."

CITES, Vietnam (2014). "Report to the CITES secretariat from the CITES Management Authority of Vietnam." SC65 Doc.38 Annex 9.

Clark, L., Van Thai, N., and Phuong, T. Q. (2009). *A long way from home: the health status of Asian Pangolins confiscated from the illegal wildlife trade in Vietnam.* Workshop on trade and conservation of pangolins native to south and southeast Asia.

Dillon, T. C., and Wikramanayake, E. D. (1997). "Parks, peace and progress: A forum for transboundary conservation in Indochina." *Parks for Peace.* **7**(3), 36–51.

Drury, R. (2009). "Reducing urban demand for wild animals in Vietnam: examining the potential of wildlife farming as a conservation tool." *Conservation Letters*, **2**(6), 263–270.

Drury, R. (2011). "Hungry for success: Urban consumer demand for wild animal products in Vietnam." *Conservation and Society*, **9**(3), 247–257.

EIA (2013). "Vietnam's Illegal Rhino Horn Trade: Undermining the Effectiveness of CITES."

Hardcastle, J., Long, B., Le Van, L. A. N. H., and Rambaldi, G. (2004). "The use of participatory three-dimensional modelling in community-based planning in Quang Nam province, Vietnam." *PLA Notes*, **49**, 70–76.

Lindsay, P., and Taylor A. (2011). "A study on the dehorning of African Rhinoceroses as a tool to reduce the risk of poaching Endangered Wildlife Trust and the South African Department of Environmental Affairs." Johannesburg, South Africa.

Mai, N. T. H. (2007). Benefit Sharing in Community Forest Management (CFM) Thua Thien Hue Province, Vietnam. *A Fair Share*, 79–88.

Nature-Vietnam, E. f. (2014). "Summary report — An evaluation of the outcomes of prosecutions for serious wildlife crimes in Vietnam."

Nghiem, L. T. P., Webb, E. L., and Carrasco, L. R. (2012). "Saving Vietnam's wildlife through social media." *Science*, **338**(6104), 192.

Ngoc, A. C., and Wyatt T. (2013). "A Green Criminological Exploration of Illegal Wildlife Trade in Vietnam." *Asian Journal of Criminology*, **8**(2), 129–142.

Nguyen, M. H., Vu, V. D., Nguyen, V. S., Hoang, V. T., Nguyen, H. D., Pham, N. T., Than, T. H., and Doan, C. (2007). *Report on the review of Vietnam's wildlife trade policy.* Hanoi, Vietnam: CRES/FPD/UNEP/CITES/IUED._

Nguyen, M. H. (2009). "The status of Vulnerable gaur Bos gaurus and Endangered banteng Bos javanicus in Ea So Nature Reserve and Yok Don and Cat Tien National Parks, Vietnam." *Oryx*, **43**(01), 129–135.

Nguyen, V. S. (2003). *Wildlife Trading in Vietnam: why it flourishes.* EEPSEA, IDRC Regional Office for Southeast and East Asia, Singapore, SG.

Nguyen, V. S. (2008). "Wildlife Trading in Vietnam Situation, Causes, and Solutions." *The Journal of Environment & Development*, **17**(2), 145–165.

Ramkumar, V., and E. Petkova (2007). "Transparency and environmental governance." *The Right to Know: Transparency for an Open World*, pp. 279–308. New York: Columbia University Press.

Sumrall, K. A. (2009). *Confronting illegal wildlife trade in Vietnam: the experience of education for Nature-Vietnam.* (Thesis), University of Michigan.

Traffic, E. A. a. P. R. S. D. and W. B. Department, Washington, DC. (2008). "What's Driving the Wildlife Trade? A Review of Expert Opinion on

Economic and Social Drivers of the Wildlife Trade and Trade Control Efforts in Cambodia, Indonesia, Lao PDR and Vietnam. East Asia and Pacific Region Sustainable Development Department, World Bank, Washington, DC." *East Asia and Pacific Region Sustainable Development Discussion Papers.*

Traffic, M. S. f. (2012). "The South Africa–Vietnam Rhino Horn Trade Nexus: A deadly combination of institutional lapses, corrupt wildlife industry professionals and Asian crime syndicates."

Transnational Crime

Ayling, J. (2013). "Harnessing Third Parties for Transnational Environmental Crime Prevention." *Transnational Environmental Law*, **2**(2), 339–362.

Ayling, J. (2013). "What Sustains Wildlife Crime? Rhino Horn Trading and the Resilience of Criminal Networks." *Journal of International Wildlife Law & Policy*, **16**(1), 57–80.

EIA (2014). *In Cold Blood: Combating organised wildlife crime.*

Elliott, L. (2012). "Fighting Transnational Environmental Crime." *Journal of International Affairs*, **66**(1), 87–104.

Etty, T., Heyvaert, V., Carlarne, C., Farber, D., Lin, J., and Scott, J. (2013). "Norms, Networks, and Markets: Navigating New Frontiers in Transnational Environmental Law." *Transnational Environmental Law*, **2**(2), 203–210.

Heyvaert, V., and Etty T. (2012). "Introducing Transnational Environmental Law." *Transnational Environmental Law*, **1**(01), 1–11.

IFAW (2005). "Caught in the web — Wildlife trade on the internet."

IFAW (2008). "Killing with Keystrokes: An Investigation of the Illegal Wildlife Trade on the World Wide Web."

IFAW (2014). "Ending Wildlife Crime Online."

IFAW, A. D. f. (2013). "Working together with INTERPOL to ensue fewer wild animals are killed needlessly."

Kheng-Lian, K. (2012). "Transboundary and Global Environmental Issues: The Role of ASEAN." *Transnational Environmental Law*, **1**(01): 67–82.

Sajeva, M., Augugliaro, C., Smith, M. J., and Oddo, E. (2013). "Regulating Internet Trade in CITES Species." *Conservation Biology*, **27**(2), 429–430.

Tagliacozzo, E. (2001). "Border Permeability and the State in Southeast Asia: Contraband and Regional Security." *Contemporary Southeast Asia*, **23**(2), 254–274.

Traffic, M. S. f. (2012). "The South Africa–Vietnam Rhino Horn Trade Nexus: A deadly combination of institutional lapses, corrupt wildlife industry professionals and Asian crime syndicates."

UNDOC (2013). "The illegal wildlife trade in East Asia and Pacific," *UNODC Transnational Organized Crime Threat Assessment.*

Warchol, G. L. (2004). "The Transnational Illegal Wildlife Trade." *Criminal Justice Studies*, **17**(1): 57–73.

White, R. D. (2011). *Transnational Environmental Crime: Toward an Eco-Global Criminology.* Abingdon, Oxon: Routledge.

Wilson-Wilde, L. (2010). "Wildlife crime: a global problem." *Forensic science, medicine, and pathology* **6**(3): 221–222.

Wright, G. (2011). "Conceptualising and combating transnational environmental crime." *Trends in Organized Crime*, **14**(4): 332–346.

Zimmerman, M. E. (2003). "The Black Market for Wildlife: Combating Transnational Organized Crime in the Illegal Wildlife Trade" *Vand. J. Transnat'l L*, **36**, 1657.

Part 2

Environmental Science

3

STATUS AND TRENDS OF PHYTOREMEDIATION IN SINGAPORE

FAM MEI-LING and SANJAY SWARUP

This chapter sets out to investigate ways in which land in Singapore can be rejuvenated or re-worked with biotic elements. This new paradigm, phytoremediation, uses plants to reduce, remove, degrade or immobilise environmental pollutants in contaminated sites. Former industrial land can also be rejuvenated to increase land availability via the introduction of biotic elements through a new paradigm of environmental management. The former Kallang Gas Works was identified as a potential site. Though Kallang Gas Works has been dormant for 20 years, she showed considerable flora development. Remediation potential was assessed against comparable sites in Venezuela with a similar petroleum derivative contaminant. Literature review showed a 80% biodegradation rate through plant-assisted bioaugmentation after 150 days. The phytoremediation plan for the former Kallang Gas Works is drawn out in phases, in which Phase (1) determines the existing site condition — clearance of subsurface structures, for example; Phase (2) determines the efficacy of existing flora and fauna; Phase (3) the removal of heavy metals and Phase (4) the removal of petroleum. The suitable tropical environment, and government interest and support can accelerate the revitalisation of the urban environment via phytoremediation.

1. Introduction

Since independence in 1965, Singapore has been ploughing her way towards becoming a concrete jungle while trying to meet housing needs, and economic growth. Gradually, be it through the "Garden City" movement or the more updated "City in a Garden" movement (Inter-Ministerial Committee on Sustainable Development, 2009), Singapore is moving towards assimilating natural processes with urban systems and flows, and even reverting canalised waterways into their meandering original state, complete with biotic input. Currently, a majority of the projects are geared towards protecting and revitalising the waterways of Singapore. However, with a projected population of 6.5 million in the next 40 to 50 years (National Population and Talent Division, 2013), large pockets of land will undergo urbanisation, further encroaching on previously protected nature/water catchment areas. This provides a more compelling reason for developing new, integrative approaches towards environmental management; there are thus more reasons to introduce biotic elements towards a new paradigm of environmental management.

Phytoremediation, loosely defined, refers to a set of technologies that uses plants *in-situ* to clean up contaminated areas (this term is further explained in **Section 2 Overview of Phytoremediation**). It is effective in selective environments and this offers a sustainable urban growth option that revitalises people's relationship with the environment; hence, there is room to extend phytoremediation beyond housing estates as Singapore has various plots of lands that were formerly industrial areas that can be given a new lease of life. Knowing this, we project that the trend of revitalizing the environment via phytoremediation will be an upward one.

2. Overview of Phytoremediation

Phytoremediation is an integrated multi-disciplinary approach involving the use of plants to reduce, remove, degrade or immobilise environmental pollutants in contaminated sites or non-contaminated sites. Contaminated sites often refer to locations in which substances exist in the groundwater or soil in amounts that exceed statutory environmental quality standards, while non-contaminated sites refer to sites that have biotic elements taking on a preventive nature — for example having biotic elements in roadside swales before water enters a stormwater catchment. As such, phytoremediation generally takes place in conjunction with the following equally important disciplines: ecology, soil chemistry, soil microbiology, environmental engineering and plant biology. Plants generally handle the contaminants without affecting the topsoil, thus conserving its utility and fertility. *Introduction to Phytoremediation* (USEPA, 2000) discussed that phytoremediation applications can be classified based on the contaminant fate in the following manner: degradation, extraction, containment or a combination of these.

2.1 *Degradation*

Degradation occurs in two major areas of the plants — primarily through the rhizosphere or through metabolism within the plant. The rhizosphere is defined as the zone around the plant roots that has significantly higher microbial numbers and activity than in the bulk soil (USEPA, 2000). Phytodegradation is the breakdown of contaminants taken up by plants through metabolic processes within the plant, or the breakdown of contaminants external to the plant through the effect of compounds (such as enzymes) produced by the plants. Additionally, degradation may occur outside the plant due to the release of compounds that cause transformation. Any degradation caused by microorganisms associated with or affected by the plant root is considered as rhizodegradation.

2.2 *Extraction*

Phytoextraction is the uptake of contaminants by plant roots and translocation within the plants. Contaminants are generally removed by harvesting the plants. This concentration technology leaves a much smaller mass to be disposed of than the excavation of the soil or other media.

2.3 *Containment*

The mechanism of containment using plants either bind the contaminants to the soil, rendering them as non-bioavailable, or immobilises them by removing the means of transport. The physical containment of contaminants by plants can take the form of binding the contaminants within a humic molecule (humification), or by root accumulation in non-harvestable plants.

3. Phytorememdiation in Singapore

3.1 *Brief History of Phytoremediation in Singapore*

The history of phytoremediation in Singapore can be said to be propelled by the desire to protect the water quality of Singapore. The Active, Beautiful, Clean (ABC) Waters programme was started in 2007 and has developed significantly thereafter. The ABC Waters programme "aims to symbiotically integrate Singapore's parks (Green), reservoirs and waterways (Blue) and recreational infrastructures and facilities (Orange)" (PUB, 2009). As such, the design features of ABC Waters include natural systems which retain and treat stormwater on site before allowing the water to naturally flow into the reservoirs. These features, over time, form ecological habitats that benefit from their proximity to the water sources.

As the programme is meant to protect the waters of Singapore, there is a natural bias towards water protection. The programme is also meant to realise PUB's larger objective of creating new spaces for the community to engage in nature and get closer to the water, so as to value water.

Table 3.1. Stormwater quality objectives for Singapore.

Pollutant	Stormwater treatment objectives
Suspended Solids	80% removal of average annual load
Total Nitrogen	45% retention of average annual load
Total Phosphorous	45% retention of average annual load

(PUB, 2009)

Table 3.2. Case studies in Singapore based on category of treatment required.

Landfill leachate	Stormwater treatment
• Constructed Wetlands at Lorong Halus Wetlands Park	• Biorentention Swales at Sengkang West Way • Bioretention Basin (Rain Garden) at Balam Estate • Sedimentation Basins

This implies that the programme is about water education rather than nature education as a whole.

Based on PUB's goals for stormwater treatment, ABC features are required to meet several stormwater quality objectives (Table 3.1). In most if not all cases, phytoremediation is introduced to achieve the desired water quality.

In the guidelines, the use of plants is mentioned in (i) "Conveyance elements" through vegetated and bioretention swales, (ii) "Detention elements" through Bioretention Basins and (iii) "Retention elements" through sedimentation basins, constructed wetlands and a cleansing biotope.

3.2 *Phytoremediation Systems and Case Studies in Singapore*

There are several case studies in Singapore, and most involve protecting waterways in Singapore. These case studies are identified by category of treatment (Table 3.2). Details on other case studies, such as the nature of pollution, the species of flora used, and the efficacy or drawbacks, are discussed in a separate report (Fam, 2015).

4. Case Studies of Phytoremediation Techniques Applicable to Singapore

In Singapore, the application of phytoremediation is currently limited to the treatment of landfill leachate and stormwater management. Other potential areas of application include:

- Total Petroleum Hydrocarbons — Kallang Gas Works, Jurong Island
- Air Pollutants — Jurong Island, Mitigation for Haze
- Heavy Metals — Jurong Island, specifically Arsenic

This essay will only delve into the treatment of petroleum hydrocarbons pollution in the former Kallang Gas Works site. More details can be found in the master thesis (Fam, 2015).

4.1 *Total Petroleum Hydrocarbons*

4.1.1 *Phytoremediation mechanism and flora species*

Total petroleum hydrocarbons (TPH) (see Table 3.3) are considered persistent hazardous pollutants, and include compounds that can bioaccumulate in food chains. Benzene and benzo[a]pyrene are examples of such recognised mutagens and carcinogens. Due to their varying physical and chemical properties, TPHs are divided into two categories: (1) Gasoline Range Organics (GRO) and (2) Diesel Range Organics (DRO).

Sites with TPH often require a combination of clean-up methods, and current techniques usually involve excavation and *ex-situ* treatment of the source material, followed by the utilization of phytoremediation to mitigate

Table 3.3. Summary of 2 categories of Total Petroleum Hydrocarbons.

Gasoline range organics	Diesel range organics
Gasoline range organics (GRO) corresponds to small chain alkanes with low boiling points (60–170) such as isopentane, 2,3-dimethyl butane, n-butane and n-pentane and volatile aromatic compounds such as the monoaromatic hydrocarbons benzene, toluene, Ethylbenzene and xylenes (BTEX)	Diesel range organics includes longer chain alkanes (C10- C40) and hydrophobic chemicals such as polycyclic aromatic hydrocarbons (PAH).
BTEX compounds are highly mobile in the environment	PAHs tend to bind strongly to soil particles.

Adapted from Kamath et al. (2004).

the residual contamination (Kamath *et al.*, 2004). The general mechanisms involved are hydraulic control, uptake, transformation, volatilisation and rhizodegradation.

Hydraulic control prevents off-site migration and/or decrease downward migration of contaminants. The flora act as a 'solar pump' removing water from soils and aquifers through transpiration. Hence, the key to forming a successful barrier against plume migration is the rooting of the flora into a shallow water table aquifer. In addition, the bottom of the aquifer has to be confined by materials of low hydraulic conductivity (e.g. clay, shale, rock; i.e. materials with hydraulic conductivity $<10 - 6\,\mathrm{cms}^{-1}$). Within the upper region of an aquifer, grasses with dense and fibrous root systems are used to transpire water and limit the percolation of contaminants through the vadose zone and to intercept rainwater that may discourage tree root penetration (Kamath *et al.*, 2004).

Moderately hydrophobic hydrocarbons (BTEX) can be removed from soil and groundwater through direct plant uptake (uptake, translocation and transformation). The rate of contaminant removal has been found to be a function of uptake efficiency, transpiration rate and the contaminant concentration in soil water. Uptake efficiency depends on plant species, age, health, and the physico-chemical properties of the root zone. Transpiration rate also varies dramatically and depends on the plant type, leaf area, nutrients, soil moisture, temperature, wind conditions and relative humidity (Kamath *et al.*, 2004). There is poor plant uptake of hydrophilic compounds, which have little affinity for root membranes. There is high uptake efficiency of moderately hydrophobic hydrocarbons such as BTEX; and poor uptake of hydrophobic hydrocarbons such as PAHs, which strongly sorb to soil and are therefore not bioavailable (Kamath *et al.*, 2004).

Once a contaminant enters the plant, reactions within the following series may occur:

- Phase 1 — Conversion
- Phase 2 — Conjugation
- Phase 3 — Compartmentation

Conversion reactions include oxidations, reductions, or hydrolysis that the plant uses to begin detoxification. Conjugation reactions chemically link the Phase I products to glutathione, sugars, or amino acids, thus altering the solubility and toxicity of the contaminant. Once conjugated, xenobiotics can be removed as waste or compartmentalised. During compartmentation, chemicals are conjugated and segregated into vacuoles or bound to the cell wall material (hemicellulose or lignin). Phase III conjugates are often described as "bound residues" because chemical extraction methods do not recover the original contaminants (Kamath *et al.*, 2004).

The natural ability of a plant to volatilise a contaminant that has been taken up through its roots can be exploited as a natural air-stripping pump system. Phytovolatilisation is most applicable to those contaminants that are treated by conventional air-stripping i.e., contaminants with a Henry's constant $KH > 10\,atm\ m^3$ water.m^{-3} air, such as BTEX, TCE, vinyl chloride and carbon tetrachloride. Chemicals with $KH < 10\,atm\ m^3$ water-m^{-3} air, such as phenol and PCP, are not suitable for the air-stripping mechanism because of their relatively high polarity.

Volatile pollutants diffuse from the plant into the atmosphere through open stomata in leaves. Radial diffusion through stem tissues has also been reported. For example, methyl-tert-butyl ether (MTBE) can escape through leaves, stems, and the bark to the atmosphere. Tree core samples of hybrid poplars exposed to TCE also showed radial diffusion from the stem rather than transpiration from leaves as the main dissipation mechanism. Generally, the concentration of VOCs in the xylem decreases with increasing distance from the roots (Kamath *et al.*, 2004).

Microbial degradation in the rhizosphere might be the most significant mechanism for removal of diesel range organics in vegetated contaminated soils. This occurs because contaminants such as PAHs are highly hydrophobic and their sorption to soil decreases their bioavailability for plant uptake and phytotransformation.

The rhizosphere of most plants promotes a wealth of microorganisms that can contribute significantly to the degradation of petroleum hydrocarbons during phytoremediation. Thus, though a plant may not directly act upon these contaminants, a plant can influence the microbial community within its root zone to a great extent. Potential rhizosphere interactions that may be important for the phytoremediation of petroleum hydrocarbons include:

1. Prolific microbial growth
2. Repression/induction of catabolic enzymes
3. Co-oxidation of contaminants
4. Changes in bioavailability
5. Chemotaxis of suitable strains

Deposition of plant-derived carbon sources through root exudation, and/or root turnover provides rhizosphere bacteria with numerous organic substrates. Rhizodeposition can account for the release of 7–27% of the total carbon fixed during plant photosynthesis and varies between plants.

Commonly reported estimates are between 10–100 mg-C g-root material, of which root exudation is reported to range between 0.4–27.7 mg-C g-root material (Kamath *et al.*, 2004). The composition and quantity of root-derived material released into the rhizosphere varies depending on the season, the age of plant and the health of the plant but generally contains sugars (15–65% total organic carbon), organic acids (9–33% total organic carbon), amino acids (2–31% total organic carbon) and phenolics (0.3–4 mg-C-g root material) (Kamath *et al.*, 2004). Plant stress and age generally increase rhizodeposition.

4.2 *Application in Singapore — Kallang Gas Works*

4.2.1 *Site history*

The Gas Works, initially formed as the Singapore Gas Company, was the predecessor to the Public Utilities Board, and primarily supplied utilities to the Singapore public. The Singapore Gas Company was formed in 1861 by the Municipal Commissioners to provide gas lighting for the town (Tan, 2005).

In 1862, Kallang Gas Works was built to supply the first piped gas in Singapore to enable street lighting. There were altogether four tanks operating in Kallang Gas Works. Tank numbers 1 and 2 were built with the original plant in 1862 and were in use until they were demolished in 1957. Tank Number 3 was built before 1900, whilst Tank Number 4 was built between 1908 to 1909. Over the years, piped gas became more important as fuel for cooking and water heating, particularly with rapid housing development in the 1960s and 1970s. Its original function for gas lighting faded as electricity took over street lighting in 1906 and the last gas lamp disappeared in 1956 (Tan, 2005).

The Singapore Gas Company serviced primarily street lighting until 1901, when the Municipal Commissioners took over its responsibilities. From 1901 to 1930, coal carbonising plants were installed and used until 1958, when the gas works was converted to use oil gasification (Tan, 2005).

On 1 May 1963, the Public Utilities Board was constituted, taking over the responsibilities of the former City Council, for gas, electricity and water supply. Naphtha, a refined petroleum product enabled the growth of the Gas department, with operations falling into two divisions, namely, Production and Marketing. However, plans to develop the Riverwalk Park along Kallang Basin led to announcements in January 1987 that the Kallang Gas Works would be removed (Tan, 2005).

Gas production continued uninterruptedly for 137 years at the Kallang Gasworks except for short breaks during the world wars. It opened to the public for the first time on 26 March 1998 before it was finally demolished (Tan, 2005).

Based on the plants' operations, the pollutants were coal tar, TPH and heavy metals. The results of the investigation showed high levels of polycyclic aromatic hydrocarbon (PAH) and total petroleum hydrocarbon (TPH) in both surface soils and groundwater. Other contaminants found at high levels include cyanide and BTEX compounds (Singapore Land Authority, 2010).

4.2.2 *Remediation goals*

Remediation goal no. 1

The Urban Redevelopment Authority revealed in the Master Plan 2014 that the plot of land (approximately 3.2 hectares) was to be redeveloped into a residential–commercial entity with green development (URA, 2014). This redevelopment ties in with remediation goal 1, which is to allow the contaminated land to be redeveloped for high-density residential use. With more access to the site (outside of scope of this thesis), potential exposures to humans and ecosystems associated with future residential land use onsite or recreational use of Kallang Riverside Park will include assessment against Dutch Target Values (DTVs) and Dutch Intervention Values (DIVs) for residential land use; and site-specific risk-based levels (RBLs) for residential land use, or using alternate sources (e.g., USEPA Regional Screening Levels) or an approach approved by the relevant authorities (National Environment Agency).

Remediation goal no. 2

The site is skirted by the adjacent Rochor Canal, which feeds into the Kallang River. The water from Kallang River flows into the Kallang Basin which is linked to the Marina reservoir, and the water within this network should be kept free of pollution. Hence, the 2nd remediation goal is to prevent contamination to the water course.

4.2.3 *Design*

The design of the phytoremediation system or plant-assisted bioremediation would depend largely on site conditions. As the site is closed to public, several details can only be retrieved through a desktop review. It is important to recognize that Kallang Gas Works operated from 1862 to 1987. During the early periods, environmental requirements were less stringent, and hence pollution levels might be so extensive that it might be necessary to first reduce phytotoxicity to a level where healthy plant growth can take place. One such measure is through *ex-situ* methods, such as the excavation of contaminated soil, before phytoremediation can take place. If the TPH level is equal to or less than 3% of soil by weight, then planting tolerant species could proceed either using an RTDF mix or high-water use plants for polluted groundwater (Collins, 2000).

4.2.4 *Location*

The site which is currently vacant State land is bounded by Kallang Road in the north, Rochor Canal in the south, Kampong Bugis Road in the east and Crawford Street in the west.

Across Kallang Road in the north, there are commercial and industrial buildings. To the east of the site beyond Kampong Bugis Road, there is a temple, a low-rise property and a large green field. At the southern tip of the site is a commercial building and the Kallang Riverside Park. Rochor Canal also skirts by the south-western end of the site. To the northwest of the site is a public housing estate which comprises high-rise public housing apartments, neighbourhood commercial centres and public amenities.

The Mass Rapid Transit (MRT) East-West Tunnels underlies Kallang Road, and a portion of the northwest corner of the site is underlain by both the second and third reserve zones of the MRT.

Hence, based on vicinity activity, the potential receptors are:

- Residents, workers at commercial places and people visiting the temple or Kallang River, who may be exposed through ingestion and dermal means or inhalation of vapour.

Site layout and corresponding potential sources of contaminants

The following areas on site and their corresponding potential sources of contaminants are shown in Fig. 3.1:

- Areas of coal shed
- Northern Gas Holders

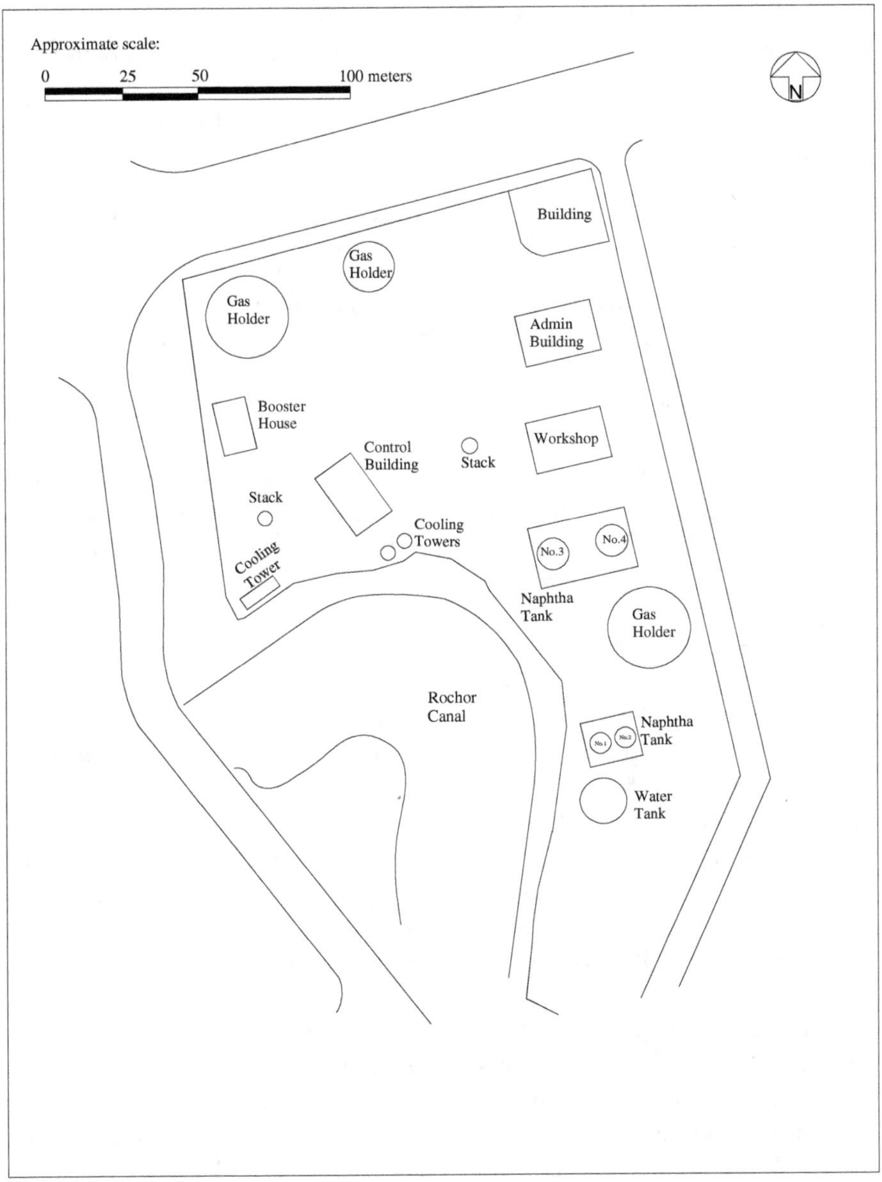

Fig. 3.1. Site layout plan of former Kallang Gas Works.

- Naphthalene tanks (1–4), Gasholders 6 and 7
- Central portions of site, condensers, scrubbers and tar/ammonia tanks

Some areas in the figures have not been identified on due to lack of information.

Soil structure, texture, and organic matter content

The design involving the species of plants would be dependent on several factors — one of which is the sub-surface make up. The Kallang area is characterised by Alluvial Member (Ka), which is "a variable terrestrial sediment ranging from pebble beds through sand, muddy sand, and clay to peat. The member is usually unconsolidated, but lightly consolidated beds may be found" (DSTA, 2009).

Soil borelogs of the general area of Kallang revealed that the site is underlain with predominantly sand, varying between silty sand and clayey sand (SLA, 2010). Based on historical investigations, the following were observed:

1. Fill material: 1 to 5 mbgs, non-uniform thickness across site, tar-like material and occasional free-flowing product found in pockets.
2. Non-cohesive soil: Dominated by sand; varying thickness and depth across site; laterally extensive that grades into marine clay; saturated and below water table.
3. Marine Clay: Starts at approximately 4 mbgs. The transition layer between sand and clay appears to act as an aquitard and tar-like material; odour and presence of hydrocarbons has been observed. Tar-like material penetrates into clay lithology.
4. Old Alluvium: Observed at 11.5 mgs. No visual impact or odour.

In terms of the influence of soil structure, contaminants may be trapped within and sorbed to the surfaces of nanopores (soil pores with diameters $<100\,\text{nm}$) that are inaccessible to organisms. In terms of soil texture, clay is capable of binding molecules more readily than silt or sand (Frick *et al.*, 1999). As the site comprises predominantly silty and clayey sand, the contaminants are considered to be bioavailable.

Soil organic matter binds lipophilic compounds, which reduces the bioavailability. High organic carbon content ($>5\%$) in soil usually leads to strong adsorption and, therefore, low availability, while moderate organic carbon content may lead to limited availability.

4.2.5 *Groundwater*

Rochor canal is no longer tidally connected to ocean. Hence tidal forcing of groundwater from the canal is not anticipated. It is anticipated that shallow groundwater flows towards Rochor Canal from the north and east site boundaries to a low point at the apex of the canal. The primary water producing aquifer is of shallow fill, consisting of non-cohesive and cohesive soils. Underlying marine clay appears to act as an aquitard in the vertical migration of groundwater.

4.2.6 *Nutrient conditions*

It has been found that the best reduction in TPH in vegetated treatment occurred when the fertiliser ratio was 100:2:0.2 (C:N:P) — which is more than the amount required for plant growth. A ratio of 100:10:1 should probably be the target to enhance microbial action. In addition, aeration is also advised, such as ploughing once the plants are established (Collins, 2000).

4.2.7 *Establishment of appropriate plants*

An important factor in establishing plants in an oil-contaminated site involves getting seeds to germinate. A good way of knowing whether the plant(s) being considered for phytoremediation will germinate successfully is to carry out germination tests in the contaminated soil prior to planting. Wiltse *et al.* (1998) found that alfalfa seed germinates in soils contaminated with up to 50 g of crude oil per kg (5% w/w).

Another approach would be to transplant seedlings or mature plants, as these plants would tolerate higher levels of contaminants. Lin and Mendelssohn (1998) were able to transplant mature *Spartina patens* into marsh soil with a crude oil content as high as 100 mg crude oil per gram of soil (0.01% w/w) and *S. alterniflora* into soils contaminated with oil concentrations as high as 250 mg crude oil per g soil (0.025% w/w) (as cited in Frick *et al.*, 1999).

Based on a case study in Venezuela, Infante *et al.* (2010) studied the seed propagation, tiller and root development, abundance, propagation easiness and the root system of 57 identified species (comprising 18 legumes, 19 grasses, 3 sedges, and 17 other herbaceous species). Infante *et al.* (2010) assessed the seed germination and biomass production of 6 commercial grasses in a silt-sandy soil contaminated with 3% light crude oil, and found that after 45 days, the hydrocarbon contamination reduced seed germination and biomass production. *Brachiaria brizantha* and *Panicum maximum* attained the highest seed germination rates in contaminated soils, whereas *B. brizantha* exhibited the highest aboveground and root production and maximum root length.

In another study, Merkl *et al.* (2005) investigated 3 types of grasses to phytoremediate a soil contaminated with 5% w/w of heavy crude oil (as cited in Infante *et al.*, 2010). Under greenhouse conditions, soils planted with *B. brizantha and C. aggregates* had a significantly lower final oil and grease concentration than the control at 1.5% (v/v) over 180 days. On the other hand, the concentration of saturated hydrocarbons was always lower in planted than unplanted soil, and *B. brizantha* induced the highest aromatic content reduction, while stimulating microbial population and activity. A positive correlation between root biomass production and oil degradation was found, as well as between oil degradation and root morphology. *B. brizantha* and *C. aggregates* showed coarser roots and *B. brizantha* presented a larger root surface area in contaminated soils. Merkl *et al.* (2005) also investigated and found that in 3% (v/v) light crude oil, the above-mentioned grasses reduced the contaminant level to 1% v/v in approximately 250 days (as cited in Infante *et al.*, 2010).

4.2.8 *Inoculation of microorganisms vs bioaugmentation*

Bossert and Bartha (1984) investigated a variety of microorganisms that are reportedly involved in the degradation of petroleum hydrocarbons. In general, the bacteria *Pseudomonas*, *Arthrobacter*, *Alcaligenes*, *Corynebacterium*, *Flavobacterium*, *Achromobacter*, *Micrococcus*, *Mycobacterium*, and *Nocardia* were reported as the most active bacterial species in the degradation of hydrocarbons in soil (as cited in Frick *et al.*, 1999).

The role of microbes also extends to a mutually-beneficial strategy for dealing with phytotoxicity, where microorganism benefit from the plant exudates while the plants benefit from the ability of microorganism to break down toxic chemicals. Radwan *et al.* (1995) found the plant *Senecio glaucus* growing along the polluted border of an oil lake in the Kuwaiti desert. The plant roots and adhering sand particles were white and clean, while the surface of the transitional zone between the root and shoot was black and polluted (as cited in Frick *et al.*, 1999). The authors suggested that microbes detoxified contaminants in the rhizosphere, which allowed the plants to survive in the oil-contaminated soils.

There are conflicting reports on the effectiveness of inoculating contaminated sites with microorganisms. Cunningham *et al.* (1993) state that it is a common experience for soil and plant inoculants to be out-competed by native microflora (as cited in Infante *et al.*, 2010). On the other hand, Radwan *et al.* (1997) found that the population of indigenous, oil degrading *Arthrobacter* species (strain KCC201) in sand treated with weathered crude oil (20% w/w) decreased dramatically following inoculation with non-indigenous, oil-degrading 28 *Arthrobacter* and *Candida* species (as cited in Infante *et al.*, 2010).

The decrease in the number of indigenous *Arthrobacter* were thought to be due to competition with the inoculated species, and the authors suggested that seeding with microbial "cocktails" irrespective of their origin may be useless or harmful if inoculated species out-competed indigenous, oil-degrading species.

Infante *et al.* (2010) found that in the tropics, a higher diversity of microorganism that are able to readily degrade hydrocarbons compounds is ubiquitously present in the soil. Since high humidity and solar radiation also favour biostimulation, bioaugmentation may be the best option for the former Kallang Gas Works site, as it has been operating for over 100 years, and the native flora would have been stimulated by the constant presence of hydrocarbon contaminants accumulated over the years. Also, Infante *et al.* (2010) indicated that it is virtually impossible to reproduce the physiology and metabolic interactions among populations of bacterial species that take place in a natural soil or a crude oil contaminated waste, as microorganisms are regulated by the microclimatic conditions of each soil, which cannot be recreated in the laboratory.

Additionally, the variety of compounds present in petroleum, comprising molecules with aromatic, aliphatic and heteroatomic moieties can only be degraded by the concerted action of different microbial communities. As a result of all the above mentioned factors, commercially available microbial consortia for field applications may turn out to be little effective for the remediation of crude oil contaminated soils.

Venezuelan case studies have shown that biostimulation is a better option than bioaugmentation (Table 3.4). In this study, an inoculum was chosen among bacteria isolated from three different sources (petroleum contaminated soil, non-contaminated soil and dry sludge from a sewage treatment plant, in which laboratory biodegradation assay results revealed that sewage plant sludge bacteria are the most effective in the removal of hydrocarbons from the contaminated soil test). The plots were fertilised

Table 3.4. Biodegradation rate, as a percentage of hydrocarbon removal (saturates plus aromatics) during biostimulation and bioaugmentation tests.

Treatment	Sampling time (days)				
	0	**60**	**90**	**150**	**180**
	Biodegradation (%)				
Soil + Oily sludge + Aeration + Fertiliser	**0**	**60**	**72**	**81**	**80**
Soil + Oily sludge + Aeration + Fertiliser + Inoculum	**0**	**68**	**71**	**75**	**78**

Source: Infante et al., 2001 as cited in Infante et al., 2010.

with urea and triple superphosphate to yield a C/N ratio of 60 and a C/P ratio of 800; moisture was maintained at 60% of the field capacity and, for aeration, the soil was thoroughly mixed three times a week with the help of a rake.

As such, for bioaugmentation to work, the soil must be amended with adequate nitrogen and phosphorus sources at proper concentrations, since an excess or deficiency will inhibit microbial activity. Interestingly, this nutritional condition may selectively affect the biodegradation rate of specific hydrocarbon families present in petroleum, such as the saturated fractions for instance.

C/ N ratios of 10:1, 20:1, 50:1, 60:1 and even 560:1 have been reported, while C /P ratios from 100:1, up to 800:1 have been recommended (Infante *et al.*, 2010).

Trials performed in Venezuela have shown that a C/N ratio of 60:1 together with a C/P ratio of 800:1 are appropriate nutritional conditions for bioremediation purposes. These proportions have been widely used for the clean-up of sludge from crude oil waste pits (Infante *et al.*, 2010).

4.2.9 *Concentration of pollution*

As there are no publicly published contaminant levels, it is assumed that NAPL, PAH, TPH and ammonia may be present in the groundwater as well, due to the expected high concentration of the respective contaminants found in the soil, and former operation processes.

Research carried out by Wiltse *et al.* (1998) gave a general indication that with a similar operation to the former Kallang Gas Works, it is possible for some species of plants to germinate, and if higher levels are encountered, seedling transplantation is a possibility.

4.2.10 *Areas of concern*

Biotransformation, bioaccumulation, biomagnification, disposal

Biotransformation is the formation of intermediate metabolites in the metabolic breakdown of chemical compounds, and this may ultimately lead to the mineralisation of toxic contaminants. The degradation of PAHs by fungi, for example, incorporates only 1 atom of oxygen, which can result in the production of carcinogenic epoxides (Sutherland, 1992). In soil conditions that favour fungal activity, early PAH metabolic products may actually increase the toxicity of the contaminant. Also, toxic intermediate metabolites may build up in the soil as a result of either the slow reactions by key bacteria or the production of "dead-end" products. Carcinogenic vinyl chloride may build up during trichloroethylene biodegradation as bacteria can convert trichloroethylene into vinyl chloride relatively quickly, but the subsequent degradation of vinyl chloride occurs slowly. In the co-metabolism of chlorinated phenols, dead-end products such as toxic chlorocatechols sometimes accumulate in the soil.

Cunningham *et al.* (1993) reported that biomagnification occurs when lipophilic organic chemicals, which concentrate in fatty tissues, increase in concentration at higher levels of the food chain. The former Kallang Gas Works is by the Kallang River, where there will be interaction between fauna and the site flora; hence in the scenario where the there is a large consumption of plant materials containing low-level contaminants, there may still be biomagnification in the food chain, unless the higher trophic level organisms themselves can efficiently degrade the contaminants (as cited in Infante *et al.*, 2010).

Complex metabolism and biological lagtime

Typical (plant-assisted) bioremediation involves a lag time before degradation begins. This implies that there may be slower rates than expected due to varying site conditions and complex metabolic reactions. There may also be a scenario where the final contaminant concentrations are higher than desired as the reaction kinetics reach a limit. Also, it is a constant competition with other mechanisms—such remediation techniques may contribute little if the rate of biodegradation is slower than other processes, such as leaching or volatilisation. Considering how the site of former Kallang Gas Works is near Kallang River and that is generally underlain with silty or clayey sand (which implies a medium to high hydraulic conductivity), leaching could potentially be an issue with respect to the proposed remediation technique.

Mixture of contaminants

Often, oil-contaminated sites are contaminated with metals, salts and/or pesticides, which will complicate phytoremediation efforts. It is likely that that there is lead and arsenic in the former gas works site and it is also known that in general, metal concentration inhibits microbial metabolism.

The Committee on In Situ Bioremediation *et al* (National Research Council, 1993) recognised that mixtures of organic contaminants may also cause difficulties in phytoremediation. Shuttleworth and Cerniglia (1996) observed that certain petroleum hydrocarbons can inhibit the growth of bacteria that degrade other petroleum contaminants. More specifically, naphthalene competitively inhibits the growth of the phenanthrene-degrading bacteria *Acidovorax delafieldii* strain TNA921 and *Sphingomonas paucimobilis* strain TNE12. In a phenomenon known as *diauxy*, microbes may selectively degrade organic compounds that are easiest to digest or provide the most energy (as cited in Frick *et al.*, 1999). Thus, if the target contaminant (e.g., petroleum hydrocarbons) occurs in combination with an adequate supply of a more preferred substrate, diauxy could result in limited microbial degradation of the target compound(s).

The Committee on In Situ Bioremediation *et al.* (1993) also recognised that mixtures of organic compounds can, however, promote the microbial degradation of petroleum hydrocarbons, particularly if one or more components of the mixture is a co-metabolite of others. Likewise, biota primarily degrading one type of organic compound may also degrade a second compound present at concentrations too low to independently support bacterial growth (as cited in Frick *et al.*, 1999).

Addition of fertilisers

In creating an environment suitable for microbial degradation, the following carbon, nitrogen and phosphorous ratio has been recommended. As such, additional fertilisers are often required to make the nutrient ratio, as reported in Section 4.2.8 **Inoculation of Microorganisms vs Bioaugmentation** :

In the case of Kallang Gas Works, as Rochor Canal is adjacent to the site, nutrient run-off will be a potential problem. Also, the action of nitrifying and denitrifying bacteria will add on to the greenhouse gas burden through the production of nitrous oxide.

4.2.11 *Costs*

The typical costs to consider for a field-scale phytoremediation project involves expenditure on design, site preparation, reporting, monitoring, and operation and maintenance. Part of site preparation would include ensuring the site is clear of debris — which is crucial in the case of the former Kallang Gas Works, as pockets of the site still consist of demolition debris. Other site preparations include preliminary greenhouse experiments along with agronomic soil testing, and potentially mathematical modelling. Also, phytoremediation projects usually require five or more years, hence it is crucial to ensure that funding for operation and maintenance is available throughout the duration of the project (Kamath *et al.*, 2004). Some of the major problems in the field have been weeds, and insect or disease infestation, and it has been estimated that at least 30% of the plants may need to be replanted in the second or third year.

A cost estimate provided by Jordahl *et al.* (2002) is summarised in Table 3.5 and Table 3.6 below:

Cunningham *et al.* (1993) estimated that the general cost of phytoremediation is US$17 to US$100 per m^3 (based on a cost of US$2,000–$15,000 per hectare to a depth of 15 cm). Further, mowing, pruning,

Table 3.5. Five-Year Cost Comparison of Phytoremediation by hybrid Poplar Trees.

Item	Cost (USD)
Design and Implementation	50,000
Monitoring Equipment	
Capital	10,000
Installation	10,000
Replacement	5,000
5-Year Monitoring	
Travel & Administration	50,000
Data Collection	50,000
Reports(annual)	25,000
Sample analysis	50,000
Total	250,000

Source: Jordahl et al. (2002) as cited in Kamath et al. (2004).

Table 3.6. Elaboration of phytoremediation tasks. Adapted from United States Environmental Protection Agency (2000).

Design	Installation	Operating

• **Site characterisation**
• **Work plan and report preparation**
• **Treatability**
	• **Site Preparation**
	  • Facilities removal
	  • Debris removal
	  • Utility line removal/relocation
	• **Soil Preparation**
	  • Physical modification: tilling
	  • Chelating agents
	  • pH control
	  • Drainage
	• **Infrastructure**
	  • Irrigation system
	  • Fencing
	• **Planting**
	  • Seeds, plants
	  • Labour
	  • Protection
		• **Maintenance**
		  • Irrigation water
		  • Fertiliser
		  • pH control
		  • chelating agent
		  • drainage water disposal
		  • Pesticides
		  • Fencing/pest control
		  • Replanting
		• **Monitoring**
		  • Soil nutrients
		  • Soil pH
		  • Soil water
		  • Plant nutrient status
		  • Plant contaminant status (roots, shoots, stems, leaves
		  • Tree sap flow monitoring
		  • Air monitoring (leaves, branches, whole tree, area)
		  • Weather monitoring

Adapted from United States Environmental Protection Agency (2000).

harvesting, monitoring vegetation for contaminants, irrigation and fertiliser costs should be included in the initial estimated costs. McIntyre and Lewis (1997) estimated that where site circumstances permit, phytoremediation may use the same equipment and materials common to agricultural practices and costs can be equated to those associated with crop production (as cited in Frick *et al.*, 1999). The cost of applying cropping systems in phytoremediation has been estimated at US\$0.02–\$1.00 per m^3 per year (or US\$200–\$10,000 per hectare with roots extending down to one meter).

Cunningham *et al.* (1993) and Bollag *et al.* (1994) estimated that costs associated with bioremediation generally range from about US\$98–\$260 per m^3. The cost of *ex situ* bioremediation of petroleum hydrocarbons in a fully contained facility ranges from US\$133–\$400 per m^3 (or US\$20,000–\$60,000

Table 3.7. Elaboration of the remediation phases and their respective actions.

Phase	Goal	Action
1	Determine efficacy of existing flora and microorganisms	• Site has been closed for 20 years, and the existing vegetation (see photo A-2 in Appendix A) is best suited for phytoremediation as it has acclimatised to contaminants.
2	Lowering Contaminants to a level suitable for Phytoremediation / Plant-assisted Bioremediation	• Excavation of impacted soil as tar was seen at 4 m below ground surface.
3	Hydraulic Control	• Site has shallow water table aquifer. Also, the bottom of the aquifer is confined by materials of low hydraulic conductivity—marine clay. This makes it ideal to prevent contaminant plume from migrating towards Rochor Canal.
4	Removal of TPH contaminant as per Remediation Goal	• Existing grasses can be considered given that site has been left on its own for over two decades, and substantial green cover (grasses/shrub/woody vegetation) was observed (see Photo A.2 in the Appendix) • Commercial tropical grasses that can be considered based on field study in Venezuela: o *B. brizantha* o *C. aggregatus* o *E. indica* • Bioaugmentation with the need to investigate on the type of native bacteria present.

per hectare to a depth of 15 cm) (Cunningham *et al.*, 1993). *In situ* bioremediation techniques are cheaper at US\$50–\$133 per m^3 or US\$7500–\$20,000 per hectare — assuming a 15 cm depth. However, these costs can quickly escalate depending on the level of monitoring, analytical costs, and degree of security and safety that are required (as cited in Frick *et al.*, 1999).

Cunningham *et al.* (1993) further emphasised that in phytoremediation, agronomic costs, including planting, tillage, fertilisation, and harvesting, can be insignificant in comparison with associated administrative costs, such as site management, regulatory reporting, and analysis of data (see Table 3.7 for considerations of agronomic and administrative cost items for the former Kallang Gas Works) (as cited in Frick *et al.*, 1999).

4.2.12 *Summary of remediation plan*

In consideration of the issues highlighted above, the remediation plan is summarised as follow:

Phase 1 — Initial field study (selecting plant species, including survey of existing plant species, determining sub-surface conditions, and microorganism selection)

Phase 2 — Implementing of *ex-situ* methods to lower contaminant to level suitable for phytoremediation

Phase 3 — Remove heavy metals and implementation of hydraulic control (Fig. 3.2) layout plan of former Kallang Gas Works and Phytoremediation Phase 3

Phase 4 — Removal of TPH contaminants (Fig. 3.3)

5. Conclusion and Areas for Further Research

This dissertation has set out to understand the extent that phytoremediation systems have been applied in Singapore, and in doing so, identify areas in Singapore that could benefit from such remediation systems — whether phytoremediation or plant-assisted bioremediation systems. To this end, it has been investigated that phytoremediation in Singapore currently extends only to stormwater treatment and a single example of landfill leachate treatment in Lorong Halus. A promising area that has been put forth in this essay is the potential rejuvenation of the former Kallang Gas Works.

With the above-mentioned case studies of phytoremediation or plant-assisted bioremediation, there also exists a short excerpt on how a naturally derived solution can assist in choreographing a civic-mindedness for natural capital (and its value), and the healthy functioning of an ecosystem. Ecology, is this sense, is expanded to include human activity and cultural flows. The following key points are thus derived.

Balancing priorities

In remediating the environment, this essay focused on natural solutions — the removal of concentrated pollutants using flora and microorganisms.

Fig. 3.2. Site Layout Plan of Former Kallang Gas Works — Phase 3.

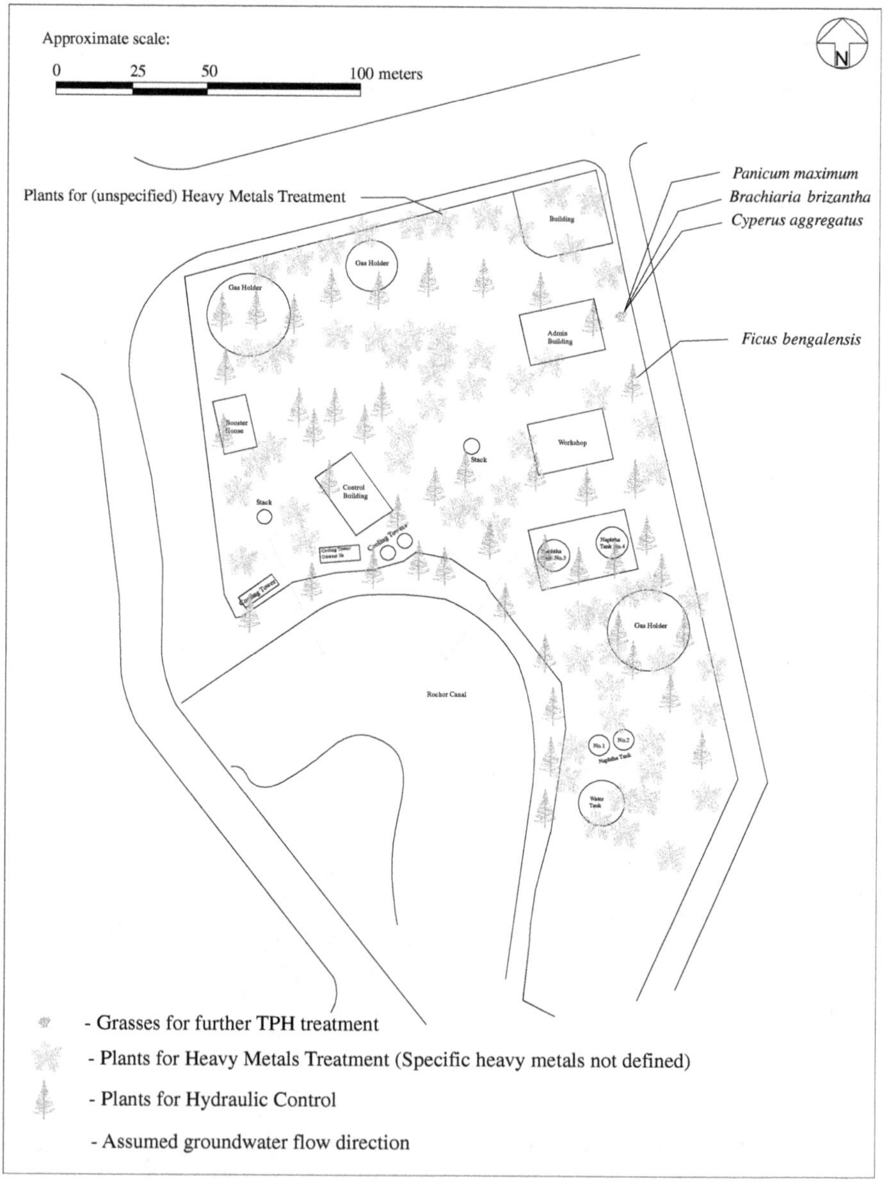

Fig. 3.3. Site Layout Plan of Former Kallang Gas Works — Phase 4.

However, in most scenarios, these natural solutions can only be feasible with not-so natural solutions that involve earth-moving equipment, such as in the case of Kallang Gas Works, where the petroleum derived waste is so concentrated that the gross pollution needs to be excavated before phytoremediation and bioremediation can respond to the lingering groundwater and sub-surface contaminants. However, effective as *ex-situ* methods are, they lack the finesse that flora and microorganisms possess in contaminant treatment.

Local solutions to a local problem

Numerous studies have been carried out largely in the United States under the Superfund Act, and also in Scandinavian countries, as these places have a major emphasis on environmental policies. However, the species of flora identified for a specific pollutant treatment (such as heavy metals or organics) is often grown only in a widely different climate in temperate regions. Singapore, being a tropical country, could boast a robust bioremediation attempt due to its warm temperatures and plethora of tropical flora that uptake a lot of water — making it efficient for remediating polluted groundwater. Kallang Gas Works, having been left in its dormant state for almost 20 years, could already be harbouring the right inoculum of microorganisms conditioned to such high concentration of pollutants. The site visit also revealed existing woody vegetation and an extensive grass network that ought to be effective for cleaning up Kallang.

On an ending note, a phytoremediation system is similar to urbanism in many ways, in that because it is a living and dynamic system of plants and microorganism, it can anticipate change, open-endedness and negotiation.

6. References

Bollag, J.-M., Mertz, T. and Otjen, L. 1994. Chapter 1 Role of Microorganisms in soil bioremediation. Bioremediation Through Rhizosphere Technology. T.A. Anderson and J.R. Coats. American Chemical Society: Washington, D.C. ACS Symposium Series 563, pp. 2–10.

Bossert, I., Kachel, W.M., and Bartha, R., 1984. Fate of Hydrocarbons during Oily Sludge Disposal in Soil. Applied and Environmental Microbiology 47, pp. 763–767.

Collins, C.D. (2000). Implementing Phytoremediation of Petroleum Hydrocarbons. *Methods in Biotechnology*, **23**, 99–107.

Cunningham, S.D. and Berti, W.R. (1993). Remediation of contaminated soils with green plants: An overview. *In Vitro Cellular & Developmental Biology Plant*, **29**(4), 207–212.

Defence Science and Technology Agency (DSTA) (2009). Geology of Singapore. Map. 1:10,000. Singapore: Government of Singapore.

Fam, M.L. (2015). The status and trends of phytoremediation in Singapore (Unpublished Master's thesis). National University of Singapore, Singapore.

Frick, C.C., Farrel, R.E., and Germida, J.J. (1999). Assessment of Phytoremediation as an In-Situ Technique for Cleaning Oil-Contaminated Sites. Petroleum Technology Alliance of Canada.

Infante, C., Morales, F, Ehrmann, E.U., Hernández-Valencia, I. and Leon, N. (2010). Hydrocarabon bioremediation and phytoremediation in tropical soils: Venezuelan study case. *Trends in Bioremediation and Phytoremediation*, **2010**, pp. 429–451.

Jordahl, J.L., Foster, L., Schnoor, J.L. and Alvarez, P.J.J. (1997). Effect of hybrid poplar trees on microbial populations important to hazardous waste bioremediation. *Environmental Toxicology and Chemistry*, **16** (6), 1318–1321.

Kamath, R., Rentz, A., Schnoor, L. and Alvarez, J.J. (2004). Phytoremediation of hydrocarbon-contaminated soils: principles and application. *Studies in Surface Science and Catalysis*, **151**, 447–478.

Lin, Q., and Mendelssohn, I.A. (1998). The combined effects of phytoremediation and biostimulation in enhancing habitat restoration and oil degradation of petroleum contaminated wetlands. *Ecological Engineering*, **10**, 263–274.

National Research Council (USA) (1993). In Situ Bioremediation: When Does it Work?. pp. 63–90. Washington, DC: The National Academies Press.

McIntyre, T., and Lewis, G.M. (1997). The advancement of phytoremediation as an innovative environmental technology for stabilization, remediation, or restoration of contaminated sites in Canada: a discussion paper. *Journal of Soil Contamination*, **6**, 227–241.

Merkl, N., Schultze-Kraft, R., and Infante, C. (2005). Phytoremediation in the tropics — influence of heavy crude oil on root morphological characteristics of graminoids. *Environmental Pollution,* **138**, 86–91.

National Population and Talent Division (2013). A Sustainable Population for A Dynamic Singapore. Prime Minister's Office, Government of Singapore. Retrieved on 9 July 2014 from: http://population.sg/whitepaper/downloads/population-white-paper.pdf.

Public Utilities Board (PUB) (2009). PUB Active, Beautiful, Clean Waters Design Guidelines.

Radwan, S.S., Sorkhoh, N.A., Fardoun, F., and Al-Hasan, R.H. (1995). Soil management enhancing hydrocarbon biodegradation in the polluted Kuwaiti desert. *Applied Microbiology Biotechnology*, **44**, 265–270.

Shuttleworth, K.L., and Cerniglia, C.E. (1995). Environmental aspects of PAH biodegradation. *Applied Biochemistry Biotechnology*, **54**, 291–302.

Singapore Land Authority (2010). Tender for Appointment and Engagement of Consultancy Services for Proposed Remediation of Former Kallang Gasworks Site at Kampong Bugis (Tender Ref No: SLA000/PT09-16/87.1.26-V2). Singapore.

Sutherland, J.B. (1992). Detoxification of polycyclic aromatic hydrocarbons by fungi. *Journal of Industrial Microbiology*, **9**, 53–62.

Tan, B. (2005). Gas Works. National Libraries Board Singapore. Retrieved 30 June 2014 from http://eresources.nlb.gov.sg/infopedia/articles/SIP_750_2005-01-11.html

United States Environmental Protection Agency (USEPA) (2000). Introduction to Phytoremediation. Retrieved on 9 May 2014 from: http://nepis.epa.gov/Exe/ZyPDF.cgi/30003T7G.PDF?Dockey=30003T7G.PDF

Urban Redevelopment Authority (URA) (2014). Island-Wide Master Plan. Retrieved on 2 May 2014 from: https://www.ura.gov.sg/maps/?service=MP.

Wiltse, C.C., Rooney, W.L., Chen, Z., Schwab, A.P., and Banks, M.K. (1998). Greenhouse evaluation of agronomic and crude oil-phytoremediation potential among alfalfa genotypes. *Journal of Environmental Quality*, **27**, 169–173.

4

MANAGING THE RISK OF NON-INDIGENOUS MARINE SPECIES TRANSFER IN SINGAPORE USING A STUDY OF VESSEL MOVEMENT[1]

CHIN-SING LIM, YI-LIN LEONG and KOH-SIANG TAN

Shipping is recognised as a major vector for the global transfer of non-indigenous marine species (NIMS). As a major transhipment port, Singapore can minimise the risk of NIMS transfer by implementing pragmatic management strategies, such as using vessel movement information to assess the risk of NIMS transfer.

Findings from vessel movement information in a major port terminal in Singapore showed that vessel residence time is short, with more than 92% of vessels spending seven days or less. There was little variation in vessel residence time to vessel arrival numbers, while the top three last ports of call were found to be from regional ports. Using two key features obtained from vessel movement records, 1) vessel residence time and 2) biogeographic origin of the vessels' last port of call, a simple risk assessment matrix was constructed and applied to assess the level of risk of NIMS transfer by transiting vessels.

Introduction

The Network of Shipping and its Implications for Marine Biosecurity

The commercial shipping network, (i.e., containerised vessel movements) are governed strongly by global markets, such as location of global manufacturing centres (e.g., labour market and factories) and economic supply and demand. Hence, shipping transport are usually serviced by fixed shipping routes aimed at optimal travel time and efficiency within international or regional ports along the coastal environment (Capineri and Randelli, 2007; Kaluza *et al.*, 2010). The confluence of increasing shipping volume and the capability of major ports to efficiently handle larger transhipment vessels has led to the development of a global shipping

[1]This paper first appeared in *Marine Pollution Bulletin*, **115**, Lim C.S., Leong Y.L., Tan K.S., Managing the risk of non-indigenous marine species transfer in Singapore using a study of vessel movement, pp. 332–344, Copyright 2017. Reproduced with permission from Elsevier.

network resembling a hub and spoke model (Notteboom and Rodrigue, 2005; Wang and Wang, 2011), forming transhipment routes to link regional ports to facilitate trading with major hub ports. Therefore, these also make existing ports and the surrounding natural marine environment the most vulnerable to impacts of NIMS introduction by shipping. Earlier studies by Kaluza *et al.* (2010) and Keller *et al.* (2011) highlighted global shipping movement patterns as a key factor in determining the risk of bio-invasion in risk assessment models. A study by Seebens *et al.*, (2013) that focused on vector-based modelling using vessel trading patterns, environmental hetereogenity (e.g., temperature and salinity differences) and biogeographic distribution to model bio-invasion probability, also indicated Singapore and the South-East Asia (SEA) region as a bio-invasion hotspot. According to Carlton (1985), world patterns of ballast water movement (over the past 100 years) have probably paralleled the changing patterns of world shipping routes. These routes or patterns are complex and are dependent on the distribution of the worlds' economies, markets, population, location of industries and resources. In addition, he found that reduced vessel duration in port led to reduced ship biota (i.e., biofouling) as a result of faster port handling (Carlton, 1985), hence it is plausible that the risk of transfer of NIMS will be increased by prolonged vessel duration in port. In some vessel management strategies (e.g., Craft Risk Management Standard, MPI, New Zealand), vessel time spent in port is also highlighted as a key component that address inspection criteria necessary to determine acceptable vessel hull cleanliness. These findings are important as they suggest that vessel movement along the shipping network, which are connected by coastal ports functioning as nodal points, have varying density and diversity (i.e., port connectivity). Port connectivity will likely drive vessel movement patterns that may directly influence risk of NIMS transfer along certain routes and biogeographic regions due to increased propagule pressure.

Challenges of Managing Bio-invasion in Singapore

In Singapore, most of the published literature on non-native species has been on terrestrial flora and fauna (Pandit *et al.*, 2006), avian (Lim *et al.*, 2003; Brook *et al.*, 2003) and freshwater aquatic flora and fauna (Ng *et al.*, 2014). According to the Global Invasive Species Database (Global Invasive Species Program), Singapore has 55 introduced non-indigenous species (not including native species that have spread to other countries).

A breakdown of the numbers revealed that a majority of the number of species belonged to plants and fishes (19 and 17 respectively), followed by insects (8), birds (5), amphibians (2), snails (2) and mammal (1). There was only one marine representative (i.e., *Mytilopsis sallei*). However, a recent review of estuarine and marine non-indigenous species (Jaafar *et al.*, 2012) listed a total of 17 species, including two bivalve molluscs, one polychaete worm, two dinoflagellates and 12 fish species, although the study did not take into consideration whether the species cause harmful impact on the environment. Overall, there is a shortage of published literature on marine invasive species and its impacts in Southeast Asia (e.g., marine plankton and invertebrates), including Singapore, where there is a focus mainly on terrestrial or freshwater aquatic environment (Peh, 2010), and in particular invasive species that have agricultural impacts. Although a recent paper (Nghiem *et al.*, 2013) estimated the total economic cost arising from invasive species impact in Southeast Asia based on published invasive species database, no marine invasive species was used in the model. In addition, it has been recognised that the loss of native biodiversity in Singapore has largely been attributed to urbanisation (e.g., land conversions) rather than from direct competition with invasive species. The failure of invasive species to establish themselves against native biodiversity (Ng *et al.*, 1993) also suggest that anthropogenic disturbances are more likely to impact the native environment than as a result of non-native species competition, although it is not certain if this is true for marine aquatic species. As a result of this lack of understanding of marine biodiversity in Singapore and in general the SEA region, the status of native and alien species in the region and its associated impacts are poorly understood and is likely to impede political urgency and development in action plans to address marine bioinvasion occurences. However, Hewitt (2002) suggested that the rich biodiversity in tropical environment do not support pattern of lower invasion risk or environmental resistance (i.e., biodiverse regions are also susceptible to invasions). Therefore, there is a need to recognise the potential impacts both in Singapore coastal environment and the country's role as a major transhipment port to manage NIMS transfer.

Managing the Risk of NIMS Transfer as a Major Transhipment Hub Port

National economic interests present a challenge for preserving native marine habitats and biodiversity. Given that global shipping requires vessels to operate in different parts of the world, abiding by international shipping guidelines like the IMO Convention makes it easier for vessels coming to Singapore to be regulated through a uniform set of rules or guidelines, including managing the risk of NIMS transfer on vessels to environmentally sensitive areas. This may involve requiring vessels to adopt ballast water or biofouling management plans. Singapore has not ratified the 2004 International convention for the control and management of ships' ballast water and sediments (BWM Convention) to date. As a major shipping hub for transhipment in the SEA region, Singapore may need to recognise and manage the risk of NIMS transfer on transoceanic vessels transiting through its ports to sensitive marine environments, if the BWM Convention comes into force.[2] This will likely require the port authorities to implement reasonable measures to ensure vessels comply with international guidelines regarding the management of ballast water. Similarly, the management of biofouling on vessels may also require Singapore to take on a proactive role in mitigating high-risk vessels transiting through the port.

The study of NIMS has fuelled the understanding of the dynamic process of distribution and spread of potential invaders, leading to the drafting and adoption of international and national environmental instruments aimed at protecting the marine environment against the risks of bio-invasion by shipping, particularly in vessel management strategies. From a management perspective, an informed decision on the overall risk of biofouling or ballast water is also important to guide policy planning, for example in the implementation of protocols (i.e., imposing frequency of or prioritising inspection) for ballast water sampling or biofouling inspection. A risk assessment model that is based on vessel movement information may be a pragmatic and timely approach to estimate the overall risk factor of a vessel likely to harbour NIMS, aiding vessel management strategies in Singapore (i.e., inspection for non-compliance).

Methods

Approximating Vessels Residence Time in Keppel Terminal, Singapore

A parameter such as vessel turnaround time has been used before in transport economic studies (Ducruet and Merk, 2013) to calculate a port

[2]The BWM Convention will enter into force on 8 September 2017, after ratification criteria have been met.

efficiency index. It is defined as the time between arrival and departure at a port so that a vessel can complete its loading, unloading and other servicing operations. Here, it is proposed as a means to estimate the vessel port residence time of containerised vessels present in a major port terminal. Vessels position in Keppel Terminal were recorded every day for an hour for three one-month periods (taken to be 30 days for this study); the first period from 29 April 2014 to 28 May 2014 (May 2014), the second period from 3 November 2014 to 2 December 2014 (November 2014) and the third period from 24 January 2015 to 22 February 2015 (February 2015). The vessels recorded in this study included instances where vessels were alongside the berths, as well as inactively moored just off the berths. This was achieved through live records of vessel positions through the open access option on www.fleetmon.com, which provided real-time ship positions in port and basic technical information about the vessel from terrestrial Automatic Identification System (AIS) receiver stations for live vessel tracking. Four classes of vessels (bulk carriers, containers, general cargo and RORO) were recorded from these data. These were chosen because the vessels generally follow fixed routes with varying degrees of patterns according to Kaluza *et al.*, (2010). Tankers tend to move less predictably between ports and the number of ports frequented is also limited as they transport mainly oil products (Kaluza *et al.*, 2010). The total number of vessels in port, vessel type, the vessel IMO registered number, the represented flag state and the deadweight tonnage of the vessels were recorded each day.

Documenting ship positions over a 24 h interval each day allowed an approximation of ship status in time, so that overall vessel traffic flow through the terminal could be observed. However, as the recording was only performed for a relatively limited time of one hour every day for a month, not all vessel arrival or departure per day may be recorded in port. One assumption made was that vessels spent > 24 h in port on each arrival. This assumption may be valid as the container vessel average turnaround time (ATT) — an index for port handling efficiency related to the amount of time a vessel could complete its operations, for Singapore was 1.16 days in 2011 (Ducruet and Merk, 2013). This provided a basic record of vessel positions in port for Keppel Terminal. By tracking individual vessels in port, vessel arrival and departure dates from Keppel Terminal could be estimated and approximated as the duration that an individual vessel spent in port over the observed periods. The duration of each vessel record in port was approximated to be 24 h, as the records taken from the vessel monitoring website was taken at every 24 h interval. This duration was taken to be the vessel residence time.

Although AIS information is available over other websites (e.g. www.marinetraffic.com), the website employs a sea boundary to define every port in the database. Given that the vessel traffic volume in port is expectedly high, having a clear zone to define the port area allows a clear record of arrival and departure in port. This option was chosen as obtaining travel records of vessels and operational data including ballast discharge and uptake volume from shipping registries was severely limited and costly.

Vessels' Last Port of Call

For two periods from November 2014 and February 2015, the last port of call of vessels prior to arrival in Singapore were recorded using an open source website (www.marinetraffic.com). Different marine eco-regions from which these ports are located in could then be identified according to Spalding *et al.* (2007) classification for the world's marine eco-regions. This information is useful in identifying shipping routes with high vessel arrivals, as well as to identify vessels that are departing from marine eco-regions with high numbers of NIMS records. Although complete information on vessels' previous port and next destination port was not available for every single detected vessel, available information were recorded for all vessels where possible.

Invasive Species Database and Geographic Distribution

The geographic distribution of marine invasive species worldwide was synthesised in an earlier work by Molnar *et al.*, (2008) from over 350 invasive species databases worldwide that reported upon the identifications of marine invasive species, including their geographic distribution, impacts on native biodiversity, and introduction pathways. The database is available at http://www.conservationgateway.org. This database is useful because it allowed probable transfer pathways and ecological impact factor for each species to be considered when deriving geographical distribution. This classification of the world's ocean provided a detailed and comprehensive biogeographic system to look at coastal and shelf areas, therefore providing a representative composition of the biodiversity in marine eco-regions to support analyses of patterns in marine biodiversity, and direct efforts in marine resource management and conservation. Using records of the vessels' last port of call recorded from November 2014 as an illustration, the geographic distribution of NIMS species found in different marine eco-regions (Spalding *et al.*, 2007) from which ports are located in could be studied.

Statistical Analysis

A distribution of the vessel residence time over the three observed periods was performed with the Kruskal–Wallis test. This distribution allowed vessel movement to be examined for similarity in duration spent in port, which could estimate the vessel turnover rate in port. The number of vessels was modelled as a linear function of the inverse of the number of days they spent in port (i.e., number of vessels $\sim$ f (1/days in port)), in an effort to elucidate any differences in their general distributions. Statistical analysis was performed with R Statistical Software (Foundation for Statistical Computing, Vienna, Austria).

Results

Length of Vessel Residence Time

From May 2014, a total of 639 vessels were recorded in Keppel Terminal. The majority of these vessels (95.8% of total vessel numbers) spent less than seven days in port. From November 2014, a total of 577 vessels were recorded. 93.6% of vessels spent less than seven days in port. From February 2015, a total of 552 vessels were recorded. 92.6% of vessels spent less than seven days in port. These results are shown in Fig. 4.1. Vessels turnaround (i.e. time between a vessel arriving and departing the port) is high in Keppel Terminal, with approximately 85% of vessels spending 1–3 days in port, for all the three observed periods (Fig. 4.1).

Total vessel residence time in port of 1,387 days were recorded from May 2014; a total of 1359 days were recorded from November 2014 and a total of 1323 days were recorded from February 2015. These numbers represented the area under the charts in Fig. 4.1. From the vessel movement records, it was also possible to deduce that some vessels made multiple departures and arrivals into Keppel Terminal in a one-month period. These vessels were possibly operating out of Keppel Terminal and within several ports in the region, given that the interval between subsequent re-arrival was short.

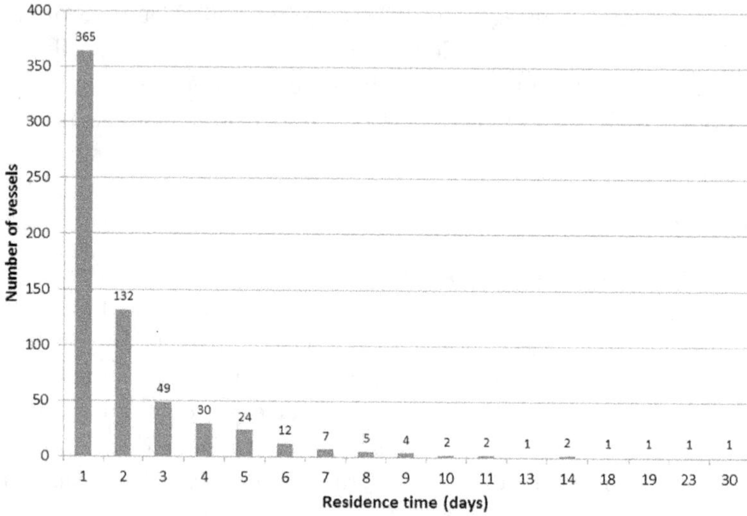

(a)

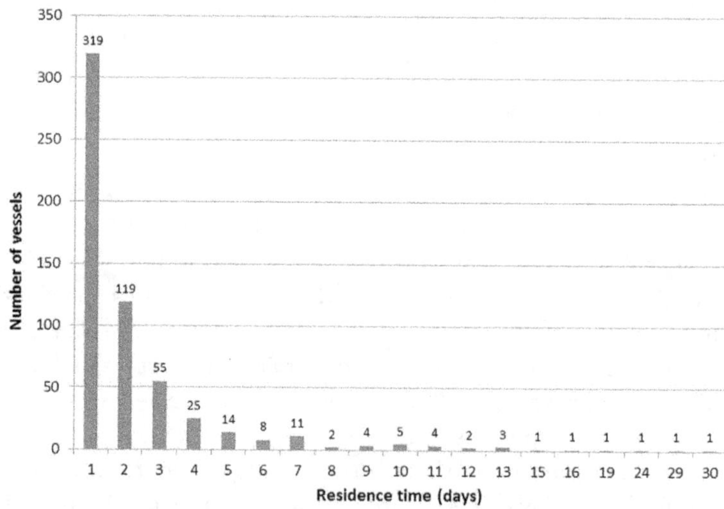

(b)

Fig. 4.1. Vessel residence time for vessel arrivals in Keppel Terminal in three observed periods: (A) May 2014; (B) November 2014; (C) February 2015. The results showed that >92% of vessels spent seven days or less in port. Approximately 85% of vessels spent 1–3 days in port.

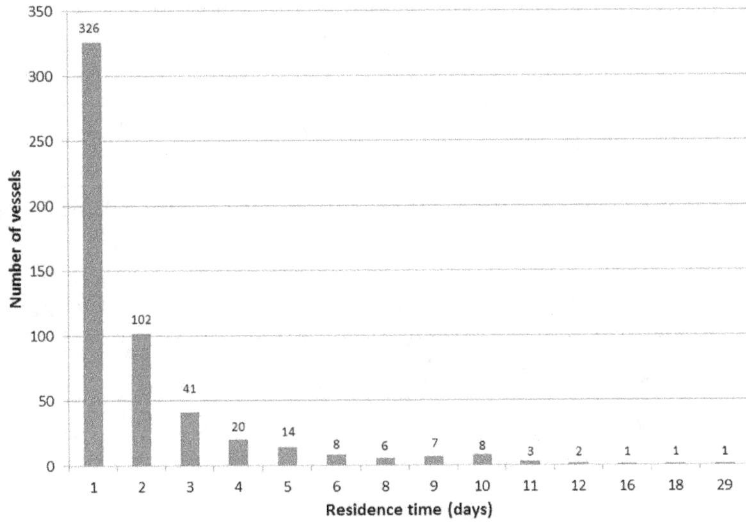

(c)

Fig. 4.1. (*Continued*)

The duration (i.e., vessel residence time) each arriving vessel spent in port varied. This variation between the number of vessels arrival each month and the vessel residence time in port for the three observed periods was similar and follow closely a hyperbolic relationship ($p < 0.01$) (Fig. 4.2).

The results of the Kruskal–Wallis test suggest no significant differences in the median number of days the vessels spent in port ($H(2) = 0.0504$, $p = 0.9751$). Further analysis also supports this, with the loosely comparable model results of each of the three individual months (Table 4.1).

Duration of Long Stay Vessels

A total of 4.2% of vessels (27 of 639 vessels) from May 2014, 5.7% of vessels (33 of 577 vessels) from November 2014 and 7.4% of vessels (41 of 552 vessels) from February 2015 were recorded to have stayed in port for seven days or more. These vessels possibly transit Keppel Terminal multiple times given the number of departure and arrival records. These long-term vessels were made up of mostly container and cargo vessels. However, only 12 of

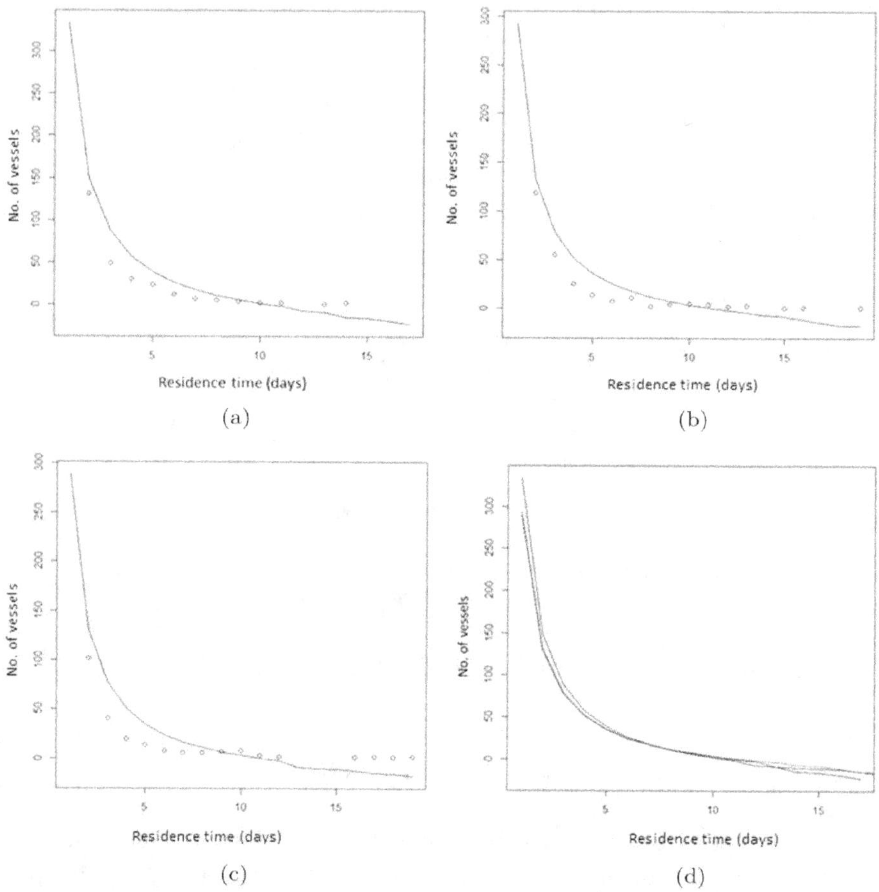

Fig. 4.2. Number of vessels modelled as a linear function of the inverse of the number of days they spent in port (i.e., number of vessels ~f (1/days in port)). A) May 2014, B) November 2014, C) February 2015, D) overlay of the fitted lines for each of the 3 months of data. There was little variation ($p < 0.01$) in the relationship between number of vessels and vessel residence time for all three periods.

the 27 vessels in May 2014 stayed in port for seven or more consecutive days (i.e. not interrupted by a departure from port). 20 of the 33 vessels from November 2014 stayed in port for seven or more consecutive days; while 20 of the 41 vessels from February 2015 stayed in port for seven or more consecutive days in port.

Table 4.1. Number of vessels modelled as a linear function of the inverse of the number of days they spent in port.

Period	Estimated (1/day)	Estimated intercept	Adjusted R-squared	P-value
May 2014	369.287	−35.253	0.9529	1.42E-11
Nov 2014	321.089	−28.297	0.9547	4.48E-13
Feb 2015	317.653	−28.666	0.9306	1.71E-11
All months	335.748	−30.591	0.9439	2.20E-16

Vessels' Last Port of Call

For two periods from November 2014 and February 2015, the vessels' last port of call, defined as ports originating prior to vessel arrival in Keppel Terminal, were recorded. From this record, it was possible to identify 1) the different shipping routes, 2) routes density based on vessel numbers that traverse these routes and 3) estimate the vessel residence time spent in Keppel Terminal after arrival. With this information, the density of different shipping routes could be established based on number of arriving vessels. Figure 4.3 showed the total vessel residence time against the number of vessels plying different shipping routes. For ease of representation, only the *last ports of call* where the total vessel residence time from all the vessel arrivals in Keppel Terminal were seven or more days, were shown in Fig. 4.3. The top three ports in these two periods, based on number of arriving vessels, were dominated by regional ports (i.e., from November 2014: Laem Chabang (Thailand), Port Klang and Tanjung Pelepas (Malaysia) and from February 2015: Port Klang and Tanjung Pelepas (Malaysia), Tanjung Priok (Indonesia)).

Geographic Distribution of Invasive Species

Using the identification of shipping ports obtained in November 2014 as a basis, the potential presence of NIMS in these ports could be identified using the geographic distribution of NIMS in the world's marine eco-regions (Molnar *et al.*, 2008). This would allow (*in principle*) the identification of the possible presence of NIMS in the ports of the corresponding marine eco-regions (David *et al.*, 2013). From the information in this period, 21 international ports (including Keppel Terminal) are situated in 14 distinct marine eco-regions (Table 4.2), defined as 'marine areas

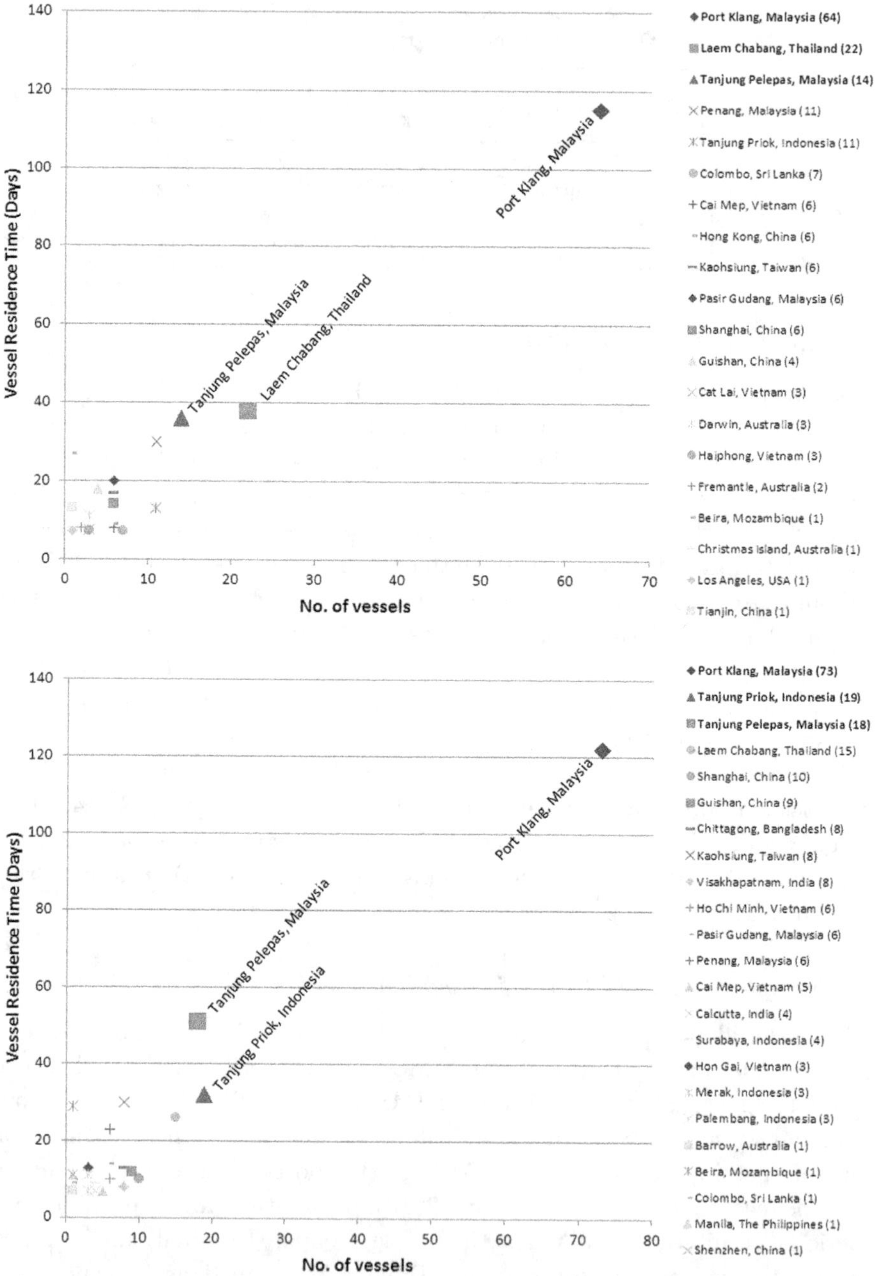

Fig. 4.3. Vessels' last port of call based on vessel residence time in port in two period: (A) November 2014; (B) February 2015. Vessels' last port of call was not collected in May 2014. Regional ports (within SEA) dominated the number of arriving vessels.

Table 4.2. Marine eco-regions where last ports of call frequented by vessels calling at Keppel Terminal are located. The number of harmful species are defined as those capable of causing significant ecological impacts. Source: Marine invasive species database (Molnar *et al.*, 2008).

Eco-region code	Eco-region	Ports located	Number of NIMS recorded in this eco-region	Number of harmful specie
50	Yellow Sea	Tianjin (China)	8	2
52	East China Sea	Shanghai (China)	11	5
59	Southern California Bight	Los Angeles (U.S.A.)	49	35
101	Bight of Sofala	Beira (Mozambique)	3	1
104	South India and Sri Lanka	Colombo (Sri Lanka)	8	4
112	Gulf of Tonkin	Haiphong (Vietnam)	7	2
113	Southern China	Guishan, Hong Kong (China) Kaohsiung (Taiwan)	12	4
115	Gulf of Thailand	Laem Chabang (Thailand)	4	2
116	Southern Vietnam	Cai Mep, Cat Lai (Vietnam)	3	2
117	Sunda Shelf/Java Sea	Tanjung Priok, Indonesia	7	5
118	Malacca Strait	Pasir Gudang, Penang, Port Klang, Tanjung Pelepas (Malaysia); Singapore	4	2
120	Cocos-Keeling/Christmas Island	Christmas Island (Australia)	2	0
141	Bonaparte Coast	Darwin (Australia)	5	3
209	Leeuwin	Fremantle (Australia)	18	9

of relatively homogeneous species composition' (Spalding *et al.*, 2007). Table 4.2 summarises the number of NIMS recorded in the 14 marine eco-regions based on the database created by Molnar *et al.*, 2008. Based on recorded impact caused on their recipient environment after introduction, the *number of harmful species* recorded is a subset from the *number of NIMS*. These were species associated with a high ecological impact score (i.e., species that had 'disrupted large ecosystem, affect multiple species and processes with wide abiotic influences').

Table 4.3 listed the description of the NIMS (including the number of harmful species), marine eco-regions, the ecological impact and the likely pathway that they were introduced through. A total of 74 distinct species (Table 4.3) were recorded from these 14 marine eco-regions. Species that appeared more than once (i.e., overlapped in the other 14 marine eco-regions) were shaded grey. In addition, species that were transported either through ballast water or ship biofouling were shaded. Only 24 species (Fig. 4.4 and Table 4.3) were recorded to 1) occur in more than one eco-region, and were transported by 2) ballast water or ship fouling (or both). Notably, for the Malacca Strait, even though there are four records, the only record so far for Singapore was the Caribbean bivalve *Mytilopsis sallei.*

These species ranged from tunicates, molluscs, dinoflagellates, arthropods, bryozoans, hydroids and annelid worms. Eight species, namely *Botryllus schlosseri* (found in four eco-regions), *Ectopleura crocea* (4), *Amathia distans* (4), *Sphaeroma walkeri* (5), *Monocorophium acherusicum* (6), *Bugula neritina* (7), *Mytilopsis sallei* (9), *Teredo navalis* (9) were found in four or more eco-regions. Many species demonstrate wide geographical distribution, occurring in both mid-temperate and tropical environments.

Discussion

Global shipping today has facilitated the transfer of NIMS to new marine environments by allowing NIMS to 'hitchhike' on vessels ballast water or biofouling, potentially increasing the environmental impacts on native recipient environments. Coastal ports in different geographical locations and marine environment are both donor and receiving centres for the transport of NIMS through shipping. These port states face different vulnerabilities and challenges when it comes to managing marine bio-invasion. Although there is general awareness and guidelines on vessel management strategies, such as those highlighted in the BWM Convention Guidelines, to reduce the risk of introduction of NIMS, the framework for implementation is arguably still in development. As a major transhipment hub-port for international vessels in the region, it is possible for Singapore to identify potential high risk shipping routes by studying vessel movements. Such information will likely assist the port state to make informed decisions on managing risk assessment matrices for vessels. Vessel management

Table 4.3. List of NIMS, their ecological impact rating, likely introduction pathways and the eco-regions where they are recorded in, according to Molnar *et al.*, 2008. This list is based on the eco-regions of the *last ports of call* of vessel arrivals in November 2014 in Keppel Terminal. Shaded rows refer to NIMS that are found in 1) more than one eco-region and 2) transported by ballast water or biofouling (or both). A dash in the cell refer to unknown impact or pathway of that particular NIMS. The eco-region locations can be found on the map in Fig. 4.4; 50: Yellow Sea; 52: East China Sea; 59: Southern California Bight; 101: Bight of Sofala; 104: South India and Sri Lanka; 112: Gulf of Tonkin; 113: Southern China; 115: Gulf of Thailand; 116: Southern Vietnam; 117: Sunda Shelf/Java Sea; 118: Malacca Strait; 120: Cocos-Keeling/Christmas Island; 141: Bonaparte Coast; 209: Leeuwin.

Species	Taxonomic group	Ecological Impact (1: Little or no disruption, 2: Disrupts single species with little or no wider ecosystem impact, 3: Disrupts multiple species, some wider ecosystem function, 4: Disrupts entire ecosystem processes with wider abiotic influences)	Impact	Pathway	Eco-regions code
Acanthogobius flavimanus	Vertebrate — fish	3	Compete for food and resources	Biofouling	59
Acar plicata	Invertebrate — Mollusc	2	—	—	113, 141, 209
Alexandrium affine	Algae	1	Generally non-toxic	Ballast Water	112
Alexandrium peruvianum	Algae	4	Produce paralytic shellfish toxins	Ballast Water	117
Alexandrium tamarense	Algae	4	Produces very potent PSP neurotoxins which can affect humans	Ballast Water	115

(*Continued*)

Table 4.3. *(Continued)*

Species	Taxonomic group	Ecological Impact (1: Little or no disruption, 2: Disrupts single species with little or no wider ecosystem impact, 3: Disrupts multiple species, some wider ecosystem function, 4: Disrupts entire ecosystem processes with wider abiotic influences)	Impact	Pathway	Eco-regions code
Alexandrium taylori	Algae	3	Cause PSP, diarrhetic shellfish poisoning, amnesic shellfish poisoning and ciguatera fish poisoning. Impact economy by impacting fisheries, as well as human health and tourist activities.	Ballast Water	117
Alitta succinea	Invertebrate - Annelid	3	Preys upon native species; changes sediment chemistry	Biofouling	59, 209
Alosa sapidissima	Vertebrate - fish	3	May change food-web interactions	-	59
Amathia distans	Bryozoan	1	Ship fouling	Ballast Water and Biofouling	50, 59, 104, 117

Ascidiella aspersa	Tunicate	2	Competes with other native filter-feeding fauna of economic importance like scallops, mussels, and oysters	Biofouling	104, 209
Asparagopsis armata	Plant	4	Dominate many algal assemblages	Biofouling	59
Balanus amphitrite	Invertebrate - Arthropod - Crustacea	1	Major fouling organism	Biofouling	59
Balanus improvisus	Invertebrate - Arthropod - Crustacea	4	Competes for food and resources with other species	Biofouling	59
Boccardiella ligerica	Invertebrate - Annelid	2	Attain high densities in non-native ecosystems	-	59
Bonnemaisonia hamifera	Algae	2	Carpets rocks, competing with native species	-	59
Botryllus schlosseri	Tunicate	3	Aquaculture	Ballast Water and Biofouling	59, 113, 141, 209

(*Continued*)

Table 4.3. (*Continued*)

Species	Taxonomic group	Ecological Impact (1: Little or no disruption, 2: Disrupts single species with little or no wider ecosystem impact, 3: Disrupts multiple species, some wider ecosystem function, 4: Disrupts entire ecosystem processes with wider abiotic influences)	Impact	Pathway	Eco-regions code
Botryllus violaceus	Tunicate	3	Reduces the available space for other fouling species,	Biofouling	59
Bugula neritina	Bryozoan	1	Ship fouling	Biofouling	50, 52, 59, 112, 113, 141, 209
Bryopsis pennata	Algae	1	Thrives in eutrophic condition; produces chemicals that are toxic to herbivores	—	118
Carcinus maenas	Invertebrate - Arthropod - Crustacean	3	Preys upon and dominates native species	Ballast Water	59, 209

Caulerpa taxifolia	Plant	4	Out-competes other Algal species, seagrasses & sessile invertebrates for food & light	Ballast Water	59
Ceratostoma inornatum	Invertebrate - Mollusc	3	Destroys populations of oysters, especially Crassostrea gigas	-	59
Ciona intestinalis	Tunicate	4	Negative affects on aquaculture and native species; destruction of habita	Biofouling	59. 209
Cladophora sericea	Algae	3	Survive in a variety of conditions and stress levels; thrives in eutrophic conditions	Biofouling	59
Codium fragile ssp tomen-tosoides	Algae	4	Nutrient depletion; causes decrease in biodiviersity; aquaculture fouling.	Biofouling	59
Cordylophora caspia	Invertebrate - Cnidarian	3	Fouling of pipes; causes changes in species compositions.	Ballast Water and Biofouling	52, 117

(*Continued*)

Table 4.3. (*Continued*)

Species	Taxonomic group	Ecological Impact (1: Little or no disruption, 2: Disrupts single species with little or no wider ecosystem impact, 3: Disrupts multiple species, some wider ecosystem function, 4: Disrupts entire ecosystem processes with wider abiotic influences)	Impact	Pathway	Eco-regions code
Crassostrea gigas	Invertebrate - Mollusc	3	Degradation of native habitat; domination over native species.	Biofouling	59
Cryptosula pallasiana	Bryozoan	2	Competitive fouling organisms in ports and harbours.	Biofouling	50, 209
Crepidula onyx	Invertebrate - Mollusc	2	Attach to shells of native mollusk	-	113
Didemnum vexillum	Ascidian	4	Changes the composition of benthic communities	Biofouling	59
Ectopleura crocea	Invertebrate - Cnidarian	1	Foulin	Biofouling	59, 104, 120, 209
Garveia Franciscana	Invertebrate - Cnidarian	4	Competition, habitat change,	Ballast Water and Biofouling	59

Ficopomatus enigmaticus	Invertebrate - Annelid	4	Nuisance fouler of power plant intake pipes and recreational boat hulls	Ballast Water and Biofouling	509
Gemma gemma	Invertebrate - Mollusc	3	Displaces and competes for resources with native species.	-	59
Geukensia demissa	Invertebrate - Mollusc	3	Fouling organism	Ballast Water and Biofouling	59
Grandidierella japonica	Invertebrate - Arthropod - Crustacea	2	Competes with native species for resources	Ballast Water	59
Godiva quadricolor	Invertebrate - Mollusc	2	Predator of native sea slugs	Ballast Water and Biofouling	209
Gymnodinium catenatum	Algae	2	Produces toxins responsible for PSP (Paralytic Shellfish Poisoning). Causes economic losses to fisheries; risk to human health.	Ballast Water	59, 115

(*Continued*)

Table 4.3. (*Continued*)

Species	Taxonomic group	Ecological Impact (1: Little or no disruption, 2: Disrupts single species with little or no wider ecosystem impact, 3: Disrupts multiple species, some wider ecosystem function, 4: Disrupts entire ecosystem processes with wider abiotic influences)	Impact	Pathway	Eco-regions code
Heterosiphonia japonica	Algae	4	May displace other Algal species	Ballast Water	59
Hypnea musciformis	Algae	3	Outcompetes native Algaey by reproducing rapidly	Biofouling	101, 104
Kappaphycus alvarezii	Algae	3	Can smother coral reefs	-	104, 117
Monocorophium acherusicum	Invertebrate - Arthropod - Other	2	Compete with natives for resources	Ballast Water and Biofouling	50, 52, 101, 104, 112, 113
Laticorophium baconi	Invertebrate - Arthropod - Other	2	Compete with natives for resources	Ballast Water and Biofouling	52, 113

Monocorophium insidiosum	Invertebrate - Arthropod - Crustacean	2	Change sediment dynamics	Biofouling	52, 113
Monocorophium sextonae	Invertebrate - Arthropod - Other	2	Modifies sediment habitats; fouling	Ballast Water and Biofouling	209
Musculista senhousia	Invertebrate - Mollusc	4	Form dense mats, affect native species due to food competition	Ballast Water	59
Myosotella myosotis	Invertebrate - Mollusc (Snail)	1	Compete with native snails for food	Biofouling	59
Mytilopsis sallei	Invertebrate - Mollusc	3	Habitat domination	Ballast Water and Biofouling	52, 104, 112, 113, 115, 116, 117, 118, 141
Oreochromis mossambicus	Vertebrate - fish	4	Impacts native species through direct predation & transmission of diseases	-	59

(*Continued*)

Table 4.3. (*Continued*)

Species	Taxonomic group	Ecological Impact (1: Little or no disruption, 2: Disrupts single species with little or no wider ecosystem impact, 3: Disrupts multiple species, some wider ecosystem function, 4: Disrupts entire ecosystem processes with wider abiotic influences)	Impact	Pathway	Eco-regions code
Phyllorhiza punctata	Invertebrate - Cnidarian	4	Threatens large commercial fisheries, compete with food source	Ballast Water and Biofouling	59
Penaeus monodon-type Baculovirus	Virus	3	Increase gill & surface fouling by epibionts; impact aquaculture	-	104, 118
Plotosus lineatu	Vertebrate - fish	3	Competes with native species for resources	-	52, 209
Polyandrocarpa zorritensis	Tunicate	3	Competes with natives for resources.	Ballast Water and Biofouling	59
Polydora cornuta	Invertebrate - Annelid	3	Compete with natives for resources	Ballast Water and Biofouling	52, 59, 113

Pseudopolydora paucibranchiata	Invertebrate - Annelid	3	Habitat domination	Ballast Water and Biofouling	59
Ruditapes philippinarum	Invertebrate - Mollusc	4	Habitat domination		59
Sabella spallanzanii	Invertebrate - Annelid	3	Affects nutrient cycling	Ballast Water and Biofouling	209
Sphaeroma quoianum	Invertebrate - Arthropod - Crustacean	4	Affects habitat by causing erosion	Biofouling	50, 59
Sargassum muticum	Algae	4	Biofouling nuisance on fisheries	Ballast Water and Biofouling	59
Sphaeroma quoyanum	Invertebrate - Arthropod - Crustacean	4	Affects habitat by causing erosion	Biofouling	59
Sphaeroma terebrans	Invertebrate - Arthropod - Crustacean	3	A threat to mangrove roots by eroding the roots	Biofouling	141, 209

(*Continued*)

Table 4.3. (*Continued*)

Species	Taxonomic group	Ecological Impact (1: Little or no disruption, 2: Disrupts single species with little or no wider ecosystem impact, 3: Disrupts multiple species, some wider ecosystem function, 4: Disrupts entire ecosystem processes with wider abiotic influences)	Impact	Pathway	Eco-regions code
Sphaeroma walkeri	Invertebrate - Arthropod - Other	2	Aquaculture	Ballast Water and Biofouling	52, 59, 101, 112, 113
Styela clava	Tunicate	4	Competes for food and resources with native and aquaculture species.		59
Styela plicata	Tunicate	2	Aquaculture	Ballast Water and Biofouling	59, 113
Suberites zeteki	Invertebrate - Porifera (Sponges)	1	Likely to compete with native specie	Biofouling	120
Tenellia adspersa	Invertebrate - Mollusc	2	Affects biocommunities	Ballast Water	59
Teredo bartschi	Invertebrate - Mollusc	3	Filter feeders that may compete with native species for resource	Biofouling	59

Teredo navalis	Invertebrate - Mollusc	1	Ship fouling (boring organism)	Biofouling	50, 52, 59, 112, 115, 116, 117, 118, 209
Theora lubrica	Invertebrate - Mollusc	3	Habitat domination	Ballast Water	59
Tridentiger trigono-cephalus	Vertebrate - fish	3	Fouling organism	Ballast Water and Biofouling	59, 209
Ulva fasciata	Algae	3	Fouling organism	Biofouling	59
Virus: WSBV	Virus	3	—	—	50, 52, 112, 113, 116
Watersipora arcuata	Ectoprocta	1	Fouling on vessels	Biofouling	50, 59, 209
Undaria pinnatifida	Plant	3	Habitat domination, impact aquaculture	Ballast Water and Biofouling	59

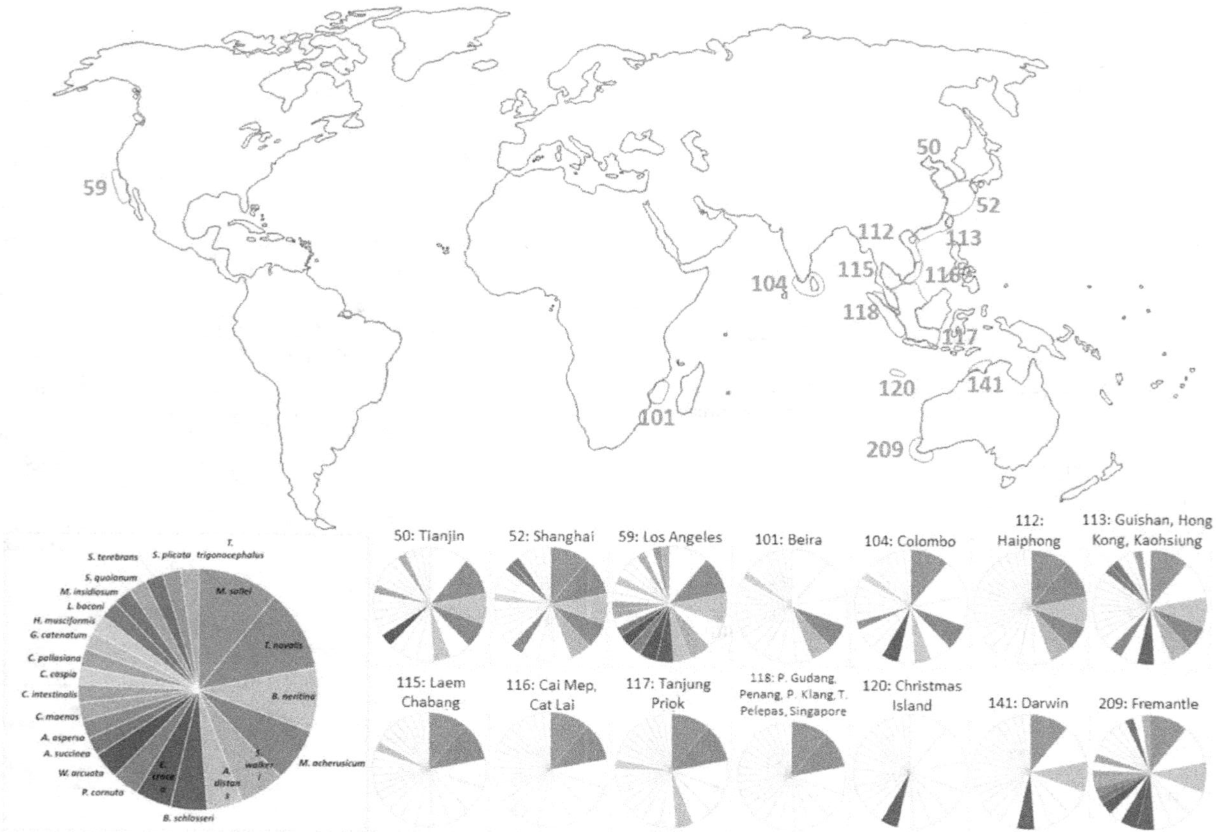

Fig. 4.4. Distribution of non-indigenous marine species based on 14 marine eco-regions (numbers referred to the marine eco-regions outlined in Table 4.3) of which the 21 international ports are located in. There is a higher number of NIMS records in temperate regions compared to tropical regions.

strategies can then be adopted effectively to manage marine bio-invasions. In this paper, a simple risk assessment matrix was constructed and applied using vessel movement information to assess the level of risk of NIMS transfer by transiting vessels in Keppel Terminal.

Vessels Residence Time in Port

There is a high turnaround time for vessels in a monthly operating cycle out of Keppel Terminal (almost 85% of vessels spent 1–3 days, while more than 92% spent seven days or less in port). The variation of the vessel arrival numbers over vessel residence time over three one-month cycles were similar in these three different periods, given that the trend lines for all three months were readily comparable by a close fit of the hyperbolic curve ($p < 0.01$) (Fig. 4.2). Given the quick turnaround time of vessels, the majority of vessels did not stay for long duration in port and only a small number of vessels were long-stay vessels. The rest would have likely made multiple port calls during the month, based on departure and arrival records.

The vessel residence time has several relevant implications. First, the number of shipping vessels spending long periods of time in local port waters is low. Therefore, the likelihood of the majority of vessels taking up biofouling is likely to be low, because of the short transit time spent in port waters. This could be supported by the correlation between vessel residence time and biofouling settlement, that reduced vessel port duration led to reduced ship biota as a result of faster port handling (Carlton, 1985). Similarly, the duration that reproducible biofouling organisms on vessels (especially in niche areas, sea chests, etc.) spent in port would likely be reduced given that vessel residence time is low for the majority of vessels. In actual circumstances, it would be operationally difficult to inspect all vessel arrivals for biofouling, given the limited time and the large area to cover on a vessel. For example, it would take 6–12 hours for two dive vessels operating concurrently to inspect and clean a VLCC at port anchorage (Hydrex White Paper No. 5). It is unlikely that underwater inspection can be enforced for all vessels given the short residence time in port. In-water inspection and cleaning by trained personnel, such as those outlined in the provisions under the *2011 Guidelines for the Control and Management of Ships' Biofouling to Minimize the Transfer of Invasive Aquatic Species* (Biofouling Guidelines), as opposed to the more costly dry-docking process, can therefore be practicable for vessels with longer scheduled time in port to manage NIMS presence on vessels, making in-water inspection and cleaning a viable solution to manage NIMS introductions for the port state. This will make the identification of long-stay vessels important for conducting biofouling inspection in-water. In addition, regular inspection of vessels ballast and biofouling management documentation could be used to generate a list of 'well maintained' and/or 'repeat offender' vessels. This list could then be used to selectively manage vessels. For example, if a regularly visiting vessel is consistently being flagged as 'high risk' but their documentation demonstrates that they are actively and effectively managing their ballast and biofouling risk, then the number of inspections that may be required could be reduced. Conversely, a vessel that repeatedly shows little regard for managing their ballast and biofouling risk could be flagged as requiring inspection at every opportunity.

Second, vessel residence time could be important when implementing measures such as vessel inspection and adopting acceptable biofouling standards for vessels intending to stay for long periods in port. A document such as the Craft Risk Management Standards (Ministry of Primary Industries, New Zealand) is useful when committing similar standards for acceptable biofouling standards to ensure minimum cleanliness before arrival or departure. The CRMS has very clear standards relating to acceptable level of biofouling on short staying (20 days and less) versus long staying vessels (21 days or more). This involves setting limits for the amount of allowable biofouling growth and general biofouling taxa on various areas of a vessel depending on how long the vessel is staying. For example, the CRMS doctrine specifically mentioned that for short-stay vessels, a further allowance is included due to the lower likelihood of introductions from vessels staying for a short time. The allowance is restricted to early stage biofouling to minimise the introduction of non-indigenous species. The doctrine also set more stringent limits for biofouling coverage on long staying vessels, including incidental coverage of less than 1–5% biofouling coverage depending on different areas of the vessel.

Thirdly, like vessel biofouling, sampling and analysis of vessel ballast water will pose a constraint when vessel operations impede ballast water to be safely accessed and sampled within a short duration of time by inspectors. A lack of standardised sampling and analysis procedures (Carney *et al.*, 2013; Frazier *et al.*, 2013), such as defining the location to sample ballast water from a vessel, is also likely to hamper inspection and affect port state inspectors to obtain a 'representative' ballast water sample for compliance testing (Regulation D2, BWM Convention). For the local port state, it is relevant to consider the risk of NIMS presence in the vessel ballast and identify/prioritise high risk vessels for re-mediation measures (e.g., discharge at reception facilities).

While it may be possible to determine the viability of organisms inside the ballast tank based on the age of the ballast water (Gollasch *et al.*, 2000, who estimates that viability of zooplankton taxa reduces significantly after three days and negligible after ten days), it could be difficult to establish the age of the ballast water taken at the primary port, as many large shipping operators practice mid-ocean ballast water exchange (Regulation D2, BWM Convention) to minimise the risk of viable propagule from the primary port to the recipient port. Moreover, given the close proximity of major ports in Southeast Asia, the success of completing a full ballast water exchange cycle according to BWM Convention standards (i.e., 95% volumetric exchange of ballast water) may be limited. Ballast water exchange is largely perceived to be an intermediate measure before numeric discharge concentration limits are formalised in the standards (Albert *et al.*, 2013). While it is possible to target vessels that have not undergone any ballast water exchanges or on-board treatment from the ballast water record book (Regulation B2, BWM Convention) for inspection, it also does not indicate that vessels that had not undergone ballast water exchanges will increase the risk of NIMS presence in the ballast water, especially when the water uptake is at the same location as the discharge. Therefore, information on the vessels' last port of call will provide a better indication of the source location of the ballast water uptake and the likelihood of NIMS presence during transit.

Geographical Distribution of Marine Invasive Species

In an earlier study, Molnar *et al.* (2008) reported the geographic distribution of NIMS worldwide, synthesising from over 350 invasive databases worldwide and reported findings of >300 marine invasive species throughout the marine eco-regions of the world. These included their geographic distribution, impacts on native biodiversity, and introduction pathways. These eco-regions are stratified by identification of unique geographical and relatively homogenous biological composition (Spalding *et al.*, 2007). This information was useful because it integrated marine invasive species data from diverse sources in a uniform manner and referenced these species information with the same geographic units (i.e., marine eco-regions), documenting the likely introduction pathways for each species and assigning a scoring threat of each NIMS to native biodiversity. Although the study was completed some years ago, it represented one of the most comprehensive set of global data to date at this scale. The regional distribution of NIMS could be illustrated by looking at the occurrence of invasive species in the marine eco-regions.

Based on the vessels' last port of call from November 2014, 21 ports (including Keppel Terminal) were located in 14 marine eco-regions. 15 of the 21 ports are located in Asia. The top three ports based on the number of arriving vessels were from regional ports (i.e., Laem Chabang (Thailand), Port Klang and Tanjung Pelepas (Malaysia)). A total of 74 distinct species were recorded from these 14 eco-regions although only 24 were reported to be distributed by 1) ballast water or biofouling (or both) and 2) known to occur in more than one eco-region (Fig. 4.4). These numbers are however, likely to be underreported. For example, the paralytic shellfish poisoning dinoflagellate, *Gymnodinium catenatum*, was described in the database as occurring in both Southern California Bight (eco-region code 59) and Gulf of Thailand (115). However, the database should also include the Malacca Strait (118), as the species was recorded in the Malacca Straits (Holmes *et al.*, 2003).

Eight species, *Botryllus schlosseri* (found in four eco-regions), *Ectopleura crocea* (4), *Amathia distans* (4), *Sphaeroma walker* (5), *Monocorophium acherusicum* (6), *Bugula neritina* (7), *Mytilopsis sallei* (9), *Teredo navalis* (9) were found in four or more eco-regions. Significantly, *B. schlosseri* and *M. sallei* have high impact on aquaculture disruption and coastal infrastructure. Many species also demonstrate wide latitudinal and longitudinal distributions, and occupy both mid-temperate to tropical environments (Fig. 4.4). It appeared that temperate regions have higher numerical records of invasive species (Fig. 4.4). This may have significance on highlighting the risk impact for NIMS transfer on long haul voyages originating from these regions. It is likely that focus on geographical distribution of high risk species and its survivability in recipient port conditions will necessitate further studies. One possibility is targeting monitoring for specific high risk species, e.g. the presence of toxins-producing algae, *Alexandrium sp.* and *Gymnodinium sp.* in SEA that causes high impact on aquaculture and human health.

By studying vessel movement between ports, it is possible to identify shipping routes across marine eco-regions documented to have high risk species. This could form the basis for a risk assessment matrix to identify vessels originating from 'hotspots' for marine invasive species. Applying the rule of precautionary principle, inspection for vessels plying shipping routes between marine eco-regions carrying high risk species could be mandated to carry out BWE or ballast water treatment. Occurrences of NIMS in a marine eco-region, for example in the event of an outbreak of harmful algal species, may be used to alert vessels which have taken in ballast. It may then be necessary to mandate ballast water treatment and sampling before discharge is allowed in the recipient port.

Conversely, vessels plying shipping routes between eco-regions which have similar documented species may possibly be exempted from detailed ballast water sampling in port. Based on the findings above, the eco-regions of Malacca Strait, of which Port Klang and Tanjung Pelepas are located in, and the Gulf of Thailand (Laem Chabang) have similar records for NIMS in these eco-regions (Molnar *et al.*, 2008), and vessels plying between these ports could possibly harbour similar NIMS composition in both biofouling and ballast water community. This may fulfil the criteria for Species' biogeographical risk assessment, and to a lesser extent, the Environmental matching risk assessment (by virtue of the proximity between the two adjacent eco-regions) under the G7 Guidelines (Regulation A4, BWM Convention), which allow vessels to be exempted from inspection. By mapping the geographical distribution of NIMS across marine eco-regions and therein coastal ports, it is possible to identify exemption zones for vessels plying ship routes between these ports; this is consistent with the G7 Guidelines which state that vessels operating only 'between specified ports and fixed routes' to be exempted from the BWM Convention.

A Simple Risk Assessment Matrix to Assess the Risk of NIMS Transfer on Vessels

From the vessel movement records, a simple risk assessment matrix using two key information in 1) vessel residence time and 2) biogeographic similarity of the vessels' last port of call, is proposed to assess the level of risk of NIMS transfer by transiting vessels in Keppel Terminal. These two key information are based on two basic principles discussed earlier, firstly, that the uptake and transfer of biofouling organisms (and its propagules) generally associated with long vessel residence time in port may increase the chances or risks of organisms transfer while the vessel is in port. Secondly, information on the last port of call of vessels provides an indication of the biogeographic distribution of organisms that a vessel may have been exposed to from that marine eco-region, especially in the uptake of ballast water from a vessel's last port of call. Operationally, this allows port inspectors to assess the overall risk level of vessel arrivals and target high risks vessels for precautionary and remediation measures, such as implementing vessel cleaning before arrival, and allow 'no de-ballasting' orders to be issued upon the vessel arrival, e.g., in the event of an algal bloom outbreak at the vessels' last port of call, preventing the discharge of such harmful algal species in Singapore. This matrix can see a design such as one shown in Fig. 4.5, where the use of vessel residence time (i.e., short duration of seven days or less or long duration of more than seven days) can be combined with biogeographic distinctiveness (i.e., similar or distinct), to ascertain the overall risk of NIMS transfer. According to Fig. 4.4, similar marine eco-regions may refer to those outlined in eco-region 115, 116 and 118, where there is an almost equal biogeographic distribution of NIMS. Distinct marine eco-regions may then refer to other eco-regions other than 115, 116 and 118. A risk rating of 1 for acceptable risk (i.e., short duration and similar biogeographic distinctiveness) versus a risk rating of 2 for non-acceptable risk (i.e., long duration and distinct biogeographic distribution) may then be used to calculate the overall risk. A three colour-code scheme can be derived to illustrate the overall risks of a vessel (e.g., green = low risk, yellow = medium risk, red = high risk). The number of vessel inspections may be targeted for vessel arrivals that attain a 'high risk' possibility. This will provide a rapid identification of high risk vessels and prioritisation of

		Vessel residence time	
		Short (≤ 7 days)	Short (> 7 days)
Biogeographic similarity	Similar	Overall risk = 1×1	Overall risk = 1×2
	Distinct	Overall risk = 2×1	Overall risk = 2×2

Fig. 4.5. Risk assessment matrix using vessel residence time in Keppel Terminal and biogeographical similarity of the last port of call. Using this matrix for vessel arrivals from November 2014, 11 vessels were found to be in a 'high risk' (i.e., red zone). Vessel arrivals from May 2014 and February 2015 were not represented here.

non-compliance inspection or other vessel management strategies for vessels likely to harbour NIMS.

The matrix is illustrated using the vessel movement information recorded from November 2014. An overall low-medium risk may be associated with Keppel Terminal based on vessel residence time and biogeographic similarity, as 97.4% (562 vessels) of total vessel arrivals were found to occupy the low and medium risk zones, while 11 vessels were found to be in a 'high risk' zone. These 11 vessels were long-stay vessels not arriving from the two adjacent eco-region 118 (i.e., Singapore, Port Klang, Pasir Gudang, Penang and Tanjung Pelepas) and eco-region 115 (i.e., Laem Chabang). This simple scoring system allows for identification of vessels to comply with, or be exempted from inspection criterion. Since this risk assessment matrix narrows the large number of vessel arrivals in port to identify high risk vessels, prioritisation of non-compliance inspection or other vessel management strategies (e.g., mandatory vessel cleaning in reception facilities) for these 'high risk' vessels can be done to facilitate vessel action plans for managing NIMS transfer in port.

Conclusion

Shipping has facilitated the transfer of NIMS across the oceans, allowing non-native species to cross water bodies that were once physically isolated by continental landmasses and shear distances. Modern shipping has overcome this barrier to introductions through shortened transit time and the creation of 'refuge spots' in vessel niche areas for these marine hitchhikers. There is now a greater resolve to deal with the introduction of NIMS through shipping, notably through the IMO resolutions on BWM Convention in 2004 and more recently through the Biofouling Guidelines in 2011. These developments signify a closer attention towards the risk and impact of shipping as a vector mechanism for NIMS transfer. By nature of global trade, the shipping industry has international maritime operations and therefore stands out as a key driver for managing NIMS transfer.

Effective marine biosecurity policies are required to provide effective management and prevention of NIMS introduction to protect native marine ecosystems. A broader approach is necessary to formulate a strategic framework to manage biosecurity, using both the understanding of marine invasive species and the dynamics of shipping network, which will enable port state controls to implement practicable solutions that are uniformly adopted. Marine biosecurity challenges transcend geopolitical boundaries and therefore these issues cannot be managed in isolation. Singapore's maritime interests and responsibilities goes beyond national boundaries, and remains vulnerable to the problems of NIMS transfer in the region, both as a potential donor and recipient environment, due to her geographical significance as a major transhipment hub port as well as having a biodiverse marine environment.

Therefore, there is a need to leverage on both the BWM Convention and the Biofouling Guidelines to manage the risk of NIMS transfer through shipping. By combining information on vessel movement, this is a key to identify high frequency shipping routes between marine eco-regions with marine species harmful to the environment. These have significance for port state controls who are drafting action plans to manage NIMS transfer through shipping, including the identification and inspection of such vessels.

The key results from this study suggested that approximately 85% of vessel arrivals in Keppel Terminal were short stay vessels, exhibiting a turnaround time of around 1–3 days while around 5–7% of vessels were observed to stay in port for more than seven days in a one-month cycle. These vessels were recorded to port multiple times in a month, possibly operating within the region. The top three ports based on number of arriving vessels, were from regional ports (i.e., from November 2014: Laem Chabang (Thailand), Port Klang and Tanjung Pelepas (Malaysia) and from February 2015: Port Klang and Tanjung Pelepas (Malaysia), Tanjung Priok (Indonesia)).

While physical inspections of biofouling and ballast water sampling on all vessels may place operational and resource limits on the port state controls, it may be possible to identify and prioritise inspection for vessels at risk of harbouring high risk NIMS. Using two key features in vessel residence time and biogeographic similarity of the vessels' last port of call, a simple risk assessment matrix may be used to improve identification of vessels harbouring high risk species. This will be key to formulating vessel action plans for managing NIMS transfer for vessels in port, especially to environmentally sensitive/dissimilar regions.

Funding

This research did not receive any specific grant from funding agencies in the public, commercial, or not-for-profit sectors.

Acknowledgement

This paper is part of an original research work submitted for the partial fulfilment for the degree of Masters of Science (Environmental Management), at the National University of Singapore. The authors gratefully thanked and acknowledged the anonymous reviewer's comments.

References

2004 International Convention for the Control and Management of Ships' Ballast Water and Sediments. International Maritime Organization, adopted in London, UK, February 2004.

2011 Guidelines for the Control and Management of Ships' Biofouling to Minimize the Transfer of Invasive Aquatic Species. International Maritime Organization. Annex 26. Resolution. *MEPC* **207**(62).

Albert, R.J., Lishman J.M., Saxena J.R. (2013). Ballast water regulations and the move toward concentration-based numeric discharge limits. Ecological Applications, **23**(2), 289–300.

Brook, B.W., Sodhi N.S., Soh M.C.K., and Lim H.C. (2003). Abundance and projected control of invasive house crows in Singapore. Journal of Wildlife Management, **67**(4), 808–817.

Capineri, C., and Randelli F. (2007). Freight transportation flows: New trade regions and trade routes. *European Journal of Transport and Infrastructure Research*, 7(2), 93–112.

Carlton, J.T. (1985). Transoceanic and interoceanic dispersal of coastal marine organisms: The biology of ballast water. *Oceanography Marine Biology Annual Review*, **22**, 313–371.

Carney, K.J., Basurko O.C., Pazouki K., Marsham S., Delany J.E., Desai D.V., Anil A.C., and Mesbahi E. (2013). Difficulties in obtaining representative samples for compliance with the Ballast Water Management Convention. *Marine Pollution Bulletin*, **68**, 99–105.

Craft Risk Management Standard: Biofouling on vessels arriving to New Zealand, 2014. Ministry for Primary Industries, New Zealand.

David, M., Gollasch S., and Leppakoski E. (2013). Risk assessment for exemptions from ballast water management — The Baltic Sea case study. *Marine Pollution Bulletin*, **75**, 205–217.

Ducruet, C., and Merk O. (2013). Examining container vessel turnaround times across the world. Port Technology International, Ed 59, pp. 19–20.

Frazier, M., Miller A.W., Lee H., and Reusser D.A. (2013). Counting at low concentrations: the statistical challenges of verifying ballast water discharge standards. *Ecological Applications*, **23**(2): 339–351.

Gollasch, S., Rosenthal H., Botnen H., Hamer J., Laing I., Leppäkoski E., Macdonald E., Minchin, D., Nauke, M., Olenin, S., Utting, S., Voigt, M., and Wallentinus, I. (2000). Fluctuations of zooplankton taxa in ballast water during short-term and long term ocean-going voyages. *International Review of Hydrobiology*, **85**, 597–608.

Hewitt, C.L. (2002). Distribution and biodiversity of Australian tropical marine bio-invasions. *Pacific Science*, **56**(2), 213–222.

Holmes, M.J., Bolch C.J.S., Gree D.H, Cembella A.D., and Teo S.L.M. (2002). Singapore Isolates of the dinoflagellate Gymnodinium catenatum (Dinophyceae) produce a unique profile of paralytic shellfish poisoning toxins. *Journal of Phycology*, **38**, 96–106.

Hydrex White Paper No. 5 (2011). Underwater ship hull cleaning: cost-effective, non-toxic fouling control.

Jaafar, Z., Yeo D.C.J., and Tan H.H. (2012). Status of estuarine and marine non-indigenous species in Singapore. *The Raffles Bulletin of Zoology, Supplement*, **25**, 79–92.

Kaluza, P., Kölzsch A., Gastner M.T., and Blasius B. (2010). The complex network of global cargo ship movements. *Journal of the Royal Society Interface*, **7**, 1093–1103.

Keller, R.P., Drake J.M., Drew M.B., and Lodge D.M. (2011). Linking environmental conditions and ship movements to estimate invasive species transport across the global shipping network. *Diversity and Distributions*, **17**, 93–102.

Lim, C.H., Sodhi N.S., Brook B.W., and Soh M.C.K. (2003). Undesirable aliens: factors determining the distribution of three invasive bird species in Singapore. *Journal of Tropical Ecology*, **19**, 685–695.

Molnar, J.L., Gamboa R.L., Revenga C., and Spalding M.D. (2008). Assessing the global threat of invasive species to marine biodiversity. *Frontiers in Ecology and the Environ*ment, **6**(9), 485–492.

Ng, H.H., Low B. W., Kwik J.T.B., and Yeo D.C.J. (2014). The tables are turned: an invasive species under potential threat. *Biological Invasions*, **16**(8), 1567–1571.

Ng, P.K.L., Chou L.M., and Lam T.J. (1993). The status and impact of introduced freshwater animals in Singapore. *Biological Conservation*, **64**, 19–24.

Nghiem, L.T.P., Soliman T., Yeo D.C.J., Tan H.T.W., Evans T.A., Mumford J.D., Keller R.P., Baker R.H.A., Corlett R.T., and Carrasco L.R. (2013). Economic and environmental impacts of harmful non-indigenous species in Southeast Asia. *PLoS ONE*, **8**(8), 1–9.

Notteboom, T.E., and Rodrigue J.P. (2005). Port regionalization: towards a new phase in port development. *Maritime Policy and Management*, **32**(3), 297–313.

Pandit, M.K., Tan H.T.W., and Bisht M.S. (2006). Polyploidy in invasive plant species of Singapore. *Botanical Journal of the Linnean Society*, **151**(3), 395–403.

Peh, K.S.H. (2010). Invasive species in Southeast Asia: the knowledge so far. *Biodiversity Conservation*, **19**, 1083–1099.

Seebens, H., Gastner M.T., and Blasius B. (2013). The risk of marine bio-invasion caused by global shipping. *Ecology Letters*, 1–9.

Spalding, M.D., Fox H.E., Allen G.R., Davidson N., Ferdaña Z.A., Finlayson M., Halpern B.S., Jorge M.A., Lombana A., Lourie S.A., Martin K.D.,

McManus, E., Molnar J., Recchia C.A., and Robertson J. (2007). Marine eco-regions of the world: A bioregionalization of coastal and shelf areas. *Bioscience*, **57**(7), 573–583.

Wang, C., and Wang J. (2011). Spatial pattern of the global shipping network and its hub-and-spoke system. *Research in Transportation Economics*, **32**, 54–6.

5

BIOGENIC VOLATILE ORGANIC COMPOUNDS (BVOCs) EMISSIONS BY SELECTED STREET TREES IN SINGAPORE

KHO YUE-MIN VERON and TAN PUAY-YOK

Ecosystem services provided by trees have been widely documented. However, they also emit biogenic volatile organic compounds (BVOCs) like isoprene and monoterpenes into the atmosphere. BVOCs can result in the formation of ground-level ozone which can potentially cause harm to human health. Yet, there is a lack of studies on BVOCs emissions in the Singapore. This study looks into the ecosystem services and disservices provided by trees and focuses particularly on the BVOCs emissions from 39 species of street trees in Singapore. An estimation of the BVOCs emissions is provided with the use of a biogenic emission model. Findings show that the annual isoprene and monoterpenes emissions in Singapore amount to about $4.2\,\mathrm{Tg\,yr^{-1}}$ and $1.1\,\mathrm{Tg\,yr^{-1}}$ respectively The high dependence of BVOCs on temperature implies that BVOCs emissions are likely to increase by at least 20% in the future due to climate change. Comparison of the normalised levels of BVOCs emissions with other countries conclude that the current emissions in Singapore are significantly lower. In order to continue to enjoy the ecosystem services provided by trees, it is recommended for planners to select the right low-emitter species trees when planning for the greening of the environment.

1. Introduction

1.1 *Overview*

Singapore — a Garden City. The Ministry of Environment and Water Resources (MEWR) and Ministry of National Development (MND), in the Singapore Sustainable Blueprint 2015 have indicated goals of achieving more green and blue spaces by 2030. Singapore is set to see, by the year 2030, '9% of land [set aside] for nature reserves, 400 km of park connectors [and] over 200ha of skyrise greenery' (MEWR and MND, 2015). Clearly, the increase in greenery allows for the environment and the community to reap the ecosystem services that trees provide.

However, trees and plants emit BVOCs. This has been classified as a disservice provided by trees (Escobedo *et al.*, 2011). There is a need to study BVOCs as despite being in lower concentrations, BVOCs have a greater impact on ozone concentrations than AVOCs (Bell and Ellis, 2005; Tsui *et al.*, 2008). With increasing propositions for biophyllic designs in urban planning, it is essential to understand the ecosystem services and disservices that trees provide in order to make informed decisions on the species of trees to select in order to create a green cityscape.

This paper thus aims to provide an estimate of BVOCs emissions in Singapore. This paper will also discuss the types of ecosystem services and disservices that trees provide, with a particular focus on BVOCs emissions The impact of factors such as temperature and Photosynthetically Active Radiation (PAR) on BVOCs emissions will also be explored. Finally, the significance of BVOCs emissions in Singapore is evaluated against the emissions recorded in other areas.

1.2 *Definition of Biogenic Volatile Organic Compounds (BVOCs)*

BVOCs are carbon-containing compounds emitted through Earth's natural vegetation into the atmosphere (Steiner and Goldstein, 2007; Williams and Koppman, 2007). BVOCs include terpenoids such as isoprene, monoterpenes and the hexenal family compounds such as hexenals, methanol, and acetone (Guenther *et al.*, 2000). Isoprene and monoterpenes are collectively known as isoprenoids. This study focuses on isoprene and monoterpenes which are the two most common types of BVOCs that are non-methane volatile organic compounds isoprene (NMVOCs).

1.2.1 *Isoprene*

Isoprene is one of the main hydrocarbons emitted mainly by deciduous, woody species (Steinbrecher, 1997; Steiner and Goldstein, 2007). Many discussions have highlighted thermotolerance as one of the functions of isoprene emitted by trees where isoprene is argued to be able to offer some protection against specific heat types for plants, such as heat flecks (Sharkey *et al.*, 2008, Singsaas *et al.*, 1999). Amongst the different types of vegetation, leaves at the top of a canopy of trees are most susceptible to heat flecks due to their exposure to large amount of solar radiation (Sharkey *et al.*, 2008). In addition, isoprene emissions also act as a defence mechanism that acts either as a physical barrier to insects and pathogens and/or as a feeding deterrent if consumed (Guenther *et al.*, 2000). Generally, the amount of isoprene emitted throughout the world annually is estimated to be at 250 to 420 $Tg\,yr^{-1}$ (Riba and Torres, 1997).

1.2.2 *Monoterpenes*

Monoterpenes, predominantly products of the secondary metabolism of plants, are derived formally from the condensation of two isoprene or isopentane units (Banthorpe *et al.*, 1972). Examples of monoterpenes include α-Pinene, β-Pinene and limonene (Peñuealas and Llusià, 2003). Monoterpenes, unlike isoprene are primarily produced by coniferous species (Riba and Torres, 1997; Steiner and Goldstein, 2007) and are stored in specialised structures of the leaves. Monoterpenes are produced as defence mechanisms which provide trees with physical protection against herbivore attacks such as those of the bark beetle (Gijzen *et al.*, 1993; Lerdau *et al.*, 1997). Generally, monoterpenes emission throughout the world annually is estimated at about 128 to 147 Tg yr^{-1} (Riba and Torres, 1997).

1.3 *Main Factors Affecting the Emission of BVOCs*

1.3.1 *Temperature*

Many studies have discussed the dependency of BVOCs emissions on temperature, particularly leaf temperature (Guenther *et al.*, 1993; Guenther *et al.*, 2000; Lamb *et al.*, 1993; Pio *et al.*, 1997; Steinbrecher *et al.*, 1997; Tsui *et al.*, 2008; Velasco, 2003). Isoprene emission was found to increase exponentially with increasing temperatures up to a high temperature (>33°C), after which it would decline gradually. Similarly, monoterpenes emission was found to increase exponentially with increasing temperatures, and diurnal monoterpenes emission variation depends mainly on leaf temperature (Guenther *et al.*, 1993).

1.3.2 *Photosynthetically active radiation*

Photosynthetically Active Radiation (PAR) belongs to the biologically significant waveband of 400–700 nm (Grant, 1997) and governs biological activities (Helas *et al.*, 1997). BVOCs emissions such as isoprene emissions are highly dependent on PAR (Guenther *et al.*, 1993; Guenther *et al.*, 2000; Lamb *et al.*, 1993; Pio *et al.*, 1997; Steinbrecher *et al.*, 1997; Tsui *et al.*, 2008; Velasco, 2003). A study in North America (Guenther *et al.*, 2000) shows that isoprene emissions were higher at lower altitudes due to higher PAR; and Velasco (2003) states that the highest emissions coincided with the highest levels of PAR. Further, the dependence of isoprene emissions on light was proven when the emission of isoprene only occurred during daytime (Tsui *et al.*, 2008).

However, similar to its behaviour with temperature, isoprene emissions' behaviour with PAR is hyperbolic. It has been observed that isoprene emissions would increase with increasing light up to a saturation point, where increasing light would result in a smaller increase in the biosynthesis of isoprene (Lerdau *et al.*, 1997; Steiner and Goldstein, 2007).

On the other hand, monoterpenes are commonly found to be light-independent, especially in the long run. Studies have shown that monoterpenes are present throughout the day (daytime and nighttime) (Velasco, 2003; Tsui *et al.*, 2008)

1.3.3 *Taxonomic group and physiology of trees*

1.3.3.1 Taxonomic group of trees

One of the factors that would determine the amount of isoprenoids emitted by land vegetation would include the taxonomy of trees that are located in the area. Previous studies have highlighted families such as Moraceae, Fagaceae, Salicaceae and Myrtaceae to contain emitting genera (Harley *et al.*, 1999; Karlik and Winer, 2001). Particularly, it has been observed that the Salicaceae and Myrtaceae families appear to contain species with high emission rates (Benjamin *et al.*, 1996). A publication on tropical trees in Okinawa Island concluded that most of the studied trees belong within the low emitting species with the exception of species from the Moraceae family (Tambunan *et al.*, 2006). However, not all species belonging to a family would take on similar emission potentials. For instance, Geron *et al.* (2001) highlight that within the oak species there was a range of emission potentials. Furthermore, there can be differences in the emission potentials of isoprene and monoterpenes for a species. It has been observed that isoprene was mainly emitted by broadleaved deciduous trees such as Oaks and Eucalyptus while monoterpenes were largely emitted by pines and other coniferous trees (Karlik and Pittenger, 2012; Steiner and Goldstein, 2007).

1.3.3.2 Physiology of trees

The physiological aspects of a tree that could affect BVOCs emission would be leaf development and the age of the leaves. It has been widely observed (Burbott and Loomis, 1967; Kuzma and Fall, 1993; Street *et al.*, 1997) that as leaves age, the potential for BVOCs emission would become lower. Further, Burbott and Loomis (1967) found that the amount of monoterpenes emitted per leaf has a positive relation to the size of the leaf. In addition, the amount of vegetation cover and foliar density also affect the BVOCs emission rates. Tsui *et al.* (2008) has shown that due to the higher amount of vegetation cover in Hong Kong than in Beijing, the emission rates in Hong Kong would show a positive relation. Apart from the individual leaf's development, age and size, it was found that BVOCs emission was higher during blooming periods of flowering trees in urban areas (Baghi *et al.*, 2012).

1.4 *Impacts of Isoprenoids on the Atmosphere*

Increasing studies have shown that BVOCs play an important role in the formation of tropospheric ozone (Duane *et al.*, 2002; Guenther *et al.*, 1994; Lerdau *et al.*, 1997). Collectively, isoprenoids can result in the formation of ozone, especially in areas with high concentrations of nitrous oxides (NO_X) (Calfapietra *et al.*, 2013). Curci *et al.* (2009) highlight that the highest amount of ozone was recorded in urban areas which were highly populated and industrialised, characterised by a high level of nitrous oxides (NO_X).

Due to its short lifetime in the troposphere, isoprene itself has no direct influence on the atmosphere. However, its by-products and relative effects on other compounds in the atmosphere can contribute to global warming and result in substantial impacts on the atmosphere. Isoprene oxidises to form carbon monoxide (CO) that has a longer lifetime in the atmosphere than the isoprene itself, which allow sample time for further oxidation to take place to form Ozone. CO can then be oxidised with other volatile organic compounds (OVOCs) in the presence of a significant concentration of nitric oxide, to produce large quantities of ozone (Fehsenfeld, 1995; Steiner and Goldstein, 2007) — a type of greenhouse gas that contributes to global warming. Further, isoprene's high reactivity can affect global temperature through its impact on methane's lifetime in the atmosphere. (Wuebbles *et al.* 1989). Due to the high reactivity nature of isoprene as compared to methane, isoprene would react with existing nitric oxide present in the atmosphere. This results in a decreased amount of nitric oxide in the atmosphere to react with methane. (Lerdau *et al.*, 1997). As such, this would result in higher level of methane concentration in the atmosphere over time and contribute to increased greenhouse gases and global temperature.

2. Literature Review

2.1 *Ecosystem Services*

Brown *et al.* (2007) refer to ecosystem services as the flows from an ecosystem that are of a relatively immediate benefit to humans and should occur naturally (i.e. existing without human intervention). However, with urbanisation and changing structure of urban areas, it is unlikely that ecosystems would remain untouched. Thus, for the purpose of this paper, ecosystem services refer to 'the components of urban forests that are directly enjoyed, consumed, or used to produce specific, measurable human benefits' (Escabedo *et al.*, 2011). The services taken into consideration would include those that are provided by semi-natural systems such as street trees. It is in this paper that street trees are considered as an urban ecosystem. Street trees are defined as 'stand-alone trees, often surrounded by paved ground' (Bolund and Hunhammar, 1999). The different types of ecosystem services provided by trees are extensively covered in literature (Bolund and Hunhammar, 1999; Brown *et al.*, 2007; Escabedo *et al.*, 2011; Nowak and Dwyer, 2007) and include but are not limited to the following:

2.1.1 *Air filtering*

Trees are capable of filtering the air to reduce gaseous air pollution caused by both anthropogenic and non-anthropogenic activities. The filtering is done through the uptake of these pollutants and particulates through the leaf stomata (Beckett *et al.*, 1998). Generally, the larger the surface area of the leaves, the larger the filtering capacity. In Lisbon, it was observed that deciduous trees with large leaves were more effective in improving air quality through the uptake of pollutants. The street trees in Lisbon were estimated to reduce the amount of air pollutants by 25.6t annually, and this was valued at about US$222,738 or US$5.80/tree (Soares *et al.*, 2011). According to Nowak and Dwyer (2007), the average annual removal of pollution per square meter of canopy cover was about 9.3 g which amounts to about US$500 per hectare of tree cover. With the increased amount of air pollutant, particularly in the urban areas due to increased use of automobiles, the lining of streets with trees due to the ability of trees to reduce the amount of air pollutants is a common practice.

2.1.2 *Regulation of micro-climate and rainwater*

Trees are known to be able to ameliorate the urban heat island (UHI) impact, reducing the temperature through evapotranspiration and providing shade to the environment which would have otherwise absorbed the incoming solar radiation (Kleerekoper *et al.*, 2012). As the temperature of the micro-climate is lowered with the presence of trees, relative humidity increases due to evapotranspiration (Jim and Chen, 2009) and this would help to regulate the micro-climate of the urban environment, making it more comfortable. Taha *et al.* (1991) concluded that trees have a high potential to regulate the micro-climate of the urban area and reduce demand for building cooling which would reduce the amount of energy consumed.

Trees can also regulate rainwater supply and quality (de Groot *et al.*, 2002). Water quality is regulated through the filtering function provided by the vegetation cover, while supply is regulated through the presence of trees which help to control the amount of surface runoff by intercepting the rainwater through their leaves. At the same time, the trunks of the trees help to regulate the rainwater that would flow into the ground, reducing the amount of runoff (Nowak and Dwyer, 2007). These would help to increase the storage of the rainwater and hence regulate water supply (de Groot *et al.*, 2002). The reduction of about 292 million m^3 of runoff by trees in Modesto, California has helped to save US$0.61 million which would otherwise have been allocated for retention and detention basins to contain the runoff (McPherson *et al.*, 1999).

2.1.3 *Recreational and cultural services*

On top of providing physical and biological benefits to humans and the physical environment, urban vegetation such as park trees and street trees also provide social and economic benefits. Green spaces can serve as public spaces that allow for the community to interact and bond in the outdoors. This has been associated with the ability to increase the quality of life of urban dwellers by improving their physical and psychological well-being. Similarly, they also allow the community to become more aware of their culture and heritage. The New York City Housing Authority (NYCHA) particularly encouraged this when public housing was undergoing the process of re-greening in an attempt to build a community that is aware of its historical and cultural past.

2.2 *Ecosystem Disservices*

Ecosystem disservices, on the contrary, are costs which negatively affect well-being. These include economic, social and environmental costs. As with ecosystem services, disservices are measured by their immediate importance to an individual's well-being, including their quality of life (Brown *et al.*,

2007). Therefore, the extent of importance of an urban forest is thus highly subjective and dependent on the individual given context, scale and heterogeneity of an urban forest (Escobedo *et al.*, 2011).

Ecosystem problems include, but are not limited to, maintenance activities that contribute to air pollution, infrastructure repair, water consumption, pollen production and chemical emissions from trees that can contribute to urban smog and ozone problems (Bolund and Hunhammar, 1999; Nowak and Dwyer, 2007).

2.2.1 *Maintenance activities that contribute to air pollution*

Urban forests, particularly park trees and street trees, have been transplanted and often require landscaping and maintenance. In cities where the urban forest is manicured, many efforts have been invested in the upkeep of the city-state. The maintenance of these urban forests range from mulching, watering, and pruning to removal and disposal which require equipment that are powered by fossil fuels (Escobedo *et al.*, 2011; Nowak *et al.* 2002). While trees are able to store and sequester CO_2, helping in the reduction of air pollutants, the need to use equipment that may emit other pollutant emissions to maintain and sustain them may override the air filter service that trees provide. Therefore, it is important to consider and evaluate if the maintenance and management of these trees emit more pollutants than the trees can filter.

Greenery in Singapore are considered as high resource input urban forests and are management intensive. The inputs required to sustain them have their respective pollutive characteristics, which may or may not be offset by the ecosystem services provided by the urban trees. Further, the need to maintain and manage these trees could be perceived by some individuals in the community as an inappropriate use of resources that could otherwise have been used to address other more critical urban community issues (Nowak and Dwyer, 2007).

2.2.2 *Chemical emissions from trees*

Extensive knowledge of ecosystem services provided by trees has resulted in the increase in programmes that initiate the planting of trees. This could result in an unintended outcome — a rise in BVOCs emitted by trees. Generally, trees are the biggest isoprene emitters and this can be exacerbated in the tropics where the leaves of trees are broad (Sharkey *et al.*, 2008) and where relative humidity is higher.

3. Methodology

3.1 *Introduction*

The emission rate from street trees in Singapore will be estimated through modelling using the biogenic emission model, which takes into account parameters such as PAR and leaf temperature. These two parameters can account for short-term variations in isoprene emissions (Guenther *et al.*, 1993). This study will be using the G93 isoprene emission rate model to predict the level of isoprene emitted by trees in Singapore. Finally, in order to evaluate the significance of the emission levels in Singapore, a comparison study will be done to compare the emissions in Singapore against previous studies in different cities.

3.2 *Biogenic Emission Model*

The BVOCs emissions estimate is based on the algorithm developed by Guenther *et al.* (1993) to normalise the effects of temperature and solar radiation on BVOCs emissions on an independent leaf level (Geron *et al.*, 1994).

The Isoprene emission rate is obtained from I in Eq. (1) obtained from Model G93 (Guenther *et al.*, 1993) where

$$I = I_S.C_L.C_T \tag{1}$$

where I is the isoprene emission rate (μgg^{-1} h^{-1}) at a temperature $T(K)$ and PAR flux L (μmol m^{-2}s^{-1}), I_s is the isoprene emission rate at a standard temperature of 30°C, and a standard PAR flux of 1000 μmol m^{-2} s^{-1}. The light correction factor, C_L is estimated as

$$C_L = \frac{\alpha C_{L1} L}{(1 + \alpha^2 L^2)^{1/2}} \tag{2}$$

where L is the PAR flux, and $\alpha(= 0.0027)$ and $C_{L1}(= 1.066)$ are empirical coefficients derived from emission rate measurements of several species as discussed by Guenther *et al.* (1993). The temperature correction factor, C_T, is given by

$$C_T = \frac{\exp[C_{T1}(T - T_S)/RT_S T]}{1 + \exp[C_{T_2}(T - T_M)/RT_S T]} \tag{3}$$

where R is the ideal gas constant (8.314 J K^{-1} mol^{-1}), T(K) is the leaf temperature, $T_s(= 303\,\text{K})$ is the standard temperature, and $T_M(= 314\,\text{K})$,

$C_{T1}(= 95000\,\mathrm{J\,mol^{-1}})$, and $C_{T2}(= 230000\,\mathrm{J\,mol^{-1}})$ are empirical coefficients derived by fit to species of several genera (Guenther *et al.*, 1993).

In the case of monoterpenes, M in Eq. (4) is given by

$$M = M_S \cdot \exp[\beta(T - T_S)] \tag{4}$$

where M is the monoterpene emission rate ($\mu\mathrm{gg^{-1}h^{-1}}$) at leaf temperature T(K), M_s is the emission rate at $T_s(= 303\,\mathrm{K})$ and $\beta(= 0.09)$ is an empirical coefficient derived from Guenther *et al.* (1993).

3.3 *Methods*

Secondary data collected for modelling include the daily temperature as well as PAR for the month of June 2015. For PAR, all values will be used despite the PAR levels dropping to minimal levels outside the time window between 0700 to 1900 hours (Tan and Ismail, 2015) in order to derive the implication of PAR on emissions. PAR in micromoles per square meter per second ($\mu\mathrm{molm^{-2}s^{-1}}$) is derived from the following equation taken from Tan and Ismail, (2015):

$$\mathrm{PAR} = 1.867.H \tag{5}$$

where PAR is photosynthetically active radiation ($\mu\mathrm{molm^{-2}s^{-1}}$), H is solar radiation ($\mathrm{Jm^{-2}s^{-1}}$) derived from $H(\mathrm{Wm^{-2}})$ from Eq (6).

$$\mathrm{E} = Pt \tag{6}$$

where E is energy in $\mathrm{Jm^{-2}s^{-1}}$, P is power in $\mathrm{Wm^{-2}}$ and t is time in seconds.

An inventory of roadside trees planted in Singapore, with details including the genus, species and quantity, will be retrieved from literature (Tan *et al.*, 2009). Due to the lack of inventory for tropical forests, the coefficients that are to be used for modelling will be taken from deciduous species in other studies (Geron *et al.*, 1994; Klinger *et al*, 2002) based on the phylogenetic approach (Benjamin *et al.*, 1996) where the coefficients would be assigned based on taxonomic relationships. The emission of a single leaf obtained from modelling would be extrapolated to the whole tree by multiplying the emission of a single leaf by the foliar mass estimates per unit crown coverage for broadleaf trees at $375\,\mathrm{gm^{-2}}$ (Geron *et al.*, 1994).

3.4 *Characteristics of Selected Street Trees*

The selection of the street trees for this study is based on the percentage of the total population each species makes up; a total of 39 street trees were

Table 5.1. Street trees in Singapore (Tan *et al.*, 2009).

S/No	Tree species	Number of trees	Percent of population
1	*Samaneasaman*	26226	9.8%
2	*Peltophorumpterocarpum*	23682	8.9%
3	*Swienteniamacrophylla*	20079	7.5%
4	*Syzygiumgrande*	18106	6.8%
5	*Khayasenegalensis*	14344	5.4%
6	*Tabebuiarosea*	9965	3.7%
7	*Mimusopselengi*	9609	3.6%
8	*Cinnamomuminers*	9371	3.5%
9	*Xanthostemonchrysanthus*	8780	3.2%
10	*Arfeuilleaarborescens*	8666	3.2%
11	*Hopeaodorata*	7022	2.6%
12	*Khayagrandifoliola*	5401	2.0%
13	*Cassia fistula*	5364	2.0%
14	*Syzygiumpolyanthum*	5094	1.9%
15	*Pterocarpusindicus*	4567	1.7%
16	*Mangiferaindica*	4206	1.6%
17	*Cratoxylumformosum*	3502	1.3%
18	*Terminalia catappa*	3231	1.2%
19	*Lagerstroemia flos-reginae*	3195	1.2%
20	*Syzygiumcampanulatum*	3097	1.2%
21	*Erythrophleumguineense*	2791	1.0%
22	*Callistemon species*	2589	1.0%
23	*Fagraeafrangrans*	2456	0.9%
24	*Millettiaatropurpurea*	2389	0.9%
25	*Syzygiumlineatum*	2214	0.8%
26	*Callistemon citrinus*	2128	0.8%
27	*Filiciumdecipiens*	1862	0.7%
28	*Kopsiaflavida*	1755	0.7%
29	*DalbergiaLatiflia*	1644	0.6%
30	*Melaleuca 'Golden gem'*	1496	0.6%
31	*Alstoniascholaris*	1451	0.5%
32	*Maniltoabrowneoides*	1319	0.5%
33	*Khayanyassica*	1267	0.5%
34	*Andirainermis*	1195	0.4%
35	*Sandoricumkoetjape*	1175	0.4%
36	*Dalbergiaoliveri*	1144	0.4%
37	*Tamarindusindica*	1117	0.4%
38	*Azadirachtaexcelsa*	1081	0.4%
39	*Pongamiapinnata*	1043	0.4%

selected. The characteristics of the trees considered include their isoprene and monoterpenes emission factors (also known as emission potentials). Table 5.1 summarises the species of the selected trees, the total population as at 2007, and the percentage of the total population each species makes up, as extracted from Tan *et al.* (2009). Table 5.2 summarises the different

Table 5.2. Summary of BVOCs emissions potentials for all species of selected street trees in Singapore (Klinger *et al.*, 2002).

S/No.	Tree species	Family	Emission factor (I)	Emission factor (M)
1	*Mangiferaindica*	Anacardiaceae – Sumac	0.1	0.1
2	*Alstoniascholaris*	Apocynaceae	0.1	0.1
3	*Kopsiaflavida*	Apocynaceae	0.1	0.1
4	*Tabebuiarosea*	Bignoniaceae	0.1	0.1
5	*Terminalia catappa*	Combretaceae	0.1	0.1
6	*Hopeaodorata*	Dipterocarpaceae	0.1	0.1
7	*Andirainermis*	Fabaceae/ Leguminosae	0.1	3
8	*Cassia fistula*	Fabaceae/ Leguminosae	0.1	3
9	*DalbergiaLatifolia*	Fabaceae/ Leguminosae	0.1	3
10	*Dalbergiaoliveri*	Fabaceae/ Leguminosae	0.1	3
11	*Erythrophleumguineense*	Fabaceae/ Leguminosae	0.1	3
12	*Maniltoabrowneoides*	Fabaceae/ Leguminosae	0.1	3
13	*Millettiaatropurpurea*	Fabaceae/ Leguminosae	0.1	3
14	*Peltophorumpterocarpum*	Fabaceae/ Leguminosae	0.1	3
15	*Pongamiapinnata*	Fabaceae/ Leguminosae	0.1	3
16	*Pterocarpusindicus*	Fabaceae/ Leguminosae	0.1	3
17	*Samaneasaman*	Fabaceae/ Leguminosae	0.1	3
18	*Tamarindusindica*	Fabaceae/ Leguminosae	0.1	3
19	*Fagraeafrangrans*	Gentianaceae	0.1	0.1
20	*Cratoxylumformosum*	Hypericaceae	70	0.1
21	*Cinnamomuminers*	Lauraceae	0.1	0.6
22	*Lagerstroemia flos-reginae*	Lythraceae	0.1	0.1
23	*Azadirachtaexcelsa*	Meliaceae	0.1	0.1
24	*Khayagrandifoliola*	Meliaceae	0.1	0.1
25	*Khayanyassica*	Meliaceae	0.1	0.1
26	*Khayasenegalensis*	Meliaceae	0.1	0.1
27	*Swienteniamacrophylla*	Meliaceae	0.1	0.1
28	*Sandoricumkoetjape*	Meliaceae – Mahogany	0.1	0.1

(*Continued*)

Table 5.2. (*Continued*)

S/No.	Tree species	Family	Emission factor (i)	Emission factor (M)
29	*Melaleuca 'Golden gem'*	Myrtaceae	70	3
30	*Syzygiumcampanulatum*	Myrtaceae	70	3
31	*Syzygiumgrande*	Myrtaceae	70	3
32	*Syzygiumlineatum*	Myrtaceae	70	3
33	*Syzygiumpolyanthum*	Myrtaceae	70	3
34	*Xanthostemonchrysanthus*	Myrtaceae	70	3
35	*Callistemon citrinus*	Myrtaceae – Myrtle family	70	3
36	*Callistemon species*	Myrtoideae	70	3
37	*Arfeuilleaarborescens*	Sapindaceae	0.1	0.1
38	*Filiciumdecipiens*	Sapindaceae – Soapberry	0.1	0.1
39	*Mimusopselengi*	Sapotaceae	0.1	0.1

BVOCs emission potentials for the 39 species of street trees selected for this study.

I: Isoprene, M: Monoterpene. L: low emitter, M: moderate emitter, H: high emitter.

4. Results

4.1 *Emissions*

4.1.1 *Isoprene emission in Singapore*

Single leaf through modelling

The isoprene emission from a single leaf is shown in Fig. X1. The range of emission for the low emitter species is from 0.01 $\mu gg^{-1}h^{-1}$ to 0.10 $\mu gg^{-1}h^{-1}$. The range of emission from high emitter species is from 7.0 $\mu gg^{-1}h^{-1}$ to 71.85 $\mu gg^{-1}h^{-1}$. Generally, the highest level of emissions for both species groups is observed during the period from 13:00h to 14:00h, while the lowest level of emission is observed during the period from 18:00h to 04:00h.

Total emission per tree per species

Figure 5.2 shows the total emissions from the 39 selected species of street trees. The emission level for low emitter species is about 0.35 mgm^{-2}. For high emitter species, the emission level is about 244.27 $mg\,m^{-2}$.

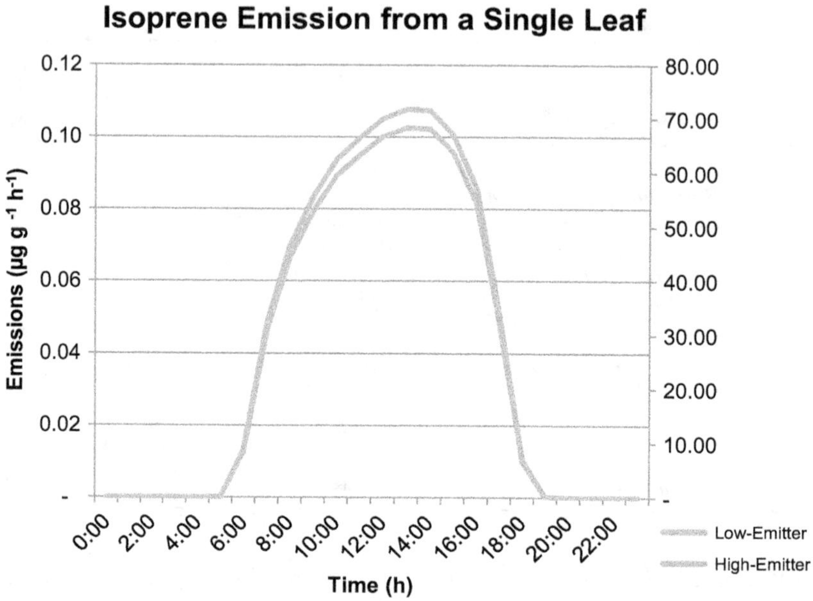

Fig. 5.1. Isoprene emissions from a single leaf of both low and high emitter species.

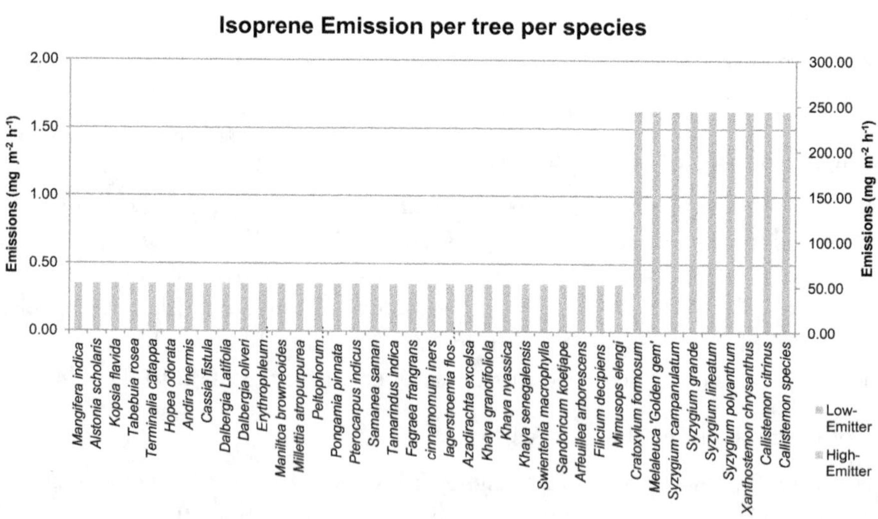

Fig. 5.2. Total isoprene emission from a whole tree per species.

Total Isoprene emission from all selected street trees in Singapore

The total isoprene emission from the selected street trees in Singapore amounts to about $11.5\,\mathrm{g\,m^{-2}h^{-1}}$. The levels of emission by the respective species of selected street trees are shown in Fig. 5.3. Of the 39 selected species of street trees, 77% belong to low emitter species while the remaining 23% are high emitter species. However, it is of note that the population of low emitter tree species collectively contribute to only 1% of the total isoprene emitted by trees in Singapore. The remaining 99% of the emission is accounted for by the 23% of high emitter street trees species. The low emitter species that accounts for a high amount of isoprene emission is the *Samanea saman*, emitting about $9151\,\mathrm{mg\,m^{-2}h^{-1}}$ of isoprene. Due to its large population, the amount of isoprene emitted is relatively higher despite the low emission potential accorded to it. For high emitter species, the isoprene emitted ranges from $0.3\,\mathrm{g\,m^{-2}h^{-1}}$ (emitted by *Melaleuca*) to $4.4\,\mathrm{g\,m^{-2}h^{-1}}$ (emitted by the *Syzygium grande* species).

4.1.2 *Monoterpenes emission in Singapore*

Single leaf through modelling

Figure 5.4 summarises the amount of monoterpenes emission from a single leaf for all emitting species. The range of emission from low emitter species ranges from $0.09\,\mu\mathrm{gg^{-1}h^{-1}}$ to $0.11\,\mu\,\mathrm{gg^{-1}h^{-1}}$. The range of emission from

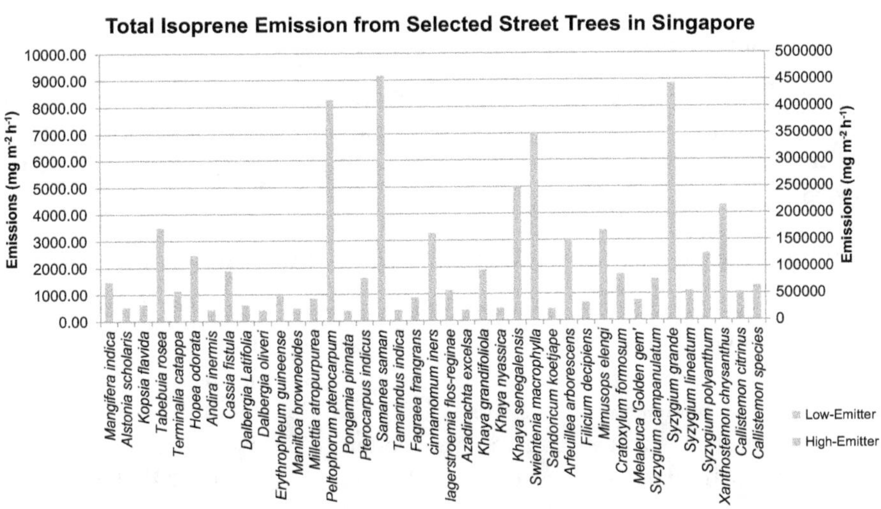

Fig. 5.3. Total isoprene emissions from selected street trees in Singapore.

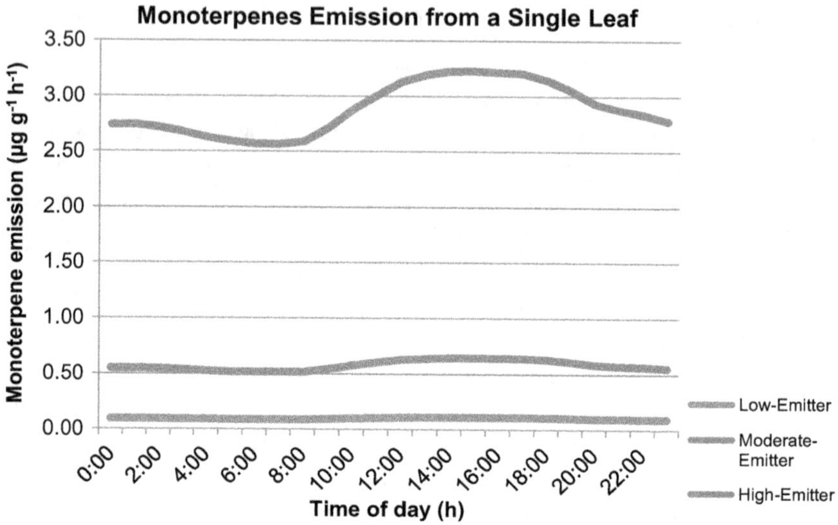

Fig. 5.4. Monoterpenes emissions from a single leaf of all emitting species.

moderate emitter species spans from $0.51\,\mu gg^{-1}h^{-1}$ to $0.65\,\mu gg^{-1}h^{-1}$. Finally, the range of emission from high emitter species ranges from $2.57\,\mu gg^{-1}h^{-1}$ to $3.23\,\mu gg^{-1}h^{-1}$ The highest emission occurs during the period from 15:00h to 16:00h, while the lowest emission is observed during the period from 06:00h to 07:00h.

Total monoterpenesemission per tree per species

Figure 5.5 shows the monoterpenes emissions from each tree of the selected 39 species of street trees. The emission levels range from about $0.87\,mg\,m^{-2}h^{-1}$ for low emitter species to $26\,mgm^{-2}h^{-1}$ for high emitter species.

Total monoterpenes emission from all selected street trees in Singapore

The total monoterpenes emission from the selected street trees in Singapore amount to about $3.1\,g\,m^{-2}h^{-1}$. The levels of emission by the respective species of selected street treesare shown in Fig. 5.6 Of the 39 selected species of street trees, 46% are low emitter species, 51% are high emitter species, and 2% are moderate emitter species. However, the low emitter tree species collectively contribute only 3% of the total monoterpenes emitted. High emitter street tree species that make up 51% of the selected street trees contribute to the remaining 96% of the emission.

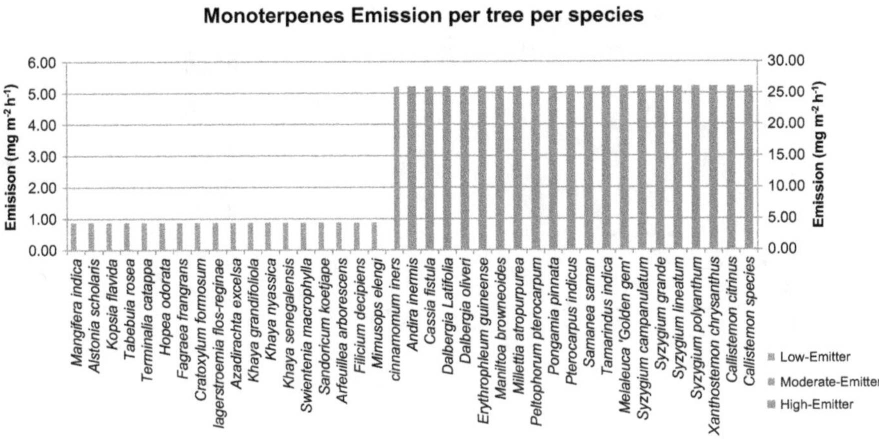

Fig. 5.5. Total monoterpenes emission from a whole tree per species.

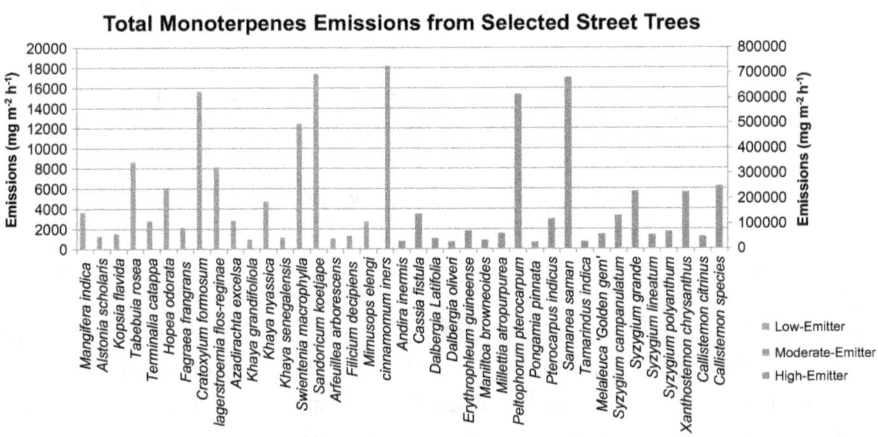

Fig. 5.6. Total monoterpenes emissions from selected street trees.

It is noted that the *Samanea saman*, while being a low emitter in terms of isoprene, is a high emitter of monoterpenes. Due to its sheer population, *Samanea saman* accounted for the highest amount of isprenoid emitted, compared to the other species.

4.2 *Temperature and Isoprene Emission*

This section of the paper explores the significance of temperature and PAR on the emission rates of the species of street trees in Singapore. There

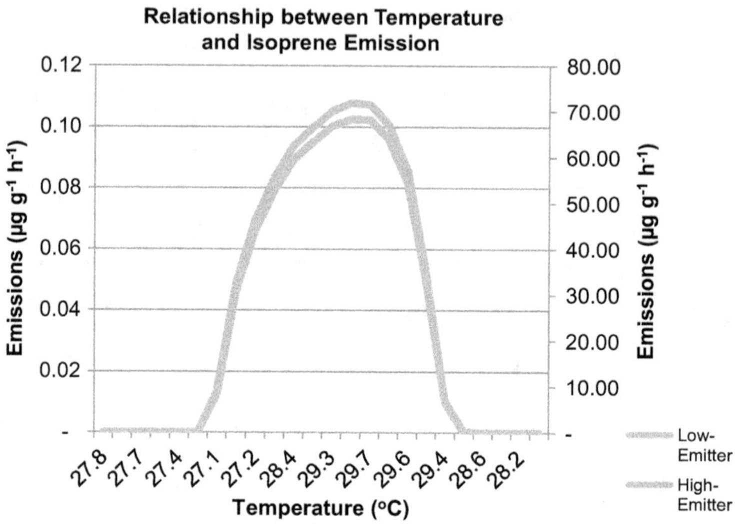

Fig. 5.7. Relationship between temperature and isoprene emission.

is a positive relationship between temperature and isoprene emission as shown in Fig. 5.7. It is observed that as temperature continues to increase to about 29.7° C, isoprene emissions begin to decrease. As temperature decreases over time, isoprene emissions also decrease to negligible levels after 19:00h. This observation is in excellent agreement with findings on the relationship between temperature and isoprene emission in other studies (Guenther *et al.*, 1993,Guenther *et al.*, 1994, Tambunan *et al.*, 2006). This could result in two implications: (i) Assuming that the average temperature in Singapore stays at around 28° C to 29° C, the isoprene emissions recorded in Singapore may be high; or (ii) with climate change and increasing surface temperature, the isoprene emission in Singapore may stay low due to the denaturation of the enzyme.

4.3 *Temperature and Monoterpenes Emission*

Findings from this study show that like isoprene, temperature has a positive impact on monoterpenes emission (Guenther *et al.*, 1993; Guenther *et al.*, 1994, Staudt *et al.*, 1997; Tingey *et al.*, 1980). As shown in Fig. 5.8, monoterpenes emissions increase exponentially with increasing temperature. However, unlike isoprene emissions which decrease to an insignificant amount after 19:00h, monoterpenes continue to be emitted throughout the day and night.

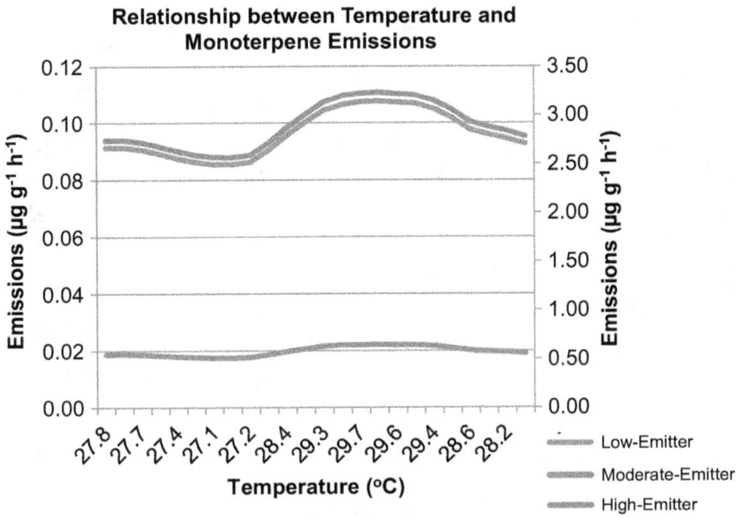

Fig. 5.8. Relationship between temperature and monoterpenes emissions.

4.4 *PAR and Isoprene Emission*

The findings from this study shows that isoprene emissions are highly dependent on PAR (Fig. 5.9). Isoprene emissions increase exponentially before peaking at PAR 873.3 μmol m^{-2}s^{-1}. The increase in emission then continues at a decreasing rate, beginning its decline when PAR reaches its peak at 888.6 μmolm^{-2}s^{-1}. Isoprene emissions subsequently decrease rapidly as PAR decreases rapidly.

4.5 *PAR and Monoterpenes Emission*

The relationship between PAR and monoterpenes emissions is weak. As shown in Fig. 5.10a and 5.10b, monoterpenes continue to be emitted when PAR was absent, particularly from 00:00h to 06:00h, and after 19:00h. The non-dependency of monoterpenes emissions on PAR is documented by Tingey *et al.* (1980) where it was found that the sum of monoterpenes emission rates did not depend on light intensity.

4.6 *Comparison of Results with Other Studies*

Table 5.3 shows the levels of emissions normalised to emissions per square km. The normalisation was based on the areas in which the individual studies were conducted in the respective cities or countries selected. It is

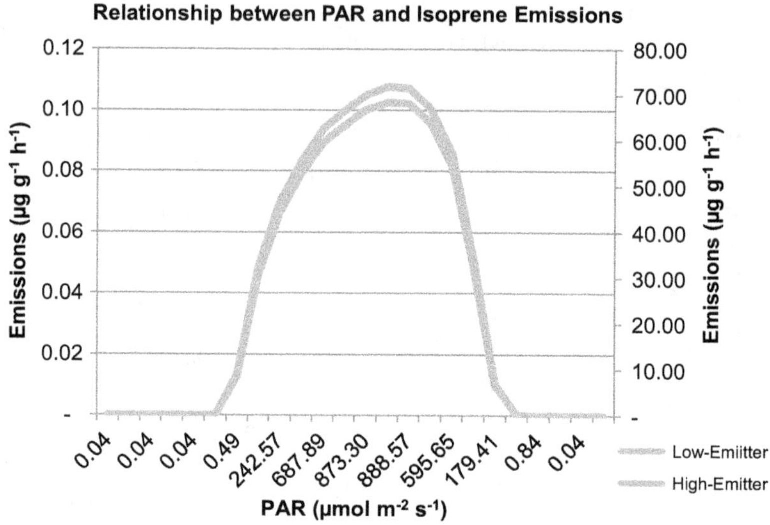

Fig. 5.9. Relationship between temperature and isoprene emission.

noted that Singapore's normalised BVOCs emissions are generally low, standing at $0.006\,\text{Tg}\,\text{yr}^{-1}\text{km}^{-2}$ for isoprene and $0.002\,\text{Tg}\,\text{yr}^{-1}\,\text{km}^{-2}$ for monoterpenes — comparable to the emissions of Mexico Valley. However, the author would like to point out that this study (and its resulting findings) has one limitation: the basis on which normalisation took place. For most of the studies selected for comparison, the areas in which the emissions were measured included forested areas; whereas in this study, only selected street trees rather than all trees were considered. As such, the area used for normalisation was the city area rather than a forested area. Therefore, the assessment of the level of normalised emissions in Singapore may not be as accurate when comparing against studies where emissions from all trees are considered.

Recent studies on selected sub-tropical trees species in India, China and Okinawa could possibly shed more light on the level of BVOCs emission in Singapore. Table 5.4 is a comparison between the emission rates of common species considered in previous studies. There are discrepancies between the findings of this study with those reported in the literature, particularly, some species that were found to be low emitters in this study, such as the *Mangifera indica* and *Dalbergia sp.* were found to be high emitters in previous studies. This is due to the emission potentials used for modelling in this study, as the data were taken from temperate tree species based on

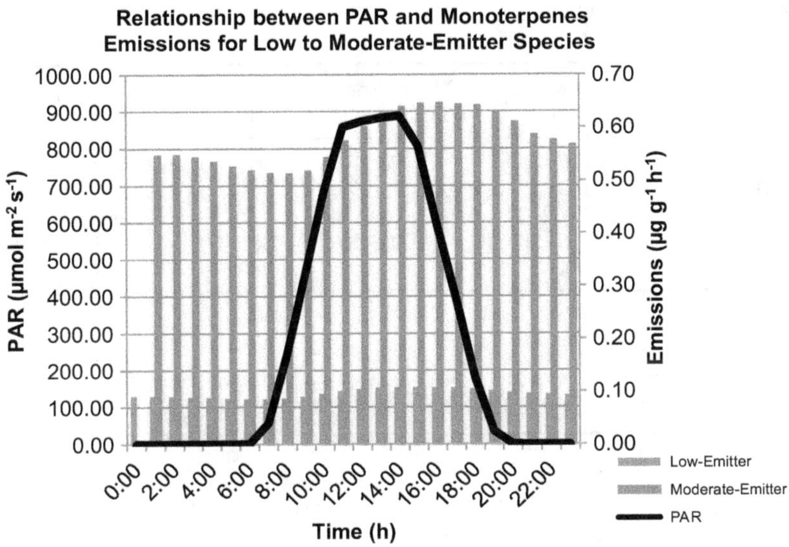

(a)

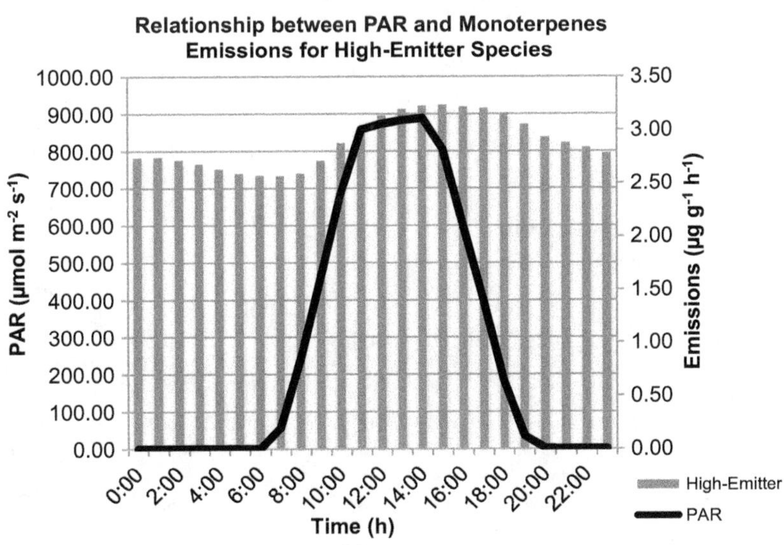

(b)

Fig. 5.10. (a) Relationship between PAR and monoterpenes emissions for low to moderate emitter species. (b) Relationship between PAR and monoterpenes emissions for high emitter species.

Table 5.3. BVOCs emissions at various countries/cities normalised at emissions per square km.

Country/ City	Isoprene ($Tg\,yr^{-1}km^{-2}$)	Monoterpenes ($Tg\,yr^{-1}km^{-2}$)	Area (km^2)	References
China	4.27E-07	3.65E-07	9.6×10^6	Klinger *et al.*, 2002
Hong Kong	0.01	0.01	447	Tsui *et al.*, 2008
Mexico Valley	0.002	0.001	6225	Velasco, 2003
North America	7.71E-06	4.71E-06	3.8×10^6	Guenther *et al.*, 2000
Singapore	0.006	0.002	695.5	This study

Table 5.4. A comparison of the observed isoprene emission rates from different tree species (based on modelling) with values reported in literature.

Scientific name	Family	Present study ($\mu g\,g^{-1}h^{-1}$)	Literature ($\mu g\,g^{-1}h^{-1}$)	Reference
(a) Species level comparison				
Alstonia scholaris	Apocynaceae	0.04	0.00	Padhy and Varshney (2005)
Mangifera indica	Anacardiaceae	0.04	13.8	Padhy Varshney (2005)
Terminalia catappa	Combretaceae	0.04	1.1	Tambunan *et al.*, (2006)
(b) Genus level comparison				
Cinnamomum japonicum	Lauracea	0.04	1.2	Tambunan *et al.*, (2006)
Dalbergia sp.	Fabaceae	0.04	6.8	Padhy and Varshney (2005)
Lagerstroemia subcostata	Lythraceae	0.04	0.9	Tambunan *et al.*, (2006)
Syzygium sp.	Myrtaceae	27.1	8.7	Padhy and Varshney (2005)

the phylogenetic approach. As such, it is important that in future studies, the coefficients used should be adapted and adjusted to fit the subtropical and tropical environment (Tambunan *et al.*, 2006).

5. Discussion

5.1 *Implications of Level of BVOCs Emissions in Singapore*

The findings of this study have shown that it is the minority of high emitter street tree species that contribute to 99% of the total BVOCs emissions. The majority of street trees considered in Singapore belong to low emitter species that contribute to only about 1% of Singapore's BVOCs emissions. As such, the selection of the right species of trees for the greening of Singapore is critical in order for the community to reap the benefits of being near to nature. Understanding the levels of BVOCs emissions in Singapore can help planners make more informed decisions when selecting the species of trees to be planted, especially in an urban environment. It is recommended to select low emitter species such as *Mangifera indica, Alstonia scholaris, Kopsia flavida, Tabebuia rosea, Terminalia catappa and Hopea odorata.* These species are low emitters of both types of BVOCs considered in this study.

However, while planting low emitter species may result in lower emissions due to their emission potential, low emitter species also have the ability to create ecosystem disservices to the environment, especially in an urban setting. Calfapietra *et al.* (2013) argue that the selection of low emitter species for an urban environment may do more harm since plants are known to emit BVOCs as a form of protection against stressful agents; the low BVOCs emissions could result in the decreased tolerance of urban trees to oxidative stress generated in an urban area. These conditions may result in increased levels of BVOCs being emitted as the trees planted have to adapt to the increased stress agents in the urban environment. Therefore, this is a consideration planners may wish to look into when selecting the species of trees to be incorporated in their urban planning.

Furthermore, the strong need to understand the situation of BVOCs emissions in Singapore is also due to the effects that BVOCs can have on human health. An increased level of BVOCs may result in an increase level of ground-level ozone (Curtis *et al.*, 2014; Lerdau *et al.*, 1997). Exposure to increased levels of ground-level ozone can result in negative impacts on the health of humans, such as decrement in lung function, the inflammation of airways, and the induction of respiratory symptoms such as increased coughing, asthma and throat irritation (United States Environmental Protection Agency (EPA), 2015). In an urban area such as Singapore, it is necessary to keep not just BVOCs low, but to also keep pollutant levels low in order to prevent negative implications on the health of the community. This is because the amount of ground-level ozone that would be formed from BVOCs is dependent on the amount of NO_X in the urban area (Curtis *et al.*, 2014). Higher levels of NO_X would result in the generation of a higher amount of ground level ozone. As such, apart from the careful selection of plant species, it is also crucial for Singapore to keep NO_Xlevels low in order to continue to reap the ecosystem services provided by the greenery. Currently, the NO_X concentration in Singapore is about $24\,\mu g\,m^{-3}$ (National Environment Agency, 2013), within the acceptable range of $40\,\mu g\,m^{-3}$ stipulated by the World Health Organisation. Therefore, it is safe to conclude that, with the exercise of caution in the selection of trees to be planted in Singapore, the community is likely to be able to enjoy the benefits of being in close proximity to nature.

5.2 *BVOCs Emissions and Climate Change*

Harley *et al.* (1999) discussed that a rise in global temperature could result in large increases in forest isoprene emissions where 'even an average warming of 2.5°C over the growing season would be expected to increase emissions by 25–40%' (Harley *et al.*, 1999).

The average surface temperature of Singapore in the long run is expected to increase by 1.4°C to 2.7°C under lowered emission scenarios, while the temperature is expected to increase by 2.9° C to 4.6°C under a business-as-usual scenario (National Climate Change Secretariat (NCCS), 2016). Taking an average of the projected increase in surface temperature, the estimated isoprene emission, based on a 2.8°C increase in the surface temperature of Singapore, in the long run would increase by about 24% from $4.21\,Tg\,yr^{-1}$ to $5.53\,Tg\,yr^{-1}$. Similarly, the monoterpenes emission is estimated to increase by about 22% from $1.12\,Tg\,yr^{-1}$ to $1.44\,Tg\,yr^{-1}$. Thus, it is important to look into the selection of trees that are planted, in order to prevent more significant increases in the amount of BVOCs emitted by trees in an urban environment.

5.3 *Limitations and Recommendations for Future Studies*

This study looks specifically at the BVOCs emissions by selected street trees in Singapore through the use of the biogenic emission model. The emission potentials accorded to the selected species of street trees were adapted from literature which largely focused on species found in the mid-latitudes and temperate regions. In addition, where measurements were not available, the emission potentials were accorded based on a phylogenetic approach suggested by Benjamin *et al.*, (1996). One of the biggest limitation of the use of this method is that generalisation cannot be made for some families such as *Leguminosae* and *Fabaceae* (Karlik and Pittenger, 2012) which some 30% of the selected species of street trees for this study belong to. As such, this could have resulted in discrepancies in emission rates and therefore painted an inaccurate picture of the situation of the BVOCs emissions by selected species of street trees in Singapore.

It is recommended for future studies to focus on establishing an inventory of emission potentials specific to species commonly found in the tropical regions, to obtain more accurate estimates of BVOCs emissions through modelling. There is also a need to keep a land cover database, which is updated annually with changes to city development. This is particularly crucial if the tropical region would like to have a better understanding of the levels of BVOCs emissions in order to make sound decisions for urban planning. In addition, it is also recommended to estimate the emissions from the street trees in Singapore through methodology such as the dynamic flow branch enclosure technique to get a more accurate picture of the level of BVOCs emissions in Singapore. It is suggested that the measurements could be taken as a point of reference for the findings in this study to be validated. Further, to allow for a more meaningful comparison of normalised levels of emission, future studies can consider estimating the BVOCs emissions from the nature reserves.

6. Conclusion

Ecosystem services provided by trees have been widely documented. This is unlike the ecosystem disservices that could also be brought about by trees. BVOCs emissions by trees is one such ecosystem disservice. In this study, estimates of isoprene and monoterpenes emissions are provided through the use of the biogenic emission model. It was found that the total annual isoprene and monoterpenes emissions from over 220,000 street trees from 39 species was $4.2\,\text{Tg}\,\text{yr}^{-1}$ and $1.1\,\text{Tg}\,\text{yr}^{-1}$ respectively. The positive influence of temperature on BVOCs emissions signify that with climate change, there may be an increase in the emissions from trees in Singapore. Therefore, with the onslaught of global warming in the next 30 years, there is a need to be critical in the selection of trees for future tree-planting programmes. Notwithstanding the differences in the emission levels in this study with previous studies, the estimates provided in this study serve as a guide to urban planners in Singapore when selecting the species of trees to be planted in order to achieve the goals set out by MEWR and MND in the Singapore Sustainable Blueprint 2015.

7. References

Baghi, R., Helmig, D., Guenther, A., Duhl, T. and Daly, R. (2012). Contribution of flowering trees to urban atmospheric biogenic volatile organic compound emissions. *Biogeosciences,* **9**, 3145–3172.

Banthorpe, D.V., Charlwood, B.V. and Francis, M.J.O. N. (1972). The Biosynthesis of Monoterpenes. *Chemical Reviews*, **72**(2), 115–155.

Beckett, K. P., Smith, P. H. and Taylor, G. (1998). Urban woodlands: their role in reducing the effects of particular pollution. *Environmental Pollution*, **99** 347–360.

Bell, M. L. and Ellis H. (2005). The impact of biogenic VOC emissions on tropospheric ozone formation in the Mid-Atlantic region of the United States. In C.A. Brebbia (Ed.) *Air pollution XIII* (pp. 89–96). Boston: WITpress Southampton.

Benjamin, M. T., Sudol, M., Bloch, L. and Winer, A. M. (1996). Low-emitting urban forests: A taxonomic methodology for assigning isoprene and monoterpene emission rates. *Atmospheric Environment*, **30**(9), 1437–1452.

Bolund, P. and Hunhammar, S. (1999). Ecosystem services in urban areas. *Ecological Economics*, **29** 293–301.

Brown, T. C., Bergstrom, J. C. and Loomis, J. B. (2007). Defining, Valuing and Providing Ecosystem Goods and Services. *Natural Resources Journal*, **47** 329–376.

Burbott, A. J. and Loomis, W. D. (1967). Effects of Light and Temperature on the Monoterpenes of Peppermint. *Plant Physiology*, **42** 20–28.

Calfapietra, C., Fares, S., Manes, F., Morani, A., Sgrigna, G. and Loreto, F. (2013). Role of Biogenic Volatile Organic Compounds (BVOC) emitted by urban trees on ozone concentration in cities: A review. *Environmental Pollution*, **183** 71–80.

Curci, G., Beekman, M., Vautard, R., Smiatek, G., Steinbrecher, R., Theloke, J. and Friedrich, R. (2009). Modelling study of the impact of isoprene and terpene biogenic emissions on European ozone levels. *Atmospheric Environment*, **43** 1444–1455.

Curtis, A. J., Helmig, D., Baroch, C., Daly, R. and Davis, S. (2014). Biogenic volatile organic compound emissions from nine tree species used in an urban tree-planting program. *Atmospheric Environment*, **95** 634–643.

De Groot, R. S., Wilson, M. A. and Boumans, R. M. J. (2002). A typology for the classification, description and valuation of ecosystem functions, goods and services. *Ecological Economics*, **41**, 394–308.

Duane, M., Poma, B., Rembges, D., Astorga, C. and Larsen, B. R. (2002). Isoprene and its degradation products as strong ozone precursors in Insubria, Northern Italy. *Atmospheric Environment*, **36**, 3867–3879.

Escobedo, F. J., Kroeger, T. and Wagner, J. E. (2011). Urban Forests and pollution mitigation: Analyzing ecosystem services and disservices. *Environmental Pollution*, **158**, 2078–2087.

Fehsenfeld, F. (1995) Measurement of chemically reactive trace gases at ambient concentrations. In P.A. Matson and R.C. Harriss (Eds.) *Biogenic Trace Gases: Measuring Emissions from Soil and Water* (pp. 206–258). Oxford: Blackwell Science.

Geron, C.D., Guenther, A. B. and Pierce, T. E. (1994). An improved model for estimating emissions of volatile organic compounds from forests in the eastern United States. *Journal of Geophysical Research*, **99**(6), 12773–12791.

Geron, C., Harley, P. and Guenther, A. (2001). Isoprene emission capacity for US tree species. *Atmospheric Environment*, **35** 3341–3352.

Gijzen, M., Lewinsohn, E., Savage, T. J. and Croteau, R. B. (1993). Conifer Monoterpenes: Biochemistry and Bark Beetle Chemical Ecology. In R. Teranishi, R.G. Buttery and H. Sugisawa (Eds.) *Bioactive Volatile Compounds from Plants* (pp. 8–22). Washington, DC: American Chemical society.

Grant, R.H. (1997). Partitioning of biologically active radiation in plant canopies. *International Journal of Biometeorology*, **40**, 26–40.

Grinde, B. and Patil, G. G. (2009). Biophilia: Does Visual Contact with Nature Impact on Health and Well-Being? *International Journal of Environmental Research and Public Health*, **6**, 2332–2343.

Guenther, A., Zimmerman, P., Harley, P., Monson. R. and Fall, R. (1993) 'Isoprene and monoterpene emission rate variability: model evaluation and sensitivity analysis', *Journal of Geophysical Research,* **98**, 12609–12617.

Guenther, A., Zimmerman, P. and Wildermuth, M. (1994). 'Natural Volatile Organic Compound Emission Rate Estimates For U.S. Woodland Landscapes', *Atmospheric Environment* **28**(6), 1197–1210.

Guenther, A., Geron, C., Pierce, T., Lamb, B., Harley, P. and Fall, R. (2000). 'Natural emissions of non-methane volatile organic compounds, carbon monoxide, and oxides of nitrogen from North America', *Atmospheric environment,* **34**, 2205–2230.

Harley, P. C., Monson, R. K. and Lerdau, M. T. (1999). Ecological and Evolutionary Aspects of Isoprene Emissions from Plants. *Oecologia*, **118**, 109–123.

Helas, G., Slanina, J. and Steinbrecher, R. (1997). Introduction: Assessment on biogenic volatile organic compounds. In G. Helas, J. Slanina and R. Steinbrecher (Eds.) *Biogenic Volatile Organic Compounds in the Atmosphere.* (pp. 1–8). Amsterdam: SPB Academic Publishing.

Karlik, J. and Pittenger, D. R. (2012). *Urban trees and ozone formation: A consideration for large-scale planting.* (pp. 1–9). California: University of California.

Karlik, J. and Winer, A. M. (2001). Measured isoprene emission rates of plants in California landscapes: comparison to estimates from taxonomic relationships. *Atmospheric Environment*, **35**, 1123–1131.

Kleerekoper, L., Esch, M. and Salcedo, T. B. (2012). How to make a city climate-proof, addressing the urban heat island effect. *Resources, Conservation and Recycling*, **64**, 30–38.

Klinger, L. F., Li, Q. J., Guenther, A. B., Greenberg, J. P., Baker, B. and Bai, J. H. (2002). Assessment of volatile organic compound emissions from ecosystems of China. *Journal of Geophysical Research*, **107**(21), 1–21. doi: 10.1029/2001JD001076.

Kuzma, J. and Fall, R. (1993). Leaf lsoprene Emission Rate 1s Dependent on Leaf Development and the Level of lsoprene Synthase. *Plant Physiology*, **101** 435–440.

Lamb, B., Gay, D., Westberg, H. and Pierce, T. (1993). 'A Biogenic Hydrocarbon Emission Inventory for the U.S.A. Using a simple Forest Canopy Model', *Atmospheric Environment*, **27A**(11), 1673–1690.

Lerdau, M., Guenther, A. and Monson, R. (1997). 'Plant Production and Emission of Volatile Organic Compounds', *Bioscience*, **47**(6), 373–382.

McPherson, E. G., Simpson, J. R., Peper, P. J. and Xiao, Q. (1999). Benefit-cost analysis of Modesto's municipal urban forest. *Journal of Aboriculture*, **25**(5), 235–248.

Ministry of Environment and Water Resources and Ministry of National Development (2015). *Singapore Sustainable Blueprint 2015.* Retrieved from http://www.mewr.gov.sg/ssb/

National Climate Change Secretariat. (2016). *Results of Second National Climate Change Study Consistent with The Intergovernmental Panel on Climate Change's 5th Assessment Report in Projecting Global Sea-Level Rise, Higher Temperatures and More Extreme Rainfall in The Region.* Retrieved from https://www.nccs.gov.sg/news/results-second-national-climate-change-study-consistent-intergovernmental-panel-climate-changes.

National Environment Agency. (2013). *Pollutant Concentrations.* Retrieved from http://www.nea.gov.sg/anti-pollution-radiation-protection/air-pollution-control/psi/pollutant-concentrations.

Nowak, D. J. and Dwyer, J. F. (2007). Understanding the Benefits and Costs of Urban Forest Ecosystems. In J. E. Kuser (ed.) *Urban and Community Forestry in the Northeast, 2nd ed.* (pp. 25–46). New York: Springer.

Nowak, D. J., Stevens, J. C., Sisini, S. M. and Luley, C. J. (2002). Effect of urban tree management and species selection on atmospheric carbon dioxide. *Journal of Aboriculture*, **28**(3), 113–122.

Peñuelas, J. and Llusià, J. (2003). BVOCs: plant defense against climate warming? *Plant Science*, **8**(3) 105–109.

Pio, C., Nunes, T. and Valente, A. (1997). Forest Emissions of Hydrocarbons. In S. Slanina (Ed.) *Biosphere-Atmosphere Exchange of Pollutants and Trace Substances*, (pp. 335–341). Heidelberg: Springer.

Riba, M.L. and Torres, L. (1997). Terpenes: Biosynthesis and transport in plants; emission and presence in the troposphere. In G. Helas, J. Slanina and R. Steinbrecher (Eds.) *Biogenic Volatile Organic Compounds in the Atmosphere.* (pp. 115–144). Amsterdam: SPB Academic Publishing.

Sharkey, T. D., Wilberley, A. E. and Donohue, A. R. (2008). Isoprene Emission from Plants: Why and How. *Annals of Botany*, **101**, 5–18.

Singsaas, E. L., Laporte, M. M., Shi, J., Monson, R. K., Bowling, D. R., Johnson, K., Sharkey, T. D. (1999). Kinetics of leaf temperature fluctuation affect isoprene emission from red oak (Quercus rubra) leaves. *Tree Physiology*, **19**, 917–924.

Soares, A. L., Rego, F. C., McPherson, E. G., Simpson, J. R., Peper, P. J. and Xiao, Q. (2011). Benefits and costs of street trees in Lisbon, Portugal. *Urban Forestry and Urban Greening*, **10**, 69–78.

Staudt, M., Bertin, N., Hasen, U., Seufert, G., Ciccioli, P., Foster P., Fujit, J. L. (1997). Seasonal and diurnal patterns of monoterpene Emissions from pinus pinea (l.) Under field conditions. *Atmospheric Environment*, **31**, 145–156.

Steinbrecher, R. (1997). Isoprene: Production by plants and ecosystem-level estimates. In G. Helas, J. Slanina and R. Steinbrecher (Eds.) *Biogenic Volatile Organic Compounds in the Atmosphere* (pp. 115–144). Amsterdam: SPB Academic Publishing.

Steinbrecher, R., Ziegler, H., Eichstädter, *et al.* (1997). Monoterpene and Isoprene Emission in Norway Spruce Forests. In S. Slanina (Ed.) *Biosphere-Atmosphere Exchange of Pollutants and Trace Substances* (pp. 352–365). Heidelberg: Springer.

Steiner, A. H. and Goldstein, A. L. (2007). Biogenic VOCs. **In** R. Koppmann (Ed.) *Volatile Organic Compounds in the Atmosphere* (pp. 83–128). Blackwell: Oxford.

Street, R. A., Duckham, S. C., Boissard, C. and Hewitt, C. N. (1997). Emissions of VOCs from Stressed and Unstressed Vegetation. In S. Slanina (Ed.) *Biosphere-Atmosphere Exchange of Pollutants and Trace Substances* (pp. 366–371). Heidelberg: Springer.

Taha, H., Akbari, H. and Rosenfeld, A. (1991). Heat island and oasis effect of vegetative canopies: Micro-meterological field-measurements. *Theoretical and Applied Climatology*, **44**, 123–138.

Tambunan, P., Baba, S., Kuniyoshi, A., Iwasaki, H., Nakamura, T., Yamasaki, H. and Oku, H. (2006). Isoprene emission from tropical trees in Okinawa Island, Japan. *Chemosphere*, **65**, 2138–2144.

Tan, P. Y. and Ismail, M. R. (2015). Photosynthetically active radiation and comparison of methods for its estimation in equatorial Singapore. *Theoretical and Applied Climatology,* 1–11, *doi:* 10.1007/s00704-015-1399-9.

Tingey, D. T., Manning, M., Grothaus, L. and Burns, W. M. (1980). Influence of Light and Temperature on Monoterpene Emission Rates from Slash Pine. *Plant Physiology*, **65**, 797–801.

Tsui, J. K. Y., Guenther, A., Yip, W. K. and Chen, F. (2008). 'A biogenic volatile organic compound emission inventory for Hong Kong', *Atmospheric Environment*, doi:10.1016/j.atmosenv.2008.01.027.

United States Environmental Protection Agency. (2015). *Health Effects of Ozone in the General Population.* Retrieved from http://www3.epa.gov/apti/ozonehealth/population.html.

Velasco, E. (2003) 'Estimates for biogenic non-methane hydrocarbons and nitric oxide emissions in the Valley of Mexic', *Atmospheric Environment*, **37**, 625–637.

Williams, J. and Koppman, R. (2007). Volatile Organic Compounds in the Atmosphere: An Overview. In R. Koppmann (Ed.) *Volatile Organic Compounds in the Atmosphere* (pp. 1–32). Oxford: Blackwell.

Wuebbles, D., Grant, K., Connell, P. and Penner, J. (1989) 'The role of atmospheric chemistry in climate change', *Journal of the Air Pollution Control Association*, **39**, 22–28.

Part 3

Environmental Governance & Management

6

SUSTAINABILITY REPORTING IN SINGAPORE

LOH ZHIYANG and AUDREY CHIA

With the increasing prevalence of corporate sustainability reporting, companies in Singapore have started to voluntarily disclose their social and environmental performance. In addition to the statutory requirements for financial reporting. The impact of this new endeavour will likely be significant especially when the Singapore Stock Exchange (SGX) makes the bold move to require corporate sustainability reporting for all listed companies in the near future. To investigate the practical issues from the application of sustainability reporting standards such as the Global Reporting Initiatives ("GRI"), this study was done to specifically clarify some of the comparability difficulties between local companies that had adopted the GRI for their corporate sustainability reports. The observations and recommendations made from this study would be useful reference points for future research.

Three issues were identified from this study. First, ambiguities were found in the disclosures of non-financial performance. While the application of the GRI standards had enabled greater transparency, it also created more uncertainties and vagueness, which made it difficult to assess the quality of the reported performance. Independent assurance of sustainability reports would therefore be important and should be made mandatory.

The second issue was on the different levels of reporting performance. For instance one company had focused its reporting of performance, for the headquarters while another company had reported the overall performance for its entire corporate group. To ensure better comparability of performance, standardised levels of reporting by business unit and geographical area would be necessary.

The last issue was on the differing extent of vertical integration between corporate sustainability reporters. Companies with different degrees of supply chain control might not be comparable in terms of non-financial performance. For better comparability, companies with shorter supply chain control, like property developers may need to include their supply chain management strategies (e.g. the policies and enforcement measures to ensure the protection of human rights by the main- and sub-contractors in the construction sector) in their reports. Guidance on the extent of supply chain disclosures should therefore be provided.

1. Introduction

1.1 *Singapore Prepares for Sustainability Reporting*

In recent years, the word "Sustainability" has been used more widely in Singapore. In 2009, a "Sustainable" Development blueprint for Singapore was announced. In May 2012, the Monetary Authority of Singapore (MAS) issued the amended Code of Corporate Governance[1] that included the consideration of "Sustainability" issues and the recognition of key stakeholder groups' perceptions in the Board's role for its conduct of affairs. In its publication on September 2013, the Institute of Singapore Chartered Accountants (ISCA) stated that more companies in Singapore will be called upon by their ever-more educated, informed and socially-conscious stakeholders to provide them with the information needed to explain their legitimacy to exist, and to create "Sustainable" value while remaining profitable.

Given that "Sustainability" has become the new norm, this study is conducted with the objective of reviewing the status of sustainability reporting practices in Singapore by performing a comparative analysis between two SGX-listed companies that had adopted the Global Reporting Initiative as their reporting framework. Before going into the analysis, this paper first examines the development of corporate social responsibility and sustainability reporting.

1.2 *An Overview of Corporate Social Responsibility and Related Initiatives*

Efforts and contributions toward Corporate Social Responsibility (CSR) are evident in emergence of numerous initiatives in the recent decades (see Table 6.1). Some of these initiatives were developed at the international level such as the Global Reporting Initiative (GRI) and the Global Compact, while others were organised at a less macro-level such as regional-based initiatives like EU's Eco-management and Audit Scheme (EMAS) and industrial-driven initiatives like the Responsible Care Program of the Chemical industry. The initiatives ranged from addressing very specific issues like the Social Accountability International's SA8000 standard for labour practices; to broader aspects such as the social responsibility focus of

[1]Source: http://www.mas.gov.sg/regulations-and-financial-stability/regulatory-and-supervisory-framework/corporate-governance/corporate-governance-of-listed-companies/code-of-corporate-governance.aspx

Table 6.1. Major CSR initiatives (Main sources: Lodhia, 2012; Hopwood *at al.*, 2010).

S/N	Name of initiative	Level of influence	Launched in	Established by	Represented by	Area of focus
1	Responsible Care Program[2]	Chemical industry	1985	Canadian Chemical Producers' Association	Chemical industry-related stakeholders	Health, safety and environmental performance
2	**AccountAbility** (AA1000) series[3]	International	1995	AccountAbility, a not-for-profit UK Company Limited by Guarantee	Multi-stakeholder groups*	Accountability, responsibility and sustainability with specific attention on governance, business models and organisational strategy
3	Global Reporting Initiative (**GRI**)	International	1997	Coalition for Environmentally Responsible Economies (CERES), a coalition of US non-profit organisations	Multi-stakeholder groups*	**Triple bottom line accountability** (reporting of sustainability-related information)
4	SA8000[4]	International	1999	Social Accountability International (SAI), a non-governmental, multi-stakeholder organisation whose	Multi-stakeholder groups*	Responsible labour practices

(*Continued*)

[2]Source: www.icca-chem.org/en/Home/Responsible-care/
[3]Source: www.accountability. org/about-us/governance/index.html
[4]Source: www.sa-intl.org/data/n0001/resources/live/2004SAIGuidanceFinal.pdf

Table 6.1. (*Continued*)

S/N	Name of initiative	Level of influence	Launched in	Established by	Represented by	Area of focus
				mission is to advance the human rights of workers around the world		
5	SIGMA[5]	European region	1999	A partnership between the British Standards Institution (the leading standards organisation), Forum for the Future (a leading sustainability charity and think-tank), and **AccountAbility** with the support of the UK Department of Trade and Industry	UK stakeholders	Integration of sustainability issues into core business processes and mainstream decision-making.
6	Global Compact[6]	International	2000	UN	UN	Responsible business conduct based on ten universally accepted principles in the areas of human rights, labour, environment and anti-corruption

7	EMAS[7]	European region	2001	The European Commission	EU	Environmental performance
8	International Council on Mining and Metals (ICMM)[8]	Mining industry	2001	World's leading companies in the mining and metals industry	22 mining and metals companies; 32 national and regional mining associations and global commodity associations	Sustainable development performance in the mining and metals industry.
9	The Equator Principles (EPs)[9]	Financial sector	2003	Global financial institutions (ABN AMRO Bank N.V., Barclays plc and Citigroup Inc. are amongst the first ten institutions to initiate the formation of the EPs)	80 Equator Principles Financial Institutions (EPFIs) in 34 countries	Social and environmental risk management for project financing.
10	The Prince's Accounting for Sustainability Project, i.e.	International	2004	HRH The Prince of Wales	Multi-stakeholder groups*	**Triple bottom line accountability** (integrative reporting of financial and

(Continued)

[5] Source: www.projectsigma.co.uk/
[6] Source: www.unglobalcompact .org/ParticipantsAndStakeholders/index.html
[7] Source: http://eur-lex.europa.eu/legal-content/EN/TXT/? qid=1405520310854&uri=CELEX:32013D0131
[8] Source: www.icmm.com/about-us/about-us
[9] Source: www.equator-principles.com/index.php/about-ep/about-ep

Table 6.1. (*Continued*)

S/N	Name of initiative	Level of influence	Launched in	Established by	Represented by	Area of focus
	Accounting for Sustainability (**A4S**)					sustainability-related information)
11	Integrated Reporting (IR)	International	2009	**A4S** and **GRI**	Multi-stakeholder groups*	**Triple bottom line accountability** (international reporting framework that supports the achievement of a sustainable economy in addition to connecting financial and sustainability outcomes at the organisational level)
12	ISO 26000	International	2010	International Organization for Standardization (ISO), an independent, non-governmental membership organisation and the world's largest developer of voluntary International Standards	Multi-stakeholder groups*	Social responsibility.

**Multi-stakeholder groups refer to a mixture of public sector, private sector and civil society groups that include regulators, investors,*

ISO26000 or the accreditation of environmental performance by EMAS; and to overarching themes like the triple bottom line perspective (i.e. economic responsibility, social equity, and environmental integrity) as exemplified by GRI's sustainability reporting framework and the Integrated Reporting (IR) framework; as well as Global Compact's principles-based standards relating to human rights, labour, environment and anti-corruption.

2. Literature Review

2.1 *The Issue with Definition: Corporate Social Responsibility, Corporate Sustainability or Corporate Responsibility?*

In his paper, Ivan Montiel conducted literature reviews of over 700 academic publications on Corporate Social Responsibility (CSR) and Corporate Sustainability (CS) from 1970 to 2005. The study objective was to identify the boundaries, differences, and similarities across CSR and CS research to cross-fertilise the research agendas (Montiel, 2008). The key differences between CSR and CS are given in Table 6.2.

Despite the differences between CSR and CS, Montiel concluded that both research fields should involve themselves in more collaborative efforts due to the significant overlap of their research interests. He further stated that in practice, companies use both CSR and CS interchangeably in their communication of sustainability performance.

Notwithstanding the conclusions of Montiel, in her book titled *The difference makers: How social and institutional entrepreneurs created the corporate responsibility movement,* Waddock cautioned that "Corporate Social Responsibility" should be distinguished from "Corporate Responsibility" or "CR". She clarified that "Corporate Social Responsibility" encompassed activities that companies have undertaken to directly benefit society like corporate philanthropy, volunteering, and community relations programs that had invited critics that such activities were *"nothing more than window dressing designed to disguise the real and sometimes negative impacts of companies' business model"* (Waddock, 2008). Waddock also differentiated the term "Corporate Responsibility" by explaining that it *"encompasses the responsibilities engendered by the firm in the course of doing business. These responsibilities arise because large corporations are recognized as powerful forces playing a critical role in fostering healthy societies because of the jobs and infrastructure of goods and services they provide. They are not separate entities from society; hence companies,*

Table 6.2. Differences identified between the definitions of CSR and CS (Source: Montiel, 2008).

Differences	CSR	CS
Research focus	Social issues	Social and environmental issues.
Research history	Began before the 1970s.	Began in the 1990s when the WCED[10] popularised the term "sustainable development" in its often-cited 1987 report, *Our Common Future*.
Research perspectives	Environmental issues were not integrated into CSR conceptualisation as environmental issues were regarded as a subset of social issues	Unlike CSR scholars, CS concepts are spilt between two camps. One prioritises the environmental responsibility dimension of business which is often termed as "ecological sustainability" and the other school of thought considers sustainability from a triple bottom line perspective, describing the dimensions of economic responsibility, social equity, and environmental integrity.
Research approach	Social and economic performance are treated as separate components.	Economic, social, and environmental pillars are interconnected.
Views on the economic dimension of business	Social responsibility supplements the primary fundamental responsibility of businesses (i.e. economic prosperity).	Social, environmental, and economic responsibilities are complementary and these three elements must be integrated to achieve perfection (i.e. sustainability).

like individuals, are subject to the social forces that create norms and expectations." (Waddock, 2008)

As can be seen, there is no generally accepted terminology for responsible business practices. For the purpose of this study, the terms "CSR", "CS" and "CR" are used interchangeably to connote "Sustainability" from

[10]The World Commission on Environment and Development (WCED) in its 1987 report titled "Our Common Future" suggested the term "Sustainable development" and it was defined as development that meets the needs of the present without compromising the ability of future generations to meet their own needs.

the business perspective and this is distinctively different from explicit philanthropic activities that benefit society as explained by Waddock above.

2.2 *What is Corporate Sustainability Reporting?*

In his paper, Schaltegger dealt with the different forms and facets of sustainability reporting. He described sustainability reporting as encompassing the formal and official corporate communication which provides information about corporate sustainability issues that are mainly related to social, environmental and economic performance; and the relationships between these aspects of corporate performance (Schaltegger, 2012). A similar explanation was also provided by the Prince's Accounting for Sustainability Project which defined the sustainability report as a report produced by an organisation to report the (positive and negative) impacts of its operations on society and the environment and its performance in this regard (Hopwood *at al.*, 2010).

Schaltegger further commented in his paper that sustainability reporting was distinct from corporate sustainability communication, and supported by the views of other scholars, Schaltegger (2012) articulated the distinction as quoted below.

> *"No matter what view is taken, sustainability reporting has to be distinguished from corporate sustainability communication which exceeds written formal communication as it includes further, direct and indirect forms of oral and written communication such as the use of labels on products, statements by board members, marketing in general, public relations, media coverage, brand communication and information communication, for example by staff in various official and unofficial contexts [...] Sustainability reporting encompasses printed and internet forms of disclosure. Thus, reporting is intentional, official, formalized and can be assumed to be purposeful."*

Schaltegger's emphasis on the words "intentional", "official" and "formalized" implied the need for a structured approach to communication and this approach has to be widely accepted and recognised as official by the users of the communicated information. Examples of this structured approach of communication (also commonly termed as "reporting framework") are the:

- Sustainability Reporting Framework by the Global Reporting Initiative (GRI);
- Connected Reporting Framework (CRF) by the Prince's Accounting for Sustainability Project (A4S);

- Integrated Reporting Framework by the International Integrated Reporting Council (IIRC); and
- AA1000 Framework by AccountAbility

2.3 *Why do Organisations Report Information on Corporate Sustainability?*

2.3.1 *Embedding sustainability in business strategies*

In the book titled *Greening Business: Research, Theory and Practice*, Worthington gathered the opinions of researchers regarding the salient factors that contributed to the growing awareness of environmental and social issues, and the consequent demands for action at both government and corporate levels. As quoted below, some of the key influences were:

- *"the publication of influential books such as Rachel Carson's Silent Spring (1962), Paul Ehrlich's Population Bomb (1968), and James Lovelocks's Gaia (1979);*
- *a series of high-profile ecological disasters including the Torrey Canyon (1967), Amoco Cadiz (1978), and Exxon Valdez (1998) oil spills; the dioxin leak at Seveso (1976); the chemical release at the Union Carbide plant in Bhopal (1984); the disaster at the nuclear installation at Chernobyl (1986);*
- *scientific discoveries and observed natural changes, for example the impact of chlorofluorocarbons (CFCs) on the ozone layer and the problem of greenhouse gas emissions (GHGs);*
- *a resurgence in interest in socially responsible business principles and practices; and*
- *media coverage of all of the above" (Worthington, 2013)*

In their book titled *Embedded Sustainability: The Next Big Competitive Advantage*, Laszlo and Zhexembayeva (2011) explained that "sustainability" in the corporate context should be *"the incorporation of environmental, health, and social value into the company's core business with no trade-off in price or quality (i.e. with no social or green premium). The goal is not green or social responsibility for its own sake. It is meeting new market expectations in ways that strengthen the company's current strategy or help it to develop a better one. At its best, it is invisible, similar to quality, yet still capable of hugely motivating employees and creating loyalty in consumers and supply chain partners."*

2.3.2 *Scholars' views on voluntary corporate sustainability reporting*

Ratnatunga and Jones (2012), in their paper on quality and comprehensiveness of sustainability information, had attempted to clarify the reasons for corporate sustainability reporting with three major theories as presented in Table 6.3.

According to the findings of Ratnatunga and Jones, there were parallels between the legitimacy and stakeholder theories as both 'sang the same tune' on the fact that any form of business groupings (described as "organisations" in the academic literatures) were inextricably related to the society that they were residing in. In addition, legitimacy theory's explanation of the need to portray an image of acting in a socially responsible manner (that is consistent with prevailing social norms and expectations) resonated with the economic self-interest reasoning from Political economy theory.

2.4 *What are the Criticisms of Sustainability Reporting?*

While sustainability reporting has become increasingly important, there are doubts on its value and reliability. Both Gray and Lehman highlighted the potential misuse of sustainability-related information which could possibly reduce such information to environmental propaganda and mask the reality of the environmental crisis and the role of business as a primary cause (Gray, 1992; Lehman, 1995).

Thomas Golden, one of the leading forensic accounting partners at PricewaterhouseCoopers, also warned that non-financial information and corporate sustainability reports provided tempting opportunities for misrepresentation because they are usually unaudited. He explained that a fraudster might no longer be able to manage earnings or misapply Generally Accepted Accounting Principles (GAAP) as easily as before the inception of the Sarbanes-Oxley Act, but he or she can find other ways to accomplish the same objectives. He added that the use of unaudited communications that contain nonfinancial and other operations data in a misrepresentative manner might be challenging at first, but will become easier with practice (Ballou *at al*, 2006).

In another paper, Gray (2010) argued against the validity and effectiveness of the frameworks by both GRI and Accounting for Sustainability (A4S) to demonstrate an organisation's relationship with

Table 6.3. Theories explaining voluntary reporting of social and environmental performance (Source: Ratnatunga and Jones, 2012).

s/n	Theory	Excerpts of theoretical construct	Research findings
1	Political economy theory	Political economy theory recognises the potential for management to tell its own story or refrain from doing so, according to its own economic self-interest (Cooper and Sherer, 1984). With regard to the economic self-interest of undertaking CSR, the theory suggests that firms which need to resolve information asymmetry for their consumers may rely on CSR activities to do so (Siegel and Vitaliano, 2006).	Studies based on this theory found that the effect of CSR on profits is stronger in competitive industries with few firms adopting this approach. This indicates that CSR may be part of the differentiation strategy.
2	Legitimacy theory	Legitimacy theory explains that the main motivation for disclosing non-financial information is due to reasons of accountability and visibility. Such disclosures relate to a way of legitimising the entity's activities and to portray the image of acting in a socially responsible manner consistent with prevailing social norms and expectations. This theory requires the existence of an implicit 'social contract' between an organisation and society.	Though evidence of the link between CSR disclosures and legitimacy theory is not conclusive, research findings suggest that government-owned enterprises and publicly-owned firms disclosed more social and environmental information than privately-owned entities.
3	Stakeholder theory	Stakeholder theory states that organisations engage in CSR mainly because they can reap some kind of benefit from such engagement. Companies are seen as having an obligation to consider society's long-run needs and wants, which implies that they engage in activities that promote benefits for society and minimise the negative effects of their actions, so long as the company is not prejudiced by engaging in such activities (Branco and Rodrigues, 2007).	Academics have discussed the significant overlap between the Stakeholder and Legitimacy theories, given that the Stakeholder theory also requires the existence of an implicit 'social contract' between an organisation and society. Further, the Legitimacy theory is founded in the Political economy theory. Hence, the convergence of these three major theories in future is likely.

ecological sustainability as it made more sense to view sustainability at the eco-systems, species and planetary levels:

> "*The Accounting for Sustainability Project seeks to* **embed** *business-centered notions of social and environmental probity within the organization as an initial step along the way toward some unspecified but clearly desirable outcome. No link with planetary or social sustainability is explored. The GRI initiative seeks to offer a means of producing narrative which can be thought to approximate a notion of sustainability through some variant on 'the Triple Bottom Line' (TBL). Once again, no attempt is made to derive substantive relationship between the exigencies of global sustainability and organizational activities and the bases of the claims to be an account or a report of sustainability are therefore rhetorical*".

Based on Gray's views, much of the sustainability-related disclosures were exaggerated and unsubstantiated claims in satisfaction of corporate interests. Gray challenged the ideals of sustainability reporting and supported the argument that the need for engaging accounts of the relationships between "un-sustainability" and corporate, economic and financial activity seemed stronger (Gray, 2010). In other words, it would make better sense to measure and report the negative impacts of business activities on the environment and compare performance based on the level of contribution to "un-sustainability".

2.5 *Summary of Background and Literature Review*

The preceding sections have covered the following:

- The definitions of corporate social responsibility and corporate sustainability
- The difference between sustainability reporting and other forms of communication
- The rationale of sustainability reporting
- The criticisms of sustainability reporting

3. Research Objectives and Methodology

3.1 *Objective*

To perform a comparative analysis of the sustainability reporting results between two selected SGX-listed Singapore companies that have adopted the GRI framework. The purpose of this analysis is to identify variations in reporting style in terms of topical coverage and emphasis that may contribute to comparability issues when assessing the non-financial performance.

3.2 *Sample Size and Sample Selection*

Although the number of companies that provide sustainability reports in Singapore has increased steadily to more than 25 as of 2014, the number is still too small for meaningful analyses such as inter-company assessments within each industry and cross-industry comparisons. Yet, because of the increasing importance of sustainability reporting which is recently reinforced by SGX in its announcement on the last quarter of 2014 to make this practice mandatory in the near-term[11], there is a necessity to understand what it means for Singapore companies and their investors. To gain a better understanding, an in-depth study of the reported content by the early adopters in Singapore is therefore relevant at this stage. For the purpose of this study, a comparative analysis between two companies would make good sense as a starting point and this type of comparative assessments can be scaled up when the number of sustainability reporters in Singapore increases.

The selected companies for the comparative research analysis are namely City Developments Limited ("CDL") and Golden Agri-Resources Limited ("GAR"). The details of their corporate background are presented in Table 6.4.

The two companies were selected as they were early adopters of the GRI framework in their respective industries and both companies have received extensive media attention on their sustainability reporting efforts. It was reported in 2010 that CDL was the first company in Singapore to disclose sustainability performance using the GRI framework in 2008[12] and CDL continued to appear on high-profile award ceremonies for its corporate responsibility achievements. In 2011, GAR was cited as a Singapore-listed company that had ramped up its sustainability reporting standards (by adopting the GRI framework) following the infamous allegations by Greenpeace that GAR's Indonesian unit was involved in illegal forest

[11]Source: http://www.eco-business.com/news/sgx-make-sustainability-reporting-mandatory/

[12]Source: http://www.eco-business.com/news/more-firms-adopting-sustainable-practices-using-gr/

Table 6.4. Companies examined.

Company name	Company background[13]
City Developments Limited ("CDL")	City Developments Limited (CDL) invests in, develops, and manages office, hotel, retail, and residential properties in Asia, Australia, Europe, and North America. It counts among its developments the 66-floor Republic Plaza—one of Singapore's tallest buildings and CDL's flagship holding. CDL boasts an extensive network of more than 400 subsidiaries. Its 55%-owned Millennium and Copthorne Hotels arm owns more than 110 hotels worldwide in major cities (New York, London, Paris, Singapore, and Sydney). The firm also owns a majority interest in City e-solutions, a hospitality technical services company engaged in reservation, insurance sales, and payroll management.
Golden Agri-Resources Ltd ("GAR")	Golden Agri-Resources is one of the largest palm oil players in the world. With over 460,000 hectares of oil palm estates across Indonesia, the company produces about 3 million tonnes of palm products annually. Its operations include harvesting oil palm trees, processing fresh fruit into crude palm oil and palm kernel oil, and refining it into such consumer products as cooking oil, margarine, and shortening. It produces two of the leading cooking oil brands in Indonesia: Filma and Kunci Mas. Additional operations in China focus on soybean oil and soybean meal production. Branded cooking products there include Grand Slam and Golden Carrier. It also owns noodle maker Florentina.

clearing which prompted the withdrawal of major customers like Unilever, Kraft and Nestle[14]

In addition, these two companies are listed on the SGX mainboard which imposes the most stringent corporate reporting standards in Singapore. Hence, the reports of the two companies can be regarded as comprehensive and reliable in the Singapore context.

Table 6.5 presents the sample size for the research analyses to perform the:

- chronological comparisons of reported content in all sustainability reports that were released by each company;
- comparisons of reported content between the two companies; and

[13]Source: Hoover's company records (database).

[14]Source: http://www.eco-business.com/news/singapore-firms-lag-in-green-reporting/

Table 6.5. Sustainability reports examined.

Company/ Reporting period	2007	2008	2009	2010	2011	2012	2013
CDL	**Sample A1**	**Sample A2**	**Sample A3**	**Sample A4**	**Sample A5**	**Sample A6**	**Sample A7**
GAR	NA	NA	NA	**Sample B1**	**Sample B2**	**Sample B3**	**Sample B4**
Benchmark for CDL (CapitaLand Limited)	NA	NA	NA	NA	NA	NA	**Sample C**
Benchmark for GAR (Wilmar International Limited)	NA	NA	NA	NA	NA	NA	**Sample D**

- comparisons of reported content between each company and a benchmark company of a similar industry for the latest reporting period

The benchmark companies were selected based on the closeness of their industry profiles with the research subjects, CDL and GAR.

3.3 *Research Method*

Thematic coding was applied for this study. According to Gibbs (2007), coding is a way of categorising the text that is under scrutiny to establish a framework of thematic ideas that can sometimes be interlinked or grouped under a general idea, hence forming a hierarchy of codes. In formulating the codes, the process will entail the intensive reading of text to identify themes which are further refined to relate with similar ideas, and eventually to link these ideas into core categories. Gibbs (2007) also explained that "the actual text is always an example of a more general phenomenon and the code title should indicate this more general idea". This way of organising the qualitative data (i.e. the reported content) will facilitate the comparative analyses across time and between reporting entities.

In addition, quantitative analysis will be included as appropriate. Gibbs (2007) clarified that "... coding in quantitative analysis is for the explicit purpose of reducing the data to a few 'types' in order that they can be counted ...." Accordingly, the quantitative analytical results will be used to complement the qualitative assessment to provide a fuller interpretation of the reported content.

3.4 *Study Limitation*

To a certain extent, the themes (or codes) are determined based on the subjective interpretation of the researcher. Hence, the coding hierarchy may not be fully representative of the themes that are embedded in the researched content. There is also a possibility that the codes are simplistic and over-generalised.

4. Results of Comparative Analyses

4.1 *Comparison of Reports by Golden Agri-Resources Ltd ("GAR")*

4.1.1 *Qualitative assessment*

Table 6.6 presents the identified codes and sub-codes (collectively termed the "coding hierarchy") of each GAR sustainability report. The codes have been consistent over the years except for the addition of "Clarity", a sub-code under "Transparency", from 2011 onwards. This additional code was triggered by the term "sustainable success", which was first used in 2011 and persisted through to the latest report. The "Clarity" sub-code was assigned as the term "sustainable success" was not defined.

Overall, GAR's strong focus on image and reputation reflected the intent to stay relevant with prevailing social norms and expectations for players within the palm oil industry. This is consistent with the "Legitimacy theory" as discussed in the literature review.

4.1.2 *Quantitative assessment*

The two common key words used in GAR's reports were "Best" and "Continue". Figure 6.1 presents the word usage frequency over time. Notably, the repetition of the word "Continue" exceeded "Best" in 2011, and remained significantly higher than the usage of "Best" in the subsequent reporting periods. This signalled the emphasis on commitment by GAR.

Table 6.6. Coding hierarchy of GAR's reports.

2010	2011	2012	2013
Characteristic • Altruistic • Collaborative • Law-abiding	Characteristic • Altruistic • Collaborative • Law-abiding	Characteristic • Altruistic • Collaborative • Law-abiding	Characteristic • Altruistic • Collaborative • Law-abiding
Leadership • Authority	Leadership • Authority	Leadership • Authority	Leadership • Authority
Reputation • Accountability • Ethical • Legitimacy • Stewardship	Reputation • Accountability • Ethical • Legitimacy • Stewardship	Reputation • Accountability • Ethical • Legitimacy • Stewardship	Reputation • Accountability • Ethical • Legitimacy • Stewardship
Strategy • Economic performance • Environmental performance • Social performance • Growth focus • Target driven	Strategy • Economic performance • Environmental performance • Social performance • Growth focus • Target driven	Strategy • Economic performance • Environmental performance • Social performance • Growth focus • Target driven	Strategy • Economic performance • Environmental performance • Social performance • Growth focus • Target driven
Transparency • Completeness • Integrity • Materiality	Transparency • **Clarity** • Completeness • Integrity • Materiality	Transparency • **Clarity** • Completeness • Integrity • Materiality	Transparency • **Clarity** • Completeness • Integrity • Materiality

4.2 *Comparison of Reports by City Developments Limited ("CDL")*

4.2.1 *Qualitative assessment*

Table 6.7 presents the codes identified in CDL reports. The coding hierarchy had been consistent over time with the addition of "Social performance" and "Integrity" codes. The "Social performance" sub-code was triggered by the coded text in 2009 stating, "*... The plan to draft a new policy on human*

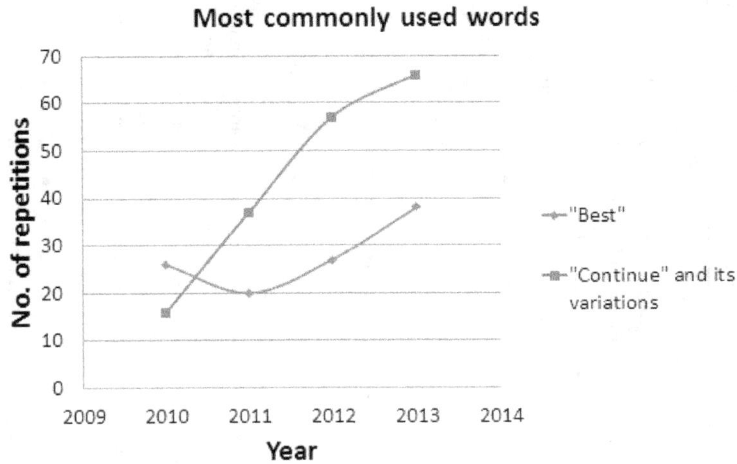

Fig. 6.1. Common key words in GAR's reports.

rights and climate change is one of the most significant signs of CDL's effort in broadening its CSR scope." The appearance of the "Integrity" sub-code in 2008 was due to the disclosure of the external auditor's statement which included the points for improvements relating to both the reported information and the underlying management processes that the information was derived.

Overall, CDL's pioneering efforts in campaigning for its green products through sustainability reports and other communication platforms accord with the intent to differentiate itself in a competitive environment and this can be explained by the "Political-economy theory" as discussed in the literature review.

4.2.2 *Quantitative assessment*

The common key words used in CDL's sustainability reports were "Green" and "First" with variations that included "Pioneer" and "Champion". According to Fig. 6.2, the trend paths of both words were closely correlated which implied that CDL's leadership in corporate responsibility was linked to its environmental performance. Notably, there was a dip in the usage of the commonly used words in the following two years after 2010 which coincided with the period when CDL focused more attention on social responsibility.

Table 6.7. Coding hierarchy of CDL's reports.

2007	2008	2009	2010	2011	2012	2013
Characteristic	Characteristic	Characteristic	Characteristic	Characteristic	Characteristic	Characteristic
• Amicable	• Amicable	• Amicable	• Amicable	• Amicable	• Amicable	• Amicable
• Charitable	• Charitable	• Charitable	• Charitable	• Charitable	• Charitable	• Charitable
• Eco-friendly	• Eco-friendly	• Eco-friendly	• Eco-friendly	• Eco-friendly	• Eco-friendly	• Eco-friendly
Leadership	Leadership	Leadership	Leadership	Leadership	Leadership	Leadership
• Pioneering	• Pioneering	• Pioneering	• Pioneering	• Pioneering	• Pioneering	• Pioneering
Reputation	Reputation	Reputation	Reputation	Reputation	Reputation	Reputation
• Ethical	• Ethical	• Ethical	• Ethical	• Ethical	• Ethical	• Ethical
• Legitimacy	• Legitimacy	• Legitimacy	• Legitimacy	• Legitimacy	• Legitimacy	• Legitimacy
• Stewardship	• Stewardship	• Stewardship	• Stewardship	• Stewardship	• Stewardship	• Stewardship
Strategy	Strategy	Strategy	Strategy	Strategy	Strategy	Strategy
• Economic performance	• Economic performance	• Economic performance	• Economic performance	• Economic performance	• Economic performance	• Economic performance
• Environmental performance	• Environmental performance	• Environmental performance	• Environmental performance	• Environmental performance	• Environmental performance	• Environmental performance
• Education	• Education	• Education	• Education	• Education	• Education	• Education
		• **Social performance**	• **Social performance**	• **Social performance**	• **Social performance**	• **Social performance**
Transparency	Transparency	Transparency	Transparency	Transparency	Transparency	Transparency
• Completeness	• Completeness	• Completeness	• Completeness	• Completeness	• Completeness	• Completeness
• Materiality	• Materiality	• Materiality	• Materiality	• Materiality	• Materiality	• Materiality
	• **Integrity**	• **Integrity**	• **Integrity**	• **Integrity**	• **Integrity**	• **Integrity**

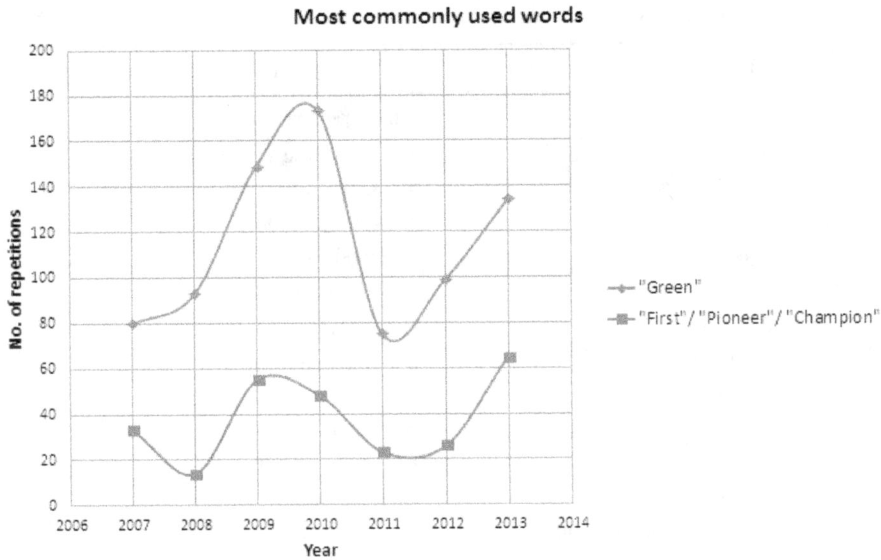

Fig. 6.2. Common key words in CDL's reports.

4.3 *Comparison of Reports by GAR and CDL*

Table 6.8 lists the thematic codes of the latest reports by CDL and GAR.

In terms of "completeness" of sustainability reporting, GAR's disclosures provided a better visibility of the policy implementation across the corporate group, while CDL's sustainability approach had been concentrated at the Singapore operations with limited disclosures of corporate responsibility efforts from its major subsidiary, the Millennium and Copthorne Hotels plc. Hence, it was unclear as to what may be the corporate group strategy for sustainability issues that concerned all other entities within CDL's ownership structure.

Although the reports of CDL and GAR contained the "integrity" element, there was a difference in the extent of transparency for reported cases of fatalities. In GAR's reports, workplace fatalities were clearly reported with details on the location, cause of accident and the types of assistance provided to the bereaved families. In CDL's report, fatal workplace injuries were not segregated from the non-fatal ones in the reported data. Unlike GAR, no further information was disclosed regarding the kind of assistance provided to the bereaved families and persons with non-fatal but serious injuries.

Table 6.8. Coding hierarchy of reports by CDL and GAR.

CDL 2013	GAR 2013
Characteristic	Characteristic
• Amicable	• Altruistic
• Charitable	• Collaborative
• Eco-friendly	• Law-abiding
Leadership	Leadership
• Pioneering	• Authority
Reputation	Reputation
• Ethical	• Accountability
• Legitimacy	• Ethical
• Stewardship	• Legitimacy
	• Stewardship
Strategy	Strategy
• Economic performance	• Economic performance
• Environmental performance	• Environmental performance
• Education	• Social performance
• Social performance	• Growth focus
	• Target driven
Transparency	Transparency
• Completeness	• Clarity
• Integrity	• Completeness
• Materiality	• Integrity
	• Materiality

Overall, the themes asserted by the reports of CDL and GAR were largely similar. Both had organisational characteristics, leadership qualities, reputational attributes, corporate strategies, and transparency in their reports.

4.4 *Comparison of Reports by Golden Agri-Resources Ltd ("GAR") and Wilmar International Limited ("Wilmar")*

Wilmar International Limited ("Wilmar) was selected as a suitable benchmark due to its compatibility with GAR's industry profile. The thematic codes from the latest reports of GAR and Wilmar are given in the Table 6.9.

Table 6.9. Coding hierarchy of reports by GAR and Wilmar.

GAR 2013	Wilmar 2013
Characteristic	Characteristic
• Altruistic	• Altruistic
• Collaborative	• Collaborative
• Law-abiding	• Law-abiding
Leadership	Leadership
• Authority	• Leading by example
Reputation	Reputation
• Accountability	• Accountability
• Ethical	• Ethical
• Legitimacy	• Legitimacy
• Stewardship	• Stewardship
Strategy	Strategy
• Economic performance	• Economic performance
• Environmental performance	• Environmental performance
• Social performance	• Social performance
• Growth focus	
• Target driven	
Transparency	Transparency
• Clarity	• Clarity
• Completeness	• Completeness
• Integrity	• Integrity
• Materiality	• Materiality

The coding hierarchies of GAR and Wilmar were fairly similar. Even the cover pages of both reports featured the same thematic images of children, the forest and palm oil seeds. These images projected the themes of future generation (i.e. local children), environmental resources (i.e. forest) and the growth potential of the palm oil industry (i.e. palm seeds)

The key differences noted were under the "Leadership", "Integrity" and "Materiality" codes. As compared to GAR, Wilmar appeared to have a softer approach to leadership. The coded texts of Wilmar's report included *"hope to convince"* and *"big corporations must take the lead"* vis-à-vis GAR's coded texts of *"leading and shaping the agenda"*; *"As a leader in the industry";* and *"... mobilizing our employees, customers, civil society organisations, investors, the Government of Indonesia, local and indigenous communities, our peers and other stakeholders".*

On management of sustainability-related issues, GAR and Wilmar were motivated differently. Under the "Materiality" sub-code, the coded text of GAR's earlier reports included *"The Greenpeace campaign against GAR was the most significant stakeholder issue"* and *"In order to meet Nestlé's Responsible Sourcing Guidelines".* These coded texts could be cross-compared against "*We did not do this due to pressure from the non-governmental organisations (NGOs)*" of Wilmar's report. The latter reflected Wilmar's self-driven motivation towards corporate sustainability.

Moreover, Wilmar did not have noticeable common words as compared to the high usage of "Best" and "Continue" by GAR. This may be due to the lesser need for Wilmar to emphasise its commitments to addressing sustainability issues.

4.5 *Comparison of Reports by City Developments Limited ("CDL") and CapitaLand Limited ("CapitaLand")*

The sustainability report of CapitaLand Limited ("CapitaLand") was selected as a benchmark for CDL's report and the codes from the latest reports are given in Table 6.10.

Under the code for "Leadership" in CapitaLand's report, there was lesser use of the words "First", "Champion" or "Pioneer" as compared to CDL. The coded texts from CapitaLand's report demonstrated a group strategy for its sustainability practice while CDL maintained a country-level approach and concentrated the reporting of sustainability performance in Singapore.

Another notable difference was in the clearer disclosure of fatalities in CapitaLand's report. CapitaLand had specifically disclosed the number of "work-related fatalities or permanent disability" while in CDL's report, similar information was reported in terms of "number of fatal and non-fatal workplace injuries per 100,000 persons employed".

5. Conclusions and Recommendations

Based on the research results, three issues on the comparability of sustainability reports were identified.

Table 6.10. Coding hierarchy of reports by CDL and CapitaLand.

CDL 2013	CapitaLand 2013
Characteristic	Characteristic
• Amicable	• Consistent
• Charitable	• Charitable
• Eco-friendly	• Eco-friendly
Leadership	Leadership
• Pioneering	• Leading by example
Reputation	Reputation
• Ethical	• Ethical
• Legitimacy	• Legitimacy
• Stewardship	• Stewardship
Strategy	Strategy
• Economic performance	• Economic performance
• Environmental performance	• Environmental performance
• Education	• Social performance
• Social performance	
Transparency	Transparency
• Completeness	• Completeness
• Materiality	• Materiality
• Integrity	• Integrity

5.1 *Transparency and Ambiguities*

Although the GRI framework has enabled better transparency, it has invited ambiguities. For instance, to highlight the positive social impacts on the surrounding communities, GAR had repeatedly pitched the palm oil business as an indispensable economic driver that will alleviate poverty. In addition, the average wage of field labourers was reported to be above the minimum national average. Yet, these 'positive' information could be limited as there was no disclosure on other important aspects such as the adverse effects on traditional economic activities that were integral to the indigenous communities; the rationale for using the minimum national wage as a benchmark to value the work of the labourers; and the chances of career advancement for these labourers. Hence, the completeness of the information disclosed may be challenged which makes it vital to validate the claims made in the sustainability reports. When sustainability reporting becomes formalised in Singapore, the regulators should require mandatory assurance of such reports.

5.2 *Level of Reporting*

Table 6.11 presents the latest sustainability report profiles of the four companies being examined in this study.

Taken at face value, it seems effortless to draw a logical conclusion that CDL is the best-performing entity with the highest GRI adherence level attained and reported content that is independently audited. Yet, the comparative analysis between CDL and GAR revealed that GAR had a more cohesive strategy for the implementation of environmentally and socially responsible practices over a wider extent of the corporate group (including entities that GAR did not have significant control over). Similarly, when compared with the report of CapitaLand, CDL's concentration of corporate sustainability efforts at the country-level stood in contrast with CapitaLand's group-wide approach across 20 countries, that translated into a sustainability-driven group management structure. Hence, this raised an important issue of comparability: whether it is reasonable to compare company-wide or localised performance with group-level performance? This may not sound like a fair question but it is a reality that companies have adopted the corporate sustainability principles differently; with some focusing on driving changes at the individual company level while others strive to achieve consistency across the corporate group. Hence, the levels

Table 6.11. Sustainability report profiles.

Entity	CDL	GAR	CapitaLand (CDL's benchmark)	Wilmar (GAR's benchmark)
GRI adherence rating[15]	A+	B	In accordance – Core	In accordance – Core
GRI version[16]	G3.1	G3	G4	G4
Externally assured[17]	Yes	No	Yes	Yes

[15]Source: http://database.globalreporting.org/search
[16]Source: http://database.globalreporting.org/search
[17]Source: Respective sustainability reports for the latest reporting period.

at which sustainability performance is reported (i.e. by company, group, business unit and/or geographical segment) should be standardised.

5.3 *Supply Chain Considerations*

The last comparability issue noted arose from the differing extent of vertical integration between companies. For social performance, both CDL and CapitaLand reported on charitable initiatives, staff welfare, compensation, developmental opportunities and gender equality. GAR and Wilmar focused a lot more attention on the fair treatment of labourers and their families (through provisions of basic needs and collaboration with local governments to alleviate poverty). Besides the difference in operating environments, one of the reasons for the difference in focus or emphasis between CDL–CapitaLand and GAR–Wilmar comparisons on social impact disclosures is the business model. CDL and CapitaLand are both involved in real-estate development that typically sub-contracts the building of properties to construction companies. On the other hand, GAR and Wilmar are in the palm oil business which usually controls a wide stretch of the supply chain; from plantation development to oil refinery and processing work. When there is a difference in the extent of vertical integration, as evidenced by CDL and CapitaLand having a significantly smaller control of the supply chain than GAR and Wilmar, the social performance would not be comparable. It would not be fair to compare the charitable efforts of CDL against GAR's contributions to their workforce in terms of health, wellness and caring for the needs of their dependents. A plausible comparison against GAR's social performance would require the inclusion (in the report) of CDL's commitment to promoting and ensuring the proper treatment of foreign construction workers in its supply chain. Therefore, the comparability of non-financial performance between companies of different industry could be compromised when the extent of supply chain integration varies significantly. Adequate guidance to sustainability reports should be provided to encourage disclosures on supply chain management and control measures.

References

Ballou, B., Heitger, D.L. and Landes, C.E. (2006). The Future of Corporate Sustainability Reporting. *Journal of Accountancy*, 65–74.

Branco, M.C. and Rodrigues, L.L. (2007). Positioning stakeholder theory within the debate on corporate social responsibility. *Electronic Journal of Business Ethics and Organization Studies*, 5–15.

Cooper, D.J. and Sherer, M.J. (1984). The value of corporate accounting reports: Arguments for a political economy of accounting. *Accounting, Organizations and Society*, 207–232.

Gibbs, G.R. (2007). *Analyzing Qualitative Data* . London, California, India, Singapore : SAGE Publications.

Gray, R. (1992). Accounting and environmentalism: An exploration of the challenge of gently accounting for accountability, transparency and sustainability. *Accounting, Organizations and Society*, 399–425

Gray, R. (2010). Is accounting for sustainability actually accounting for sustainability... and how would we know? An exploration of narratives of organizations and the planet. *Accounting, Organizations and Society*, 47–62.

Hopwood, A. Unerman, J. and Fries, J. (2010). *Accounting for sustainability: Practical insights (with a foreword by HRH The Prince of Wales)* Earthscan publications.

KPMG. (2013). *The KPMG Survey of Corporate Responsibility Reporting.* KPMG.

Laszlo, C. and Zhexembayeva, N. (2011). *Embedded sustainability: The next big competitive advantage.* California, USA : Greenleaf Publishing Limited

Lehman, G. (1995). A legitimate concern for environmental reporting. *Critical Perspectives on Accounting*, 393–412.

Lodhia, S.K. (2012). The Need for Effective Corporate Social Responsibility/Sustainability Regulation. In Jones, S. and Ratnatunga, J., *Contemporary Issues in Sustainability Accounting, Assurance and Reporting* (pp. 139–152). UK: Emerald Group Publishing Limited.

Montiel, I. (2008). Corporate social responsibility and Corporate sustainability: Separate pasts, Common futures. *SAGE journals: Organization and Environment*, 245–269.

Ratnatunga, J. and Jones, S. (2012). A Methodology to Rank the Quality and Comprehensiveness of Sustainability Information Provided in Publicly Listed Company Reports. In Jones, S. and. Ratnatunga, J. *Contemporary Issues in Sustainability, Accounting, Assurance and Reporting* (pp. 227–266). UK: Emerald Group Publishing Limited

Schaltegger, S. (2012). Sustainability Reporting Beyond Rhetoric: Linking Strategy, Accounting and Communication. In Jones, S. and Ratnatunga, J., *Contemporary Issues in Sustainability, Accounting, Assurance and Reporting* (pp. 183–195). UK: Emerald Group Publishing Limited

Siegel, D. and Vitaliano D. (2006). *An empirical analysis of the strategic use of corporate social responsibility.* Social Science Research Network.

Sustainability Disclosure Database. (2015). Retrieved from Global Reporting Initiative: http://database.globalreporting.org/search

Waddock, S. (2008). *The Difference Makers: How social and institutional entrepreneurs created the corporate responsibility movement.* UK: Greenleaf Publishing Limited.

Worthington, I. (2013). *Greening business: Research, Theory, and Practice.* Oxford University Press.

7

AIR QUALITY IMPROVEMENT AND GREENHOUSE GAS MITIGATION IN PAKISTAN: AN INTEGRATED APPROACH

KALEEM ANWAR MIR and RAJASEKHAR BALASUBRAMANIAN

The study aims to establish a linkage between both the energy (Pakistan Integrated Energy Model, i.e. Pak-IEM) and environment (Greenhouse gas and Air pollution INteractions and Synergies, i.e. GAINS) models. With this framework, the energy outputs of Pak-IEM could be fed into GAINS for exploring different emission and control strategy scenarios. Prior to this analysis, the activity data projected until 2030 under the business-as-usual scenario for different sectors (energy, mobile, agriculture, industry, etc.) implemented in the current version of the GAINS model for Pakistan were inconsistent. A strong basis for energy projections now exists in Pakistan as a result of the development of the Pak-IEM; that can provide energy projections for all sectors as a fully integrated systems model, and could underpin the GAINS analysis for Pakistan. This work focuses primarily on developing a reference emission scenario in GAINS based on the reference energy scenario generated by Pak-IEM; mainly to assess emissions of three air pollutants (SO_2, NO_x and $PM_{2.5}$) and one greenhouse gas (CO_2), as well as the health and vegetation impacts of increased emissions with an aim to develop a framework for identifying cost-effective strategies to mitigate air pollution and GHG emissions in Pakistan under the co-benefits approach. It also analyses a technology-based alternative emission control scenario that takes into account the uniform application of advanced end-of-pipe emission control technologies in comparison with a baseline case.

Introduction: Air Pollution and Climate Change Linkages

There exists a very close linkage between air pollution and climate change that has mostly been kept separate in applied scientific study as well as in political debates. However, this is changing due to modern scientific assessments (IPCC, 2001) and advanced integrated research (EEA, 2003). The most significant association is that a number of customary air pollutants and greenhouse gases arise from common burning sources and act together in the air both physically and chemically. Eventually both types of gaseous pollutants execute a range of environmental impacts at local, regional and global scales. Previously, the scientific investigation of climate change impacts, causes and dynamics are mainly concentrated on the atmospheric role of greenhouse gases. These greenhouse gases include CO_2 (Carbon dioxide), CH_4 (Methane), N_2O (Nitrous oxide), HFCs (Hydrofluorocarbons), PFCs (Perfluorocarbons) and SF_6 (Sulfur hexafluoride) that are regularly debated in worldwide negotiations on climate change. There is less research available that clarifies the relevance of certain short-lived air pollutants, particularly PM (Particulate Matter) and O_3 (Ozone), in climate change science apart from their role in radiative forcing and atmospheric chemistry. Only the most recent scientific assessments begin to decouple connections concerning these synergies and as a result are opening the doors for integrated research and action. At present, climate change researchers agree on the finding that the radiation balance of the earth is affected by GHG emissions. We can consider both air pollution and climate change as two sides of the same coin because some toxic and short-lived compounds usually known as air pollutants can also disturb climate. For example, particulate matters and ozone are both known to have great impact on Earth's radiation balance; as a result both are included in climate change considerations. Hence it is not possible to explicitly separate several compounds into mutually exclusive groups of either greenhouse gases or air pollutants and particles.

Besides the chemical and physical interaction of air pollution and climate change in the atmosphere, the impacts on human health and ecosystems are also of concern, although there are variations in exposure and sensitivity. For example, in the case of increased radiation, there will be growth in the development of ground level ozone which in turn will cause harm to vegetation as well as human health. Moreover, there will be effects on the sensitivity of the ecosystem towards acid deposition (Posch, 2002). The capacity of plants to store carbon dioxide (CO_2) could be altered due to augmented deposition and weakening of plant growth caused by air pollution (Abrol and Ingram, 1996). Therefore, both air pollution and climate change can adversely affect biodiversity (Bakkenes *et al.*, 2002; Ertsen *et al.*, 1998).

Mitigation measures for various well-known greenhouse gases may lead to an immediate decrease in the emissions of air pollutants. These mitigation measures may consist of fuel switching from oil and coal to natural gas, which is a cleaner fuel, the co-generation of heat and power (CHP), and energy efficiency enhancements. As a result of a decrease in the carbon contents of the fuel, there will also be a decrease in PM, NO_x and SO_2 emissions. In addition, several other measures for mitigating CO_2, such as IGCC (Integrated Gasification Combined Cycle) plants for producing electricity, eliminate PM and SO_2 emissions and decrease NO_x emissions. However, there are also cases where GHG mitigation measures result in an increase in air pollutant emissions. For example, replacing gasoline cars with diesel ones will definitely result in decreasing CO_2 emissions; but this course of action will also, at the same time, increase PM emissions unless no particular technical control measure is applied in the car. Similarly, the increased utilisation of biomass in small combustion processes will result in a significant increase in PM and NO_x emissions. On the contrary, strategies for controlling air pollution do not necessarily result in decreasing GHG emission, particularly when the emissions are controlled by applying end-of-pipe measures. In this kind of cases, air pollution control strategies produce higher GHG emissions due to additional use of fuel or energy. Nevertheless, the decrease in GHG emissions could be achieved only if air pollution control strategies are applied through carbonaceous fuel substitution or energy efficiency/energy saving measures (Pleijel *et al.*, 2009). We consider these links very important because the key sources for air pollution and climate change are energy use and agricultural practices, so the mitigation methods to cut air pollution correspondingly has advantages for climate change by reducing GHG emissions and vice versa; eventually resulting in co-benefits where the actions benefit both. The co-control of GHG emissions and air pollution results in extra benefits on health and environmental impacts that actually fulfil the core objective of an emission control strategy.

Air Pollution and GHG's Status in Pakistan

Pakistan is a low-income developing nation. It is separated into four provincial regions: Punjab, Sindh, Khyber-Pakhtunkhwa (KPK) and Balochistan. Pakistan shares its borders with China in the north, India in the east, Afghanistan in the north-west and Iran in the west. The total area (land) of Pakistan is 796,096 square kilometres. Pakistan has a complex terrain, ranging from the plain areas of Punjab and Sindh to the high Himalayan mountain range in the KPK and Balochistan. Figure 7.1's left and right panels display maps of Pakistan within Asia, and a map of the separate provinces within Pakistan, respectively.

Pakistan is a very low GHG emitter. According to the Carbon Dioxide Information Analysis Centre (CDIAC) of the U.S. Department of Energy (DOE), considering the year 2000's data, the Pakistan's share in global

Fig. 7.1. Pakistan location within Asia (top panel) and Pakistan provinces Map (bottom panel).

GHG emissions is merely 0.8%. Likewise, per capita GHG emissions in Pakistan are approximately one third the world average. Based on per capita GHG emissions, excluding land use change, Pakistan holds 135th place worldwide. When land use change is included, it stands at 149th place. On the other hand, if we look only at CO_2 emissions resulting from fuel burning, Pakistan's global share is almost 0.44%, and its per capita emissions are nearly one sixth of the world average (MoCC, 2012).

Air pollution is becoming an increasing concern in large cities in Pakistan (Alam *et al.* 2007). The suspended particles that include smoke and dust in Pakistan are usually two times the world average and five times greater than that of developed countries. Although a certain amount of air pollution originates from natural causes, the majority comes from human activities where the principal sources are power plants fired with fossil fuels, movement of vehicles on roads, open burning of solid waste, biomass burning and extremely inefficient energy use (Tahir *et al.* 2010; ADB, 2006). In Pakistan, the most severe problem relating to air pollution is the presence of high levels of particulate matter (PM) suspended in ambient air. One of the main reasons for the rapid rise in air pollution is the rising rate of urbanisation that causes growth of infrastructure, and rise in urban transport and cutting down of trees (deforestation), which contribute to more pollution in the air. Hence, it is imperative to decrease emissions of air pollutants that deteriorate local air quality in Pakistan, which at the same time will help to curtail GHG emissions.

The Pakistan Environmental Protection Act (PEPA) of 1997 is the central environmental law, succeeding the Pakistan Environmental Protection Ordinance, 1983. The 1997 Act launched the wide-ranging conditions, bans, and execution for pollution stoppage and control, with the goal of promoting sustainable development. The National Environmental Action Plan (NEAP) of 1997 ratified by the Pakistan Environmental Protection Council (PEPC), included air pollution in its fundamental programs. Certain significant goals, such as the unleaded gasoline initiation, decrease of sulphur in diesel oil, use of compressed natural gas (CNG) driven cars and buses, were realized. These actions may already be having co-benefits for GHG reduction. Nevertheless, several actions of NEAP 1997 have still yet to commence (Pak-SCEA, 2006).

GAINS Modelling Framework for Pakistan

GAINS model covers the entire Pakistan's energy system by dividing it into four regions:

- Region 1—Punjab
- Region II—Sindh excluding Karachi
- Region III—Karachi
- Region IV—Baluchistan & KPK

This division is based on the recognition that significant regional differences in energy supply and demand exist. In view of this database, where possible, the model has captured important variances in terms of emission projections of GHGs and air pollutants. Emissions being tracked in GAINS Pakistan use Intergovernmental Panel on Climate Change (IPCC) default emission factors. For the estimation of emissions for all the pollutants monitored, the GAINS model takes into account the relevant activity data, emission factors and removal efficiency represented by the formula below:

$$E_{i,p} = \sum_{k}\sum_{m} A_i,\ kef_{i,k,m,p}x_{i,k,m,p}$$

where i, k, m, p denotes country, activity type, mitigation measure and pollutant correspondingly. $\mathrm{E}_{i,p}$ is the pollutant emissions in country i, while $\mathrm{A}_{i,k}$indicates activity level of type k (like consumption of coal in power plants) in country i. Likewise, $ef_{i,k,m,p}$ specifies emission factor of pollutant p for activity k in country i after application of control measure m, and $X_{i,k,m,p}$ is the share of total activity of type k in country i to which a control measure m for pollutant p is applied (Amann *et al.* 2008). Table 7.1 explains the approach employed in GAINS model.

Furthermore, most data sources for the existing baseline scenario in the GAINS model were taken from the Pakistan Energy Yearbook (prepared by Hydrocarbon Development Institute of Pakistan (HDIP)), Statistical Bureau of Pakistan, National Transport Research Centre (NTRC), Punjab Statistical Bureau, Sindh Statistical Bureau, KPK Statistical Bureau, Baluchistan Statistical Bureau, Population Census Organization, Ministry of Food, Agriculture & Livestock, and Ministry of Industries.

Pak-IEM Modeling Framework for Pakistan

In Pak-IEM the entire energy system of Pakistan is considered, hence it can be used for analysing national policies. In the first version of Pak-IEM,

Table 7.1. Multi-Impact plus Multi-pollutant approach in GAINS (Amann *et al.* 2008).

	PM	SO_2	NO_x	VOC	NH_3	CO_2	CH_4	N_2O	HFCs	PFCs	SF_6
Effects on Health:											
Particulate matter	√	√	√	√	√						
Ground level O_3			√	√			√				
Damage to vegetation:											
Ground level O_3			√	√			√				
Acidification		√	√		√						
Eutrophication			√		√						
Radiative forcing:											
direct						√	√	√	√	√	√
through aerosols	√	√	√	√	√						
through OH			√	√			√				

the whole country is taken as a single region due to the nature of the available data, and this is considered the most appropriate approach at this developing stage of Pak-IEM local capacity. However, it is acknowledged that there exist substantial regional differences in energy demand and supply; hence regionalisation needs to be done in the future versions of Pak-IEM. The planning horizon of Pak-IEM runs from the base year (starting year) to the final year using a series of periods of varying duration in between. The base year chosen for the Pak-IEM model is 2006/07 (i.e. the period reported as 2007 actually runs from July 2006 to the end of June 2007), and the length of the first period is kept as a single year for the purpose of calibrating the model data to the energy balance for that year. 2007 was chosen as the base year because of the availability of most of the data for various sectors of energy systems. Presently, the model runs are extended to 2030, though Pak-IEM can be run for any desired time horizon, as the model allows for the separation of data years and model years (Pak-IEM, 2010).

Pak-IEM employs the TIMES (The Integrated MARKAL/EFOM System) model generator, the successor to the MARKAL modelling structure, which has been envisioned, developed, and constantly advanced by the International Energy Agency–Energy Technology Systems Analysis Programme (IEA-ETSAP). MARKAL/TIMES is in extensive use in several countries around the world as an energy systems optimisation model. Around 200 organisations in 70 countries are deploying MARKAL/TIMES. It has a track record of 30 years of success and growth. The group members of International Resources Group (IRG) have been engaged in capacity

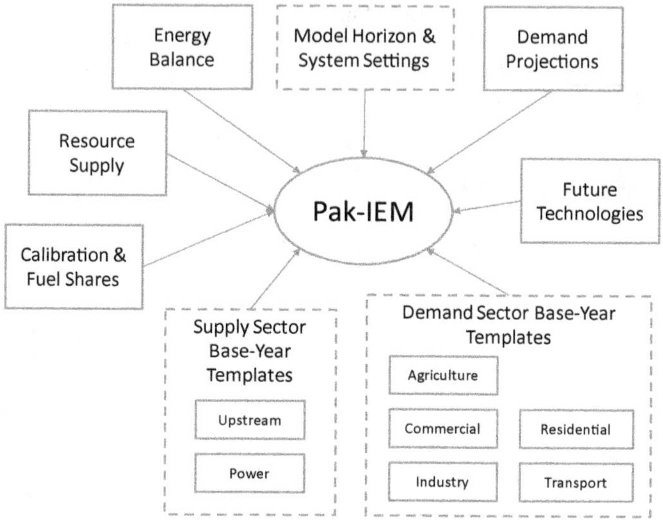

Fig. 7.2. Structure of the Pak-IEM data inputs (Pak-IEM, 2010).

building and transfer, development of model, and policy analysis concerning MARKAL/TIMES during most of that time.

Pak-IEM uses VEDA (VErsatile Data Analyst), a model management tool that oversees all aspect of working with the model. The input data for the model is structured on excel templates and is managed by VEDA-FE (VEDA-Front End), where the results/outcomes of the model are handled by VEDA-BE (VEDA-Back End). The base year for the model is adjusted to 2007 and the planning horizon runs till 2030. Figure 7.2 shows the basic structure of Pak-IEM data inputs (Pak-IEM, 2010).

Linking of Pak-IEM with GAINS Model

There are two key challenges to linking Pak-IEM with GAINS-South Asia: (1) Mapping Pak-IEM outputs to GAINS categories, ensuring no double counting of mitigation potential and attributing abatement potential to more or less aggregated Pak-IEM technology groups; (2) Pak-IEM considers Pakistan as a single region while the GAINS structure splits Pakistan into four key regions on the basis of economic activities.

The essence of the methodology is to first use Pak-IEM to output energy data (reference energy scenario — energy consumption projections till 2030) that can then be used as an input in the GAINS model to develop a reference emission scenario indicating emissions projections till 2030. A key part of the approach is formatting Pak-IEM reference scenario energy outputs to make them consistent with the GAINS model templates. The approach is significantly helped by the use of VEDA-BE software (Pak-IEM model results analysis tool used with TIMES models) that allows model outputs to be formatted according to user-defined sets and tables. In effect, VEDA-BE can reproduce the GAINS model template structure.

The more challenging aspect of the approach is to ensure that the mapping of sectors/technologies/fuels is consistent. In Pak-IEM, Pakistan is considered as single region while in the GAINS model, Pakistan is divided into four regions (Punjab, Karachi, KPK and Balochistan, and Sindh). These divisions are developed based upon the availability of data and a high concentration of emission sources. The mega city of Karachi has been kept as one region, independent of the rest of Sindh, due to high economic activity in the city, its large population, the huge number of industries located around the city, and heavy traffic load. The city of Karachi contributes to approximately 25–30% of the national Gross Domestic Product (GDP). KPK and Balochistan are combined into a single region due to their lower level of economic activities and low population density. The province of Punjab is treated as an independent region as it contributes to about 50–55% of the GDP of Pakistan. Hence, Pak-IEM output data is disaggregated into the four key regions defined in the GAINS model. All current and projected energy data from 2007 (base year) to 2030 required for the GAINS reference emissions scenario was obtained from the Pak-IEM reference energy scenario.

Pak-IEM Reference Energy Scenario Results

Figure 7.3 shows that natural gas and electricity account for most of the near term growth in final energy use. However, in the longer term, domestic supplies of natural gas are declining and imports are required to both maintain and increase supplies. By 2030, the most significant growth in final energy is based on electricity (which grows by a factor of 5) and oil products (growing by a factor of 4), driven particularly by industry demand. Natural gas use increases by 50% while coal use increases significantly, from 4 to 36 Mtoe due to use by industry. Renewable energy use, which is primarily traditional biomass (e.g. wood, agricultural residues, and dung), remains broadly flat due to resource constraints. Per capita electricity consumption, which increases from 457 kWh/yr to 1,450 kWh/yr, remains lower than the 2007 world average of 2,752 kWh/yr (Pak-IEM, 2011).

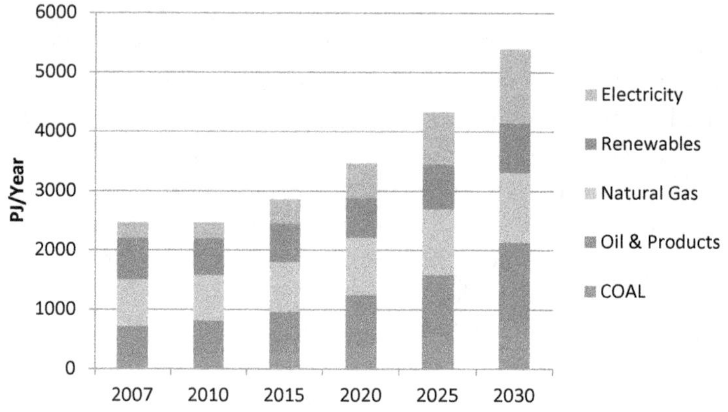

Fig. 7.3. Final energy consumption by fuel.

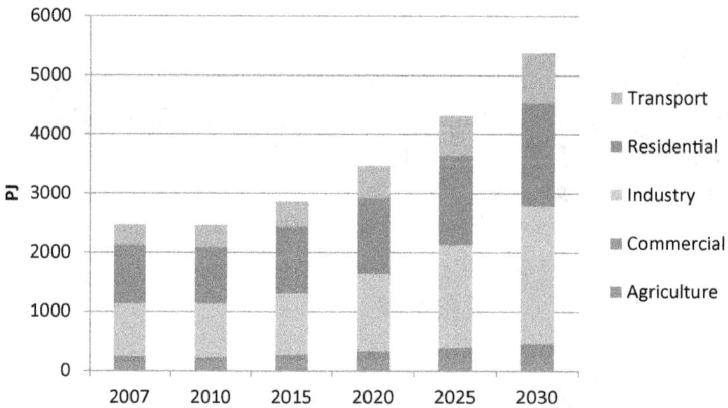

Fig. 7.4. Final energy consumption by sector.

Figure 7.4 shows that by 2030, the overall, final energy use more than doubles, while industrial energy use increases around a factor of 3 (driven by strong predicted GDP growth) and transport energy use almost quadruples. Residential energy use doubles, and continues to be the second largest component of final energy use in 2030. Agricultural and commercial energy use grows, but they remain small contributors to the overall final energy consumption. This figure helps to identify where policies that can promote moderation of energy growth relative to GDP growth should be examined (Pak-IEM, 2011).

Reference Emission Scenario: An Inventory of Emissions for 2007

Based on fuel quantity figures, energy data and specific combustion characteristics, the GAINS model approximates that in 2007, Pakistan released the emissions reported in Table 7.2 into the atmosphere.

For the year 2007, power plants and energy combustion in the industrial sector caused almost 77% of the total SO_2 emissions. For NO_x, the transport sector contributed around 66% of the total emissions while the industrial and power sectors added 16% and 11% respectively. In the case of $PM_{2.5}$ (fine particles), the largest sources were the industrial and domestic sectors; contributing 57% and 35% respectively. For CO_2 emissions, the power sector accounted for roughly 30% of the total anthropogenic CO_2 emissions in Pakistan for the year, while the industrial sector contributed 32%.

As indicated in Fig. 7.5, per-capita emissions for SO_2 and NO_x are significantly lower in Pakistan than in the United States, European Union, China and India. On the other hand, its per capita emissions for $PM_{2.5}$ are higher than the per-capita emissions of the United States, European Union and India for $PM_{2.5}$. Figure 7.6 indicates that when compared on a per-capita basis, estimates of CO_2 emissions for Pakistan are much lower when compared with numbers for the European Union, the United States, India and China. It is important to examine the future growth in levels of air pollutants to determine which of these have major implications for the region over time.

Air Quality and Impacts for 2007

Based on the emission inventory described above for the year 2007, air quality and subsequent impacts on ecosystems as well as on human health are estimated by the GAINS model.

Table 7.2. Estimates of Emissions of Air Pollutants (kilo tons) and GHG (million tons) in Pakistan for 2007.

Sector	SO_2	NO_x	$PM_{2.5}$	CO_2
Power Plant	321.55	60.46	27.25	33.42
Industry	301.62	84.305	547.786	35.61
Domestic	68.63	40.706	337.45	16.14
Transport	120.66	355.53	44.40	25.49
Total	**812.46**	**541.00**	**956.89**	**110.66**

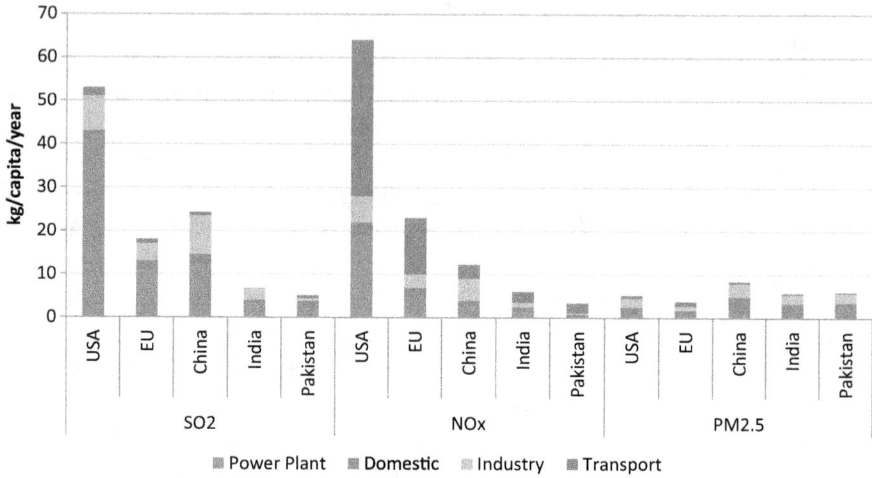

Fig. 7.5. Air pollutants emissions comparison on per capita basis by sector in different countries for 2007.

Annual mean concentrations of $PM_{2.5}$ in ambient air calculated by GAINS for the year 2007 were in the range of 50–100 $\mu g/m^3$ in large parts of Pakistan. GAINS also computes annual mean ground level ozone concentrations. GAINS computed the annual mean ozone concentration in Pakistan for year 2007 in a range between 40 and 60 ppb in most portions of Pakistan. Higher concentrations, exceeding 60 ppb, are estimated for the northern areas.

For the year 2007, estimates of GAINS for the loss in statistical life expectancy of people of Pakistan attributable to outdoor exposure to $PM_{2.5}$, fell in the range of 30–60 months. It must be kept in mind that the approach employed in GAINS for estimating health impacts only takes into account exposure to $PM_{2.5}$ from anthropogenic sources with negative health impacts. It does not, thus, link the particles emitted from natural sources such as sea salt and soil dust with loss in statistical life expectancy.

Besides estimating health impacts, GAINS also gives estimates of possible crop losses of various economically significant crops, such as rice, wheat, corn and soybean, that are attributable to ground level ozone. In Pakistan, there exist some areas where the ground level ozone concentration is significant and can cause substantial losses (ranging from 5–10%) in agricultural yield.

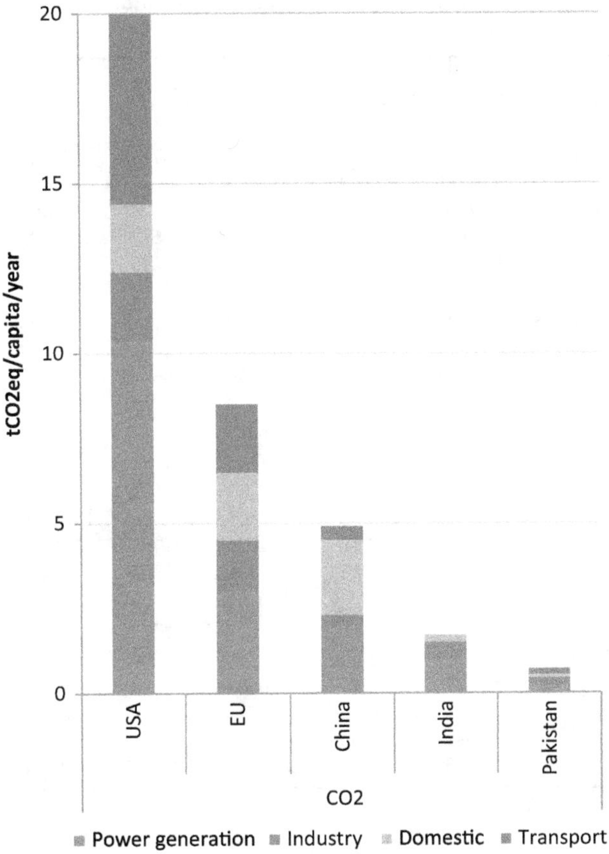

Fig. 7.6. CO_2 emissions comparison on per capita basis by sector in different countries for 2007.

Baseline Projections of Emissions up to 2030

The steep increase in energy use for power generation, industry, transport, and households will be paralleled by a drastic growth in emissions to the atmosphere. Table 7.3 presents the baseline projection of emissions of air pollutants in Pakistan by sector. Following the projected increases in economic activities, sulphur dioxide (SO_2) emissions are estimated to grow by a factor of 2.4 between 2007 and 2030, nitrogen oxides (NO_x) emissions by a factor of 2.2, fine particles ($PM_{2.5}$) emissions by a factor of 2.5, and CO_2 emissions by a factor of 2.3.

Table 7.3. Baseline projection of emissions of air pollutants & CO_2 in Pakistan by sector (kt).

	2007	2010	2015	2020	2025	2030
SO_2						
Power Plant	321.55	322.67	376.67	457.87	571.66	711.81
Industry	301.62	326.51	391.54	486.17	613.18	798.31
Domestic	68.63	66.54	75.76	92.25	114.05	143.61
Transport	120.66	119.27	137.17	165.26	205.21	253.85
Total	**812.46**	**834.98**	**981.13**	**1201.55**	**1504.10**	**1907.59**
NO_x						
Power Genration	60.46	61.26	71.73	87.70	110.06	139.62
Industry	84.31	96.38	117.15	144.05	178.52	222.03
Domestic	40.71	40.34	46.59	56.48	70.26	87.44
Transport	355.53	351.58	404.50	487.33	605.06	747.60
Total	**541.00**	**549.56**	**639.97**	**775.56**	**963.90**	**1196.69**
$PM_{2.5}$						
Power Genration	27.26	27.93	33.39	41.67	53.18	70.62
Industry	547.79	681.90	847.47	1043.21	1285.39	1593.28
Domestic	337.45	340.13	382.69	445.71	533.33	643.01
Transport	44.40	44.81	52.44	63.38	78.17	95.77
Total	**956.90**	**1094.77**	**1315.99**	**1593.97**	**1950.06**	**2402.66**
CO_2(Mt)						
Power Genration	33.42	33.50	38.95	47.32	59.12	74.15
Industry	35.61	38.80	46.70	57.47	71.84	90.10
Domestic	16.14	16.11	18.83	22.90	28.49	35.06
Transport	25.49	24.90	28.49	34.06	41.86	50.75
Total	**110.66**	**113.31**	**132.97**	**161.75**	**201.31**	**250.06**

Baseline Projections of Air Quality and Health Impacts

The increase in emissions would lead to profound deteriorations of air quality in Pakistan. In spite of current efforts in Pakistan to control air pollution, the emissions pathway as depicted in baseline projections would result in deteriorating air quality in the coming years. Most pertinent for health impacts, annual mean concentrations of fine particulate matters ($PM_{2.5}$) already surpass the World Health Organization acceptable level of 10 $\mu g/m^3$ (WHO, 2003) almost throughout Pakistan, and typically reach 50 to 100 $\mu g/m^3$ in 2007, in the Punjab region. By 2030 such levels would extend over most of Pakistan, while in some parts of the Punjab region concentrations may increase to more than 150 $\mu g/m^3$.

Consequently, health impacts from air pollution would sharply rise. For outdoor pollution, loss in statistical life expectancy for the Pakistani population on average is calculated to increase from 30–60 months in 2007, to 60–100 months in 2030, and exceeding to 100 months in a few parts of Pakistan. The total number of years of life lost because of outdoor air pollution would grow by a factor of two, from 4 years in 2007, to 8 years lost/year in 2030. This steeper growth is caused by the expected increase in total population and the aging of the people that will raise the number of older people. In comparison to air pollutant emissions, fast economic growth accompanied with boosted energy consumption will lead to a significant rise in the emission of greenhouse gases, i.e., CO_2 emissions. In the baseline scenario, CO_2 emissions will increase by a factor of 2.3 from 120 Mt in 2007 to 270 Mt in 2030.

Alternative Emission Control Scenario for Pakistan

As presented earlier, taking into account the currently anticipated economic development, the present plan for implementing additional emission control strategies seems to be inappropriate for decreasing air pollutant emissions in Pakistan. The countries in the west, mostly industrialized countries, have demonstrated that emissions of air pollutants can be effectively decoupled from economic development by uniformly employing advanced end-of pipe emission control measures. These end-of pipe emission control measures include, Selective Catalytic Reduction (SCR) for lowering NO_x emissions, desulphurization of flue gases for lowering SO_2 emissions, use of high efficiency devices for regulating PM emissions from various industrial processes, and innovative emission control measures to control emissions from vehicles, among others. Such emission control measures are extensively deployed in developed countries.

For this purpose, a hypothetical scenario assuming that the above stated advanced end-of pipe emission control measures will be implemented to all related emission sources in Pakistan till 2030 is examined using GAINS model. Such widespread implementation of advanced end-of pipe emission control technologies would considerably reduce future air pollution emissions in Pakistan to levels lower than the emissions projected in the baseline scenario (Table 7.4).

Based on the result generated by the GAINS model for the Advanced Emission Control Technology scenario (AECT) scenario, SO_2 emissions could be lowered by 59% in 2030 compared to 2007 estimates in the Baseline Emission Projection (BEP) scenario, and 25% when compared with the 2030 emission estimates. Similarly, NO_x emissions could be lowered by 83% compared to 2007 estimates in the BEP scenario, and 37% when

Table 7.4. Emissions (in Mt) of Advanced Emission Control Technology (AECT) Scenario in 2030 Compared to the Baseline Emission Projections (BEP) Scenario and the Year 2007 Emission Estimates.

Sectors	SO_2			NO_x			$PM_{2.5}$		
	2007	2030		2007	2030		2007	2030	
		BEP	AECT		BEP	AECT		BEP	AECT
Power Plant	0.32	0.71	0.08	0.06	0.14	0.06	0.03	0.07	0.00
Industry	0.30	0.80	0.26	0.08	0.22	0.13	0.55	1.59	0.57
Domestic	0.07	0.14	0.14	0.04	0.09	0.07	0.34	0.64	0.26
Transport	0.12	0.25	0.01	0.36	0.75	0.18	0.04	0.10	0.02
Total	**0.81**	**1.91**	**0.48**	**0.54**	**1.20**	**0.45**	**0.96**	**2.40**	**0.85**

compared with the 2030 emission estimates. Likewise, $PM_{2.5}$ emissions could be lowered by 89% compared to 2007 estimates in the BEP scenario, and 35% when compared with the 2030 emission estimates.

Considerable environmental and health benefits could also result. According to the AECT scenario in 2030, ambient concentrations of $PM_{2.5}$ would reduce from more than 150 $\mu g/m^3$ in the BEP scenario to around 50–100 $\mu g/m^3$. In this way, by 2030, cleaner air would reduce the loss in statistical life expectancy attributable to fine particles, from the range of 60–100 months in the baseline case to 30–60 months. It must be noted that, as the strategy only takes into account the control of emissions from large sources, health impacts due to indoor air pollution would not be influenced.

Declines in precursor emissions would also result in reducing the concentrations of ground level ozone, and thus reduce damage on crops and health. With reference to the AECT scenario in 2030, concentrations of ground level ozone would reduce from a range of 50–60 ppb in the baseline case to 40–50 ppb. In the AECT scenario, by 2030, potential wheat crop losses would decline from a range of 10–20% in the baseline case to 5–10%. In summary, maximum use of advanced end-of-pipe emission control measures for large sources might significantly lower future air pollution emissions, ultimately resulting in substantially better air quality and reducing impacts on health, vegetation and the ecosystem.

Limitations of the Study

The activity data in Pak-IEM for the base year (2007) were five years old at the time when the reference energy scenario was developed (2012). For that reason, the starting point does not necessarily represent the current situation in Pakistan. It is assumed that core aspects of the energy sector have not changed much in Pakistan to date.

In the model output data (activity data) of reference energy scenarios generated from Pak-IEM, Pakistan was considered as a single region and it was not in the disaggregated form of four regions as desired by the GAINS model. Therefore, the process of regionalising Pak-IEM from the single national framework is based on current demographics and energy use patterns in the corresponding GAINS regions and therefore may not fully reflect the evolution of the energy and, thereby, future emissions picture.

At present, only a single scenario, the reference energy scenario, was examined to develop a reference emission scenario. As a result, it may be possible that some emissions sources that may come into play under alternate assumptions might not be captured in the initial approach demonstrated.

Key sectors of particular importance to air quality are also considered, including information on brick kilns, back-up generators in urban areas, and kerosene lighting. However, some additional work may well be required to ensure further robustness of these data estimates, and to factor in spatial issues, in order to determine levels of energy use in rural vs. urban areas, and across different regions.

Most of the emission factors employed in the GAINS model for the estimation of emissions of air pollutants and greenhouse gases are IPCC default emission factors due to the unavailability of region/country specific emission factors for Pakistan. Hence, implementation of regional emission factors/fuel characteristics is also one of the general shortcomings of the study.

The health impact calculation in GAINS does not quantify impacts from emissions from natural sources and secondary organic aerosols. The current GAINS calculation assumes a linear exposure–response curve, up to the concentrations calculated for future years (i.e., up to 200 $\mu g/m^3$) based on the findings of the PAPA study (HEI, 2004). Obviously, such calculations include numerous uncertainties, which are impossible to eliminate at the present time.

The estimation of crop losses due to ozone requires actual crop production data in addition to the calculated relative yield losses. For the GAINS model, such information was provided by IIASA's LUC programme. However, the GAINS model resolution of 1×1 degree appears to be poor for the crop production estimates at country level.

The GAINS model applies yield response functions derived for European and American conditions to Asian crops. These relationships may, however, underestimate the ozone sensitivity of equivalent crops and varieties grown under Asian conditions (Emberson *et al.*, 2009). The scarcity of surface ozone measurement data in Asia for comparison with the GAINS estimates is another barrier to adequate validation of the GAINS approach.

The GAINS model health impacts assessment in the urban population are correlated with annual mean urban background concentrations, and health impacts of PM are calculated for exposed population older than 30 years. No effects are estimated for children. The calculation assumes that individuals will remain exposed to the exposure level calculated for 2020, for the rest of their life time.

Summing Up

The ongoing and anticipated economic development in Pakistan will certainly put additional pressure on current efforts to enhance air quality by way of controlling emissions of Particulates ($PM_{2.5}$) from large stationary sources, and nitrogen oxide (NO_x) from transport sector. As per the Baseline Emission Projections (BEP) scenario, SO_2 emissions are likely to increase by a factor of 2.4 in the next two decades if well-designed air pollution control strategies are not implemented. This is due to the increase in coal use by industry and for electricity generation. NO_x emissions would grow by a factor of 2.2 in 2030 owing to the deficiency of emission control policies. As a result, in the absence of implementing additional effective air pollution control strategies, in the coming decades, Pakistan will not be able to overcome the adverse impacts of air pollution on human health and vegetation, as has already been seen in most parts of the country.

For illustration (and the most relevant for health impacts), annual mean concentrations of fine particles ($PM_{2.5}$) already exceed the World Health Organization guideline of $10\,\mu g/m^3$ virtually throughout Pakistan, and typically reaches an annual average of 50–100 $\mu g/m^3$ in Punjab. By 2030, such levels would extend over most of Pakistan, while in some parts of Punjab, concentrations would increase to more than $150\,\mu g/m^3$. Consequently, health impacts from air pollution will sharply rise. For outdoor pollution, loss in statistical life expectancy is calculated to increase from 30–60 months in 2007, to 60–100 months in 2030, for the Pakistani population (on average), and exceeding 100 months in some parts of the country. The number of total life years lost due to outdoor pollution is expected to grow by a factor of two, from 4 years in 2007, to 8 years in 2030. At the same time, CO_2 emissions will increase by a factor of 2.3 from 120 Mt in 2007 to 270 Mt in 2030.

However, there are advanced end-of pipe emission control technologies available for maintaining adequate air quality levels regardless of pressure due to growth in economic development. To test this premise, an alternative technology-based emission control scenario, called the Advanced Emission Control Technology (AECT) that applies control strategies similar to those in industrialized countries, is developed. These best-practice available technologies include numerous end-of-pipe air pollution control technologies, such as Selective Catalytic Reduction (SCR) for lowering NO_x emissions, desulphurization of flue gases for lowering SO_2 emissions, use of high-efficiency devices for regulating PM emissions from various industrial processes, and innovative emission control measures to control emissions from vehicles and other sources. These can provide Pakistan with a maximum control strategy that would be on par with Europe.

According to the AECT scenario, complete application of currently available advanced end-of-pipe measures will significantly improve air quality in Pakistan. It is estimated that negative health impacts from air pollution could be reduced by 56% in 2030 by applying such advanced emission control technology to all large sources in Pakistan. Moreover, this also results in considerable environmental and health benefits. Ambient concentration of $PM_{2.5}$ would reduce from $150\,\mu g/m^3$ to around 50–$100\,\mu g/m^3$. In this way, cleaner air would reduce the loss in statistical life expectancy attributable to fine particles, from the range of 60–100 months in 2007 to 30–60 months in 2030.

Nevertheless, the wide-ranging implementation of advanced emission control measures happens at a cost and would require substantial fiscal resources. Economic resources and political will would be the governing factors affecting the implementation of such a control strategy throughout Pakistan.

In comparison to all the examples above, as a way forward, a cost-effective emission control strategy could be developed with the help of optimisation mode of the GAINS model. Although this is not the part of the present work it could be analysed in detail in the future. This tool selectively assigns particular mitigation measures across all economic sectors, regions and pollutants, and could identify strategies that improve air quality at almost half of cost compared to the all end-of pipe measures scenario. An essential part of such an emission control strategy is that it also includes measures for eliminating indoor air pollution resulting from the burning of solid fuels.

At the same time, well-designed pollution control strategies also result in reducing the GHG emission simultaneously as a co-benefit. It is seen in the GAINS optimised scenario, that low carbon policies cause reductions in SO_2, NO_x and $PM_{2.5}$emissions at no extra cost, and that each percent reduction in CO_2 emissions on average cuts health impacts from particles pollution by 1%.

It also shows that, in order to accomplish particular targets on improving ambient air quality, the related cost can be lowered by implementing selected low carbon measures. A scenario developed using the GAINS model validates this point; that extra costs for adopting low carbon or climate-friendly measures (such as the co-generation of heat and power (CHP), energy-efficiency improvements, integrated coal gasification combined cycle (IGCC) plants, fuel substitution, etc.) are largely counterbalanced by savings in investment for installing air pollution control equipment. Therefore, a smart mix of mitigation measures lower both air pollution and GHG emissions at the same time, at reduced costs; compared to conventional wide ranging end-of pipe control measures.

It should be noted that in this study F, an attempt was also made to highlight the scope for strategies that actually expand air pollution control and GHG mitigation co-benefits. By doing so, this study presents an initial step in quantitative analyses necessary to encourage rational policies that favourably address Pakistan's air quality and health circumstances.

References

Abrol, Y.P., and Ingram, K.T. (1996). Effects of elevated CO2 and temperature on growth and yields of some crop plants. In: Bazzaz F, Sombroek W (eds) Global climate change and agricultural production. Wiley, New York, pp. 123–140.

ADB, (2006). Country Synthesis Report on Urban Air Quality Management: Pakistan, Asian Development Bank and the Clean Air Initiative for Asian Cities (CAI-Asia). Available from: http://cleanairasia.org/wp-content/uploads/portal/files/documents/pakistan_0.pdf.

Alam, S., Fatima, A., and Butt, M. S. (2007). Sustainable development in Pakistan in the context of energy consumption demand and environmental degradation. *Journal of Asian Economics*, **18**(5), 825–837.

Amann, M., Bertok, I., Borken-Kleefeld, J., Chambers, A., Cofala, J., Dentener, F., Heyes, C., Hoeglund-Isaksson, L., Klimont, Z., Purohit, P., and Rafaj, P., (2008). GAINS Asia. A tool to combat air pollution and climate change simultaneously. Methodology.

Bakkenes, M., Alkemade, J. R. M., Ihle, F., Leemans, R., and Latour, J. B. (2002). Assessing effects of forecasted climate change on the diversity and distribution of European higher plants for 2050. *Global Change Biology*, **8**(4), 390–407.

EEA, (2003). Europe's Environment: The Third Assessment, Environmental assessment report No 10, European Environment Agency, Luxembourg: Office for Official Publications of the European Communities, 341 pp.

Emberson, L. D., Ashmore, M. R., Cambridge, H. M., Simpson, D., and Tuovinen, J. P. (2000). Modelling stomatal ozone flux across Europe. *Environmental Pollution*, **109**(3), 403–413.

Ertsen, A. C. D., Alkemade, J. R. M., and Wassen, M. J. (1998). Calibrating Ellenberg indicator values for moisture, acidity, nutrient availability and salinity in the Netherlands. *Plant Ecology*, **135**(1), 113–124.

HEI, (2004). Health Effects of Outdoor Air Pollution in Developing Countries of Asia: A Literature Review. Special Report #15, International Scientific Oversight Committee, Health Effects Institute, Boston.

IPCC, (2001). Climate Change 2001: The Scientific Basis. Contribution of Working Group I to the Third Assessment Report of the Intergovernmental Panel on Climate Change [Houghton, J.T.,Y. Ding, D.J. Griggs, M. Noguer, P.J. van der Linden, X. Dai, K. Maskell, and C.A. Johnson (eds.)]. Cambridge University Press, Cambridge, United Kingdom and New York, NY, USA, 881pp.

MoCC, (2012). National Climate Change Policy, Ministry of Climate Change (MoCC), Government of Pakistan, Islamabad.

Pak-IEM, (2010). Pakistan Integrated Energy (Pak-IEM): model design report. International Resources Group for Asian Development Bank (ADB) and Ministry of Planning and Development (MOPD), Government of Pakistan, Islamabad.

Pak-IEM, (2011). Pakistan Integrated Energy (Pak-IEM): model final report volume I. International Resources Group for Asian Development Bank (ADB) and Ministry of Planning and Development (MOPD), Government of Pakistan, Islamabad.

Pak-SCEA, (2006). Pakistan Strategic Country Environmental Assessment, Report No. 36946-PK, South Asia Environment and Social Development Unit, World Bank Available from: http://siteresources.worldbank.org/SOUTHASIAEXT/Resources/Publications/448813-1188777211460/pakceavolume1.pdf.

Pleijel, H. (Ed.), (2009). Air Pollution and Climate Change — Two Sides of the Same Coin? Swedish Environmental Protection Agency, Stockholm, Sweden.

Posch, M. (2002). Impacts of climate change on critical loads and their exceedances in Europe. *Environmental Science and Policy*, **5**, 307–317.

Tahir, S. N. A., Rafique, M., and Alaamer, A. S. (2010). Biomass fuel burning and its implications: deforestation and greenhouse gases emissions in Pakistan. *Environmental Pollution*, **158**(7), 2490–2495.

WHO, (2003). Health aspects of air pollution with particulate matter, ozone and nitrogen dioxide: report on a World Health Organization working group, Bonn, Germany 13–15 January 2003.

8

SELF-MANAGEMENT IN ENVIRONMENTAL MANAGEMENT PHILOSOPHY

ZHANG YUZHE and CECILIA LIM

Before the large-scale development of the Earth by humans, or human's damage to the environment, the natural environment, itself, was harmonious and sustainable. Who damaged the environment? Human beings. Nature itself does not have "trash", since all of the elements involved in the natural cycle can maintain its balance. Human activities break up this natural balance and lead to many conflicts between the environment and mankind. Thus, there is now a need to manage the environment. It should be noted that environmental management is directed at managing the activities of humans rather than managing the natural "environment" itself, a tree or a piece of land. Therefore, we need to understand what are the main human activities related to the environment and what the main motivations for these activities are.

There are numerous human activities in the world, but the main categories relating to environment include the following: economic activities, the construction of living facilities, commercial activities, transportation, and recreational activities. These activities are guided by our dominant needs: development, economic growth, the quest for a better life, and the enjoyment of life. These needs dominate our lives worldwide leading to the imbalances between human demands and the assimilative capacity of the environment, causing environmental degradation. In the context of our needs and sustainable development, what are the hidden motivations and how can we reach sustainability? All of these questions involve a fundamental problem: what is the relationship between human beings and the environment on earth, and what should this relationship be?

These issues are ultimately issues of philosophy and ethics. The relationship between man and nature determines the roles of both man and nature, and further affects the purpose and manner of human activity. Environmental philosophical and ethical studies and theories exist both in western and eastern perspectives, and these studies and theories have affected our policies and decision-making in relation to the environment at various levels. This paper discusses some important philosophical theories in both western and eastern fields. In western philosophy, the main environmental theories include Anthropocentrism (Deontology), Animal Liberation Theory and Biocentrism; in oriental philosophy, the main environmental theories embody Chinese Taoist Environmental Philosophy and Confucian Ecological Philosophy. Of course there are many other environmental theories, but these represent the main ideologies for both western and eastern philosophies and have profound impacts on environmental management theory and practice. This paper also examines some main environmental issues, including economic growth and environment, biodiversity conservation, science and environment, ethical concerns relating to modern technology, and environment valuation and sustainability. Finally, after a series of analysis, this paper offers a suggestion relating to environmental management in theory — i.e. the need for self-management and the reasons supporting this suggestion.

Introduction

Background

In the past few decades, especially in the last 30 years, environmental issues have increasingly intensified. Population growth, increased demands for energy (especially fossil fuels), urban expansion, an increase in solid waste, deforestation, air pollution, water pollution, marine pollution, loss of biodiversity, etc. — all of the above issues have contributed to global climate change. It needs all of the countries around the world to unite to address these problems. And even just a single environmental problem, like air pollution, has many stakeholders. Hence, we need a comprehensive understanding of environmental issues and a systematic solution to respond to them.

However, how can we get a comprehensive understanding and work out a systematic solution? The first step is to make the issues simple and find out the root cause. Think about this question — who pollutes the environment? It is easy to answer: humans. But we are not aiming at polluting our environment on purpose, while actually we do pollute it. Then what is wrong? Simply speaking, this pollution, in fact, is the imbalance between what we extract from nature, what we discharge to nature and the environment's assimilative capacity. Why is it unbalanced? Because we ask nature for much more than it can offer, and we dump much more into nature than it can bear. This then, is related to our issues of philosophy: How do we treat nature and how should we treat nature.

Actually, there are already a lot of discussions and theories about environmental philosophy or ethics both in western and eastern thought. And these discussions and theories have played a very important role in environmental management.

Basic Concepts — Philosophy, Environmental Philosophy; Ethics, Environmental Ethics

In the study of environmental philosophy, there are some basic concepts that need to be introduced.

Both in the western and eastern worlds, philosophy has a long history. Basically, Philosophy comes from humans seeking the origin of all things, or we should say, philosophy is "humans seeking the origin of all things". Then comes a series of discussions: Who are we? Where do we come from? What should we do and not do?

Philosophy has been divided into many sub-fields. Environmental philosophy is a branch of philosophy that is concerned with the natural environment and human's place within it. It asks crucial questions about human environmental relations such as "What do we mean when we talk about nature?" "What is the value of nature?" "How should we respond to environmental challenges such as environmental degradation, pollution and climate change?" "How can we best understand the relationship between the natural world and human technology and development?" and "What is our place in the natural world?" As such, it uniquely positions itself as a field set to deal with the challenges of the 21st Century.

We think the fundamental question is the relationship between human and nature, namely our attitude towards nature.

Ethics or moral philosophy is the branch of philosophy that involves systematising, defending, and recommending concepts of right and wrong conduct. As a branch of philosophy, ethics investigates the questions "What is the best way for people to live?" and "What actions are right or wrong in particular circumstances?" In practice, ethics seeks to resolve questions of human morality, by defining concepts such as good and evil, right and wrong, virtue and vice, crime and punishment.

In general, environmental ethics are a systematic account of the moral relations between human beings and their natural environment. Environmental ethics assumes that moral norms can and do govern human behaviour toward the natural world. A theory of environmental ethics, then, must go on to explain what these norms are, to whom or to what do humans have responsibilities, and show how these responsibilities are justified.

Here we speak about environmental philosophy rather than environmental ethics, because ethics basically focus on problems of morality, which mainly concern "shoulds" and "values". However, note that not all value questions concern moral value. Environmental philosophy offers a more comprehensive and wider perspective for us to discuss environmental issues, which will provide more useful discussions and theories for environmental management.

Main Theories of Western Environmental Philosophy

Anthropocentrism

'Anthropocentrism is the belief that human beings are the central or most significant species on the planet (in the sense that they are considered to have a moral status or value higher than that of all other organisms), or the assessment of reality through an exclusively human perspective.'[1] 'Anthropocentrism is always used as values or a value measure, which treat the interests of humanity as the basis of all values and moral evaluation and considers only humans as the value judgement body.'[2] In the ethical relationship between man and nature, it focuses on Man. Thus, the thinking of "Mankind/people is the goal".This theory of anthropocentrism was first proposed by Kant. The core ideas of anthropocentrism include the following: 1) In the relationship between man and nature, only the human (who has consciousness) is the subject; nature is the object. At any time when it comes to the "value", it means " significance for the human." 2) All human activities are aimed to meet the needs of their own survival and development. If this goal cannot be achieved, those activities have no meaning. Thus, everything should be in the interests of humanity as a start and as its final destination.

In daily life, "anthropocentrism" is very common. We tend to judge an action or an article by using words like "valuable, worthy to do, useful to society"; or we ask "what are the benefits to me" before we do something. Concepts, like these words, imply the hidden "anthropocentrism" tendency in our ideas. The value of natural objects is not objective, but subjectively defined by human needs: is it valuable or not — for human beings?

Animal Liberation Theory and Animal Rights

Animal liberation theory is one of the important theories in animal ethics. Since this theory talks about animal rights and the relationship between human and animals, it is also an environmental philosophical argument.

The classical claims of animal liberation are found in Peter Singer's book. Thus, Peter Singer is always regarded as the representative of animal

[1] Anthropocentrism — *Merriam-Webster Dictionary.*
[2] Baidu Encyclopedia, interpretation of the word Anthropocentrism. http://baike.baidu.com/link?url=0QjcTA2TazZpQ9UOPITqDeEL8iJdYhC0uouOFYAjjX-F4KiTK5E55rGj4QFkWMUS2r-TB9jkW7mZw1afxEHfC_

liberation theory. The basic idea of this theory is that we should expand our moral caring to animals, admit they also have moral standing, treat their interests and demands equally, and stop hurting them. Peter Singer and other animal protectionists hold that animals have the same sentience as human beings. Thus, they should have the same right as humans to enjoy their life and should not be harmed by humans. This theory challenges many human activities: the farming of animals for food, the wearing of fur, the use of products made from animals e.g., leather furniture, the use of animals for medical research etc.

Animals also feel pain when they get hurt. But some critics will ask — how do you know that animals can feel pain and they have emotions like us? This is a thorny question. For ourselves, pain is a direct experience; but for another, either human or animal, you can never feel their pain directly. Pain or hurt is a kind of "mental event", and, personal. But some neurologists have found that higher animals and humans have very similar neurological and physiological reactions when faced with hurt, including mammals, birds and so on. Peter Singer thinks this is enough to prove that animals have the same sentience as human beings.

This theory provides a fundamental thinking for animal protection, and also considers that animals have equal rights to enjoy life and be free from pain or hurt like human beings.

Many animal protectionists and environmentalists suggest and encourage vegetarianism. What is the relationship between vegetarianism and the environment? From the perspective of the impact of the raising of animals, without thinking of it at the moral level, the production of meat adds to environment's burden. Firstly, raising animals requires massive amounts of resources, like land, food, energy, and water. Secondly, raising animals pollutes the air, especially greenhouse gases like carbon dioxide, methane, and nitrous oxide. (See Figs. 8.1 and 8.2)[3] Thirdly, raising animals causes a lot of water pollution.

The above scientific data proves that vegetarianism is good for the environment. We should reset our values and lifestyle. When involving moral issues or personal values, the laws or regulations are ineffective. The only way to achieve a high level of moral standard and social general values is through self-reflection and self-management.

---

[3]Meat Eater's Guide: Get to Know the Carbon Footprint of Your Diet. http://www.treehugger.com/green-food/meat-eaters-guide-get-to-know-the-carbon-footprint-of-your-diet-lamb-beef-cheese-are-the-worst.html.

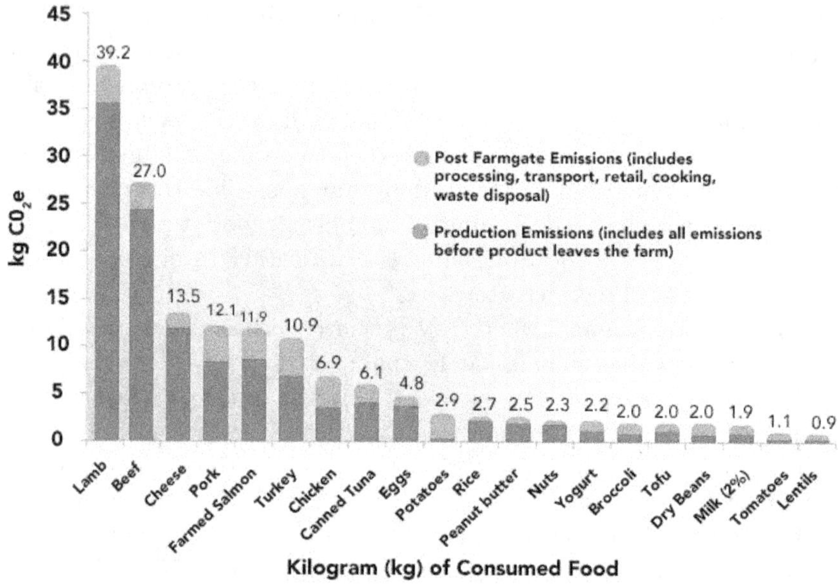

Fig. 8.1. Full Lifecycle Greenhouse Gas Emissions from Common Proteins and Vegetables. *Source*: *Environmental Working Group*.[4]

Biocentrism

Biocentrism is an evolution of animal liberation and animal rights. The biocentric thought is nature based, not human based. Biocentrists believe that all species have inherent value and that humans are not "superior" to other species in a moral or ethical sense. Advocates of biocentrism promote the preservation of biodiversity, animal rights, and environmental protection.[5] Biocentric ethics calls for a rethinking of the relationship between humans and nature. It states that nature does not exist simply to be used or consumed by humans, but that humans are simply one species amongst many, and that because we are part of an ecosystem, any actions which negatively affect the living systems of which we are a part, adversely affects us as well, whether or not we maintain a biocentric worldview (Bari, 1995; Ingram and Parks, 2002).

[4]Meat Eater's Guide to Climate Change + Health: Climate and Environmental Impacts. http://breakingnews.ewg.org/meateatersguide/a-meat-eaters-guide-to-climate-change-health-what-you-eat-matters/climate-and-environmental-impacts/.
[5]Wikipedia, Biocentrism (ethics). https://en.wikipedia.org/wiki/Biocentrism_(ethics).

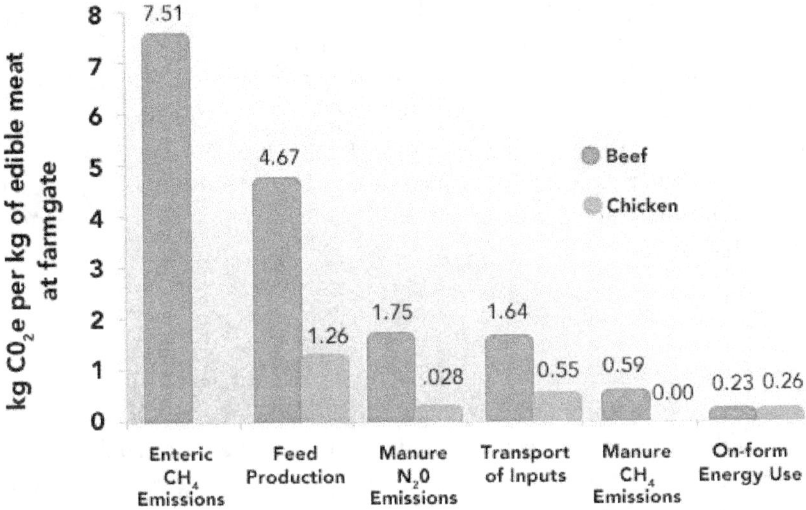

Fig. 8.2. Sources of Emissions from Beef and Chicken Production. *Source*: *Environmental Working Group*[6]

The four main pillars of a biocentric outlook are:

- Humans and all other species are members of Earth's community.
- All species are part of a system of interdependence.
- All living organisms pursue their own "good" in their own way.
- Human beings are not inherently superior to other living things.[7]

The word 'biocentric' means 'life-centred' (Des Jardins, 2000). Biocentric ethics is one theory of "non-anthropocentrism" in environmental ethics. It was firstly put forward by Albert Schweitzer (1875–1965) in his book "*Philosophy of civilisation: Civilisation and Ethics*" in 1923. He believed that all life have "Survival-Consciousness", so humans should have reverence for all life just like we revere ourselves. This is the "reverence of life" principle. When a person regards all life as important as ourselves, he is a real moral man.

However, since biocentrism holds that all life is equally to be revered, then what is our choice when there is a conflict between two lives? When a

[6]Meat Eater's Guide to Climate Change + Health: Climate and Environmental Impacts. http://breakingnews.ewg.org/meateatersguide/a-meat-eaters-guide-to-climate-change-health-what-you-eat-matters/climate-and-environmental-impacts/.

[7]Wikipedia (n.d.). "Biocentrism ethics". Retrieved from: https://en.wikipedia.org/wiki/Biocentrism_(ethics).

mosquito is biting you, will you will kill it or let it go? Especially when the mosquito is carrying the dengue virus, what is your choice? Furthermore, how should we value and compare the values of different lives? Is a human's value higher than that of a mosquito and if so, why? Based on existing discussions, there is no correct answer to these questions. But, at least, this theory and "reverence of life" principle provide a thinking of moral virtue rather than only a fixed rule of action. From this perspective, we can have further discussions.

Socrates

The thoughts of Socrates are not ideas related to environmental philosophy, but we can be inspired by them. Socrates believed that man should be on an endless quest to seek the natural truth. The external world and human sensations (internal world) are always changing, so the knowledge you have is also uncertain. Socrates wanted to aspire to seek an unchanged, certain, and eternal truth, which should not rely on the external world but return to ourselves and study ourselves. Thus, this answers the question why we need to focus more on human's own mind, thinking and activities rather than the outside world when we try to solve environmental issues, and this provides supports for the later suggestion: self-management in environmental management.

Main Theories of Oriental Environmental Philosophy

We will now focus on oriental environmental philosophy, and here, "oriental" means ancient Chinese. Chinese ancient philosophy is profound, with many branches and factions; but the main theories on the environment are the philosophies of Taoism and Confucianism. These are the basis of Chinese environmental philosophy.

Chinese Taoist Environmental Philosophy

Lao Tzu (sixth to fifth century B.C.) and Chuang Tzu (fourth to third century B.C.) are representatives of Taoism. The basic idea of Taoism related to the environment is its "Nature and Humanity" ideology, which is also the basis of Taoism. Lao Tzu said "Tao (the law of the universe) generates one, one generates two, two generates three, and three generates everything" in his "Tao Te Ching". Tao is the fundamental law of the universe. It is inscrutable, difficult to describe, and seemingly invisible, but omnipresent, and it is the fundamental force of all changes in the infinite universe. In "Equality of Things", Chuang Tzu put forward his view that "The World and I coexist, and all things and I are One". Nature is actually an incarnation of Tao, and to revere Tao is to revere nature. Lao Tzu and Chuang Tzu regarded the reverence of nature, respecting Tao and valuing virtue, as guidelines of life and behaviour.

Thus, to seek harmony between humanity and nature should be our environmental values. Lao Tzu realised the importance of protecting the ecosystem. He warned people not to think that humans are above nature and do what humans want to do without considering nature. Chuang Tzu held that we should comply with the nature of all things, and follow their respective laws. The real morality is to let all things grow and develop naturally in their own way, and maintain their natural endowments; otherwise, we will bring chaos and even a recipe for disaster to the originally harmonious nature.

Then, what should we do to achieve harmony between humans and nature? For this, Lao Tzu proposed a "Wisely Suspending" thinking ("知止"的思想), and Chuang Tzu proposed a "Communion with Nature" thinking ("天和"的思想). These are the essence of their thoughts — respect for nature as well as seeking ways to obtain harmony between humans and nature. "Wisely Suspending" means that if we know enough is enough and do things in moderation, then we will not encounter danger. This is a wise and useful suggestion today because of our over-consumption. Chuang Tzu regarded "Communion with Nature" as the best condition of the relationship between human and nature, and as a beauty of balance. Here, in the thoughts of Taoism, humanity must obey the Tao of nature, namely the rules and principles of human's activities must obey natural laws.

Confucian Ecological Philosophy

Ancient Confucianism advocated "Periodic bans." It means that you cannot go logging or hunting at any time. In "T*he Book of Rites*", the chapter "Festival Righteousness" it is recorded that Zeng Zi said: "Cutting trees and hunting animals, must be at a reasonable time." This is aimed at obeying the natural laws of growth. The Confucian ecological concern is to maintain the balance of the whole society's way of life with how an individual lives his life. In the "Book of Rites · Months and seasons", it states that in spring and summer, people are prohibited from doing any harmful action to natural growth, especially destroying nests for birds and lairs for animals, and killing or taking birds' eggs, nestlings, young animals, cubs and so forth. This is because spring and summer are the seasons for all living things to breed and grow, in China. This is also because these constraints are also for the benefit of humans, as allowing animals to breed naturally during their breeding seasons will result in more for the common man. In Xun Zi's "*Kingships*", it also lets all life survive and grow naturally, admitting that all life has its value and significance. This can be regarded as the earliest consideration of sustainable development.

Xun Zi developed and enriched the thought "reverence of natural laws" into "make use of natural laws". Thus, the development of Confucian ecological thought is "understand natural laws" — "reverence of natural laws" — "make use of natural laws". Xun Zi thought that humans should understand and make use of natural laws, by respecting and obeying them. Xun Zi argued that humans and nature have their own roles, so humans cannot take over the place of nature. Thus, humans simply cannot conquer nature, but can transform and utilise nature within certain constraints, and the premise must be to adapt and use natural law and rules.

Conclusion and Discussion

The roots of environmental philosophy described in Taoism and Confucianism are the same. Their basic ideas include three points: 1) "Nature and Humanity" ideology; 2) Humans should respect and comply with the laws of nature; 3) Complying with the laws of nature is the moral standard and the harmony between humans and nature is the final target. Confucianism put forward the "Restraint use" ecological idea, requiring timely and right amount of access to natural resources, which is a form of obeying the laws of nature. The "Restraint use" ecological perspective can also be regarded as one way of self-management.

We human beings cannot change or avoid natural laws, even if we have advanced technologies. How we should treat nature is not a personal question, it is a world question for all human beings. Thus, we should not violate nature's laws. Natural laws are regarded as the moral standard, which means the standard is not based on human's preferences and values. Thus, it is fairer and more justified.

Main Discussions of Environmental Issues

After a relatively systematic introduction and discussion about environmental philosophy from both western and eastern views, we can discuss some actual problems. The hottest controversies are focused on the conflicts between economic growth and the environment, and between human activities and animal protection. Nature plays roles as giver (source), and receiver (sink). Thus, the main contradiction is reflected in these two aspects: giving — human's demand is more than what nature can offer; receiving — humans discharge more than what nature can absorb. What causes these two imbalances? Mainly economic growth and development. Then, a series of questions need to be asked and analysed: 1) Why do we need economic growth and development? 2) What is development? 3) If the answer is "pursuit of a better life", then what is "a better life"? Now we will discuss these related issues.

Economic Growth and the Environment

As discussed above, nature has two roles: giver and receiver; the imbalance between human's demand and what nature can provide, is mainly reflected as resource depletion; the imbalance between human's discharging and the environment's assimilative capacity, is always reflected as pollution. Resource depletion and pollution, these two issues are mainly caused by economic activities.

After the industrial revolution, in more than a hundred years of industrialisation, "economy" and "development" have become more and more valued and worthy of pursuit in human society. At the beginning, human beings did not feel the pressure of environmental problems, or we can say that there were less environmental problems. But along with industrial development came pollution — this was the first environmental issue, rather than resource depletion. The famous cases include the London smog incident of December 1952 and the Los Angeles photochemical smog event of September 1955, causing the death of some 400 elderly people (above 65 years) within two days, because of respiratory system failure. These events attracted public attention, and related laws were passed. However, people living elsewhere are still seeking economic growth. Thus came the serious Beijing Smog of 2013. On January 12 in 2014, the Beijing Meteorological Bureau issued the following reading: "01-12-2013 20:00; PM2.5; 886.0; 755; Beyond Index." (See Table 8.1)

The above case study relates to air pollution, but there are many cases of pollution in other fields, such as surface water pollution, ocean pollution, soil pollution and so forth. Thus, if we do not treat these issues seriously, this pollution will be continuous and aggravated.

Table 8.1. Air Quality Index and levels of health concern of PM2.5 (Global Sherpa, 2013).

Air Quality Index (AQI) Values	Levels of Health Concern	Colors
When the AQI is in this range:	.. *air quality conditions are*	... *as symbolized by this color:*
0–50	Good	Green
51–100	Moderate	yellow
101–150	Unhealthy for Sensitive Groups	Orange
151 to 200	Unhealthy	Red
201 to 300	Very Unhealthy	Purple
301 to 500	Hazardous	Maroon

Credit: Beijing Air Pollution Exposes China's Health & Environment Risks (http://globalsherpa.org/china-air-pollution-health-environment/)

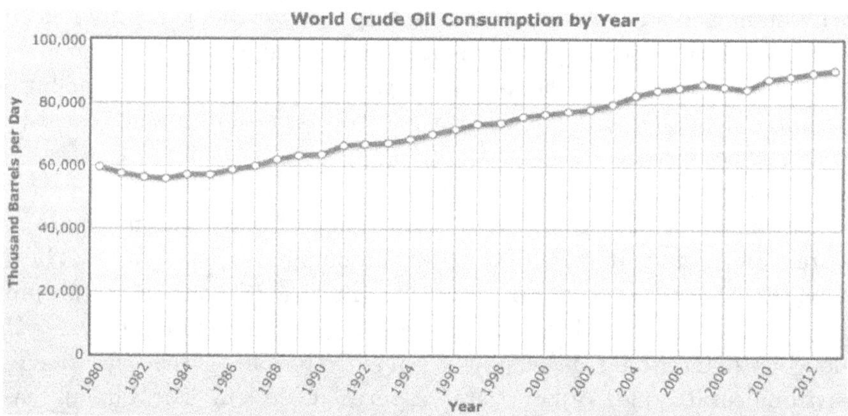

Fig. 8.3. World Crude Oil Consumption by year. *Credit: United States Energy Information Administration.*

Apart from pollution problems, natural resources are used to develop the economy, so resource depletion is getting more and more serious. (See Figs. 8.3, 8.4, 8.5). However, even if environmental issues have become worldwide concerns, most states still regard economic growth as the nation's first goal.

"Economic growth," "development," "better life," "higher GDP," etc., these values are created by humans, but are they necessary to our life? What is the meaning of human development and a better life? Why do we need a "better" life rather than a "good" life? If we keep pursuing a "better" life,

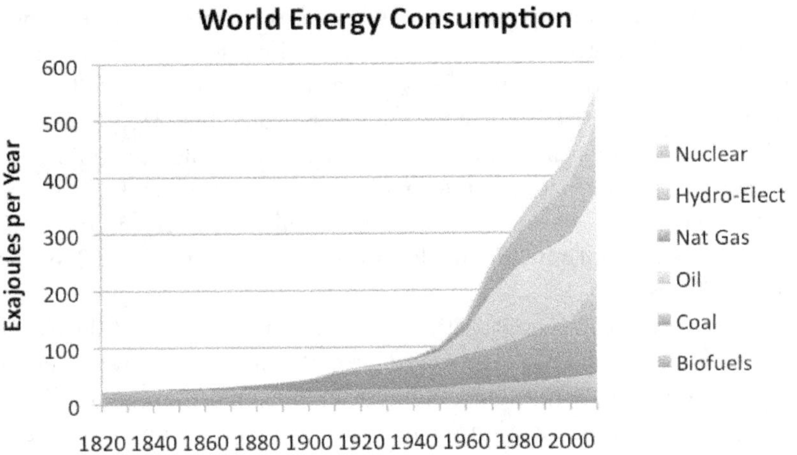

Fig. 8.4. World Energy Consumption by Source (Based on Vaclav Smil estimates from *Energy Transitions: History, Requirements and Prospects* together with BP Statistical Data for 1965 and subsequent). (Tverberg, 2012).

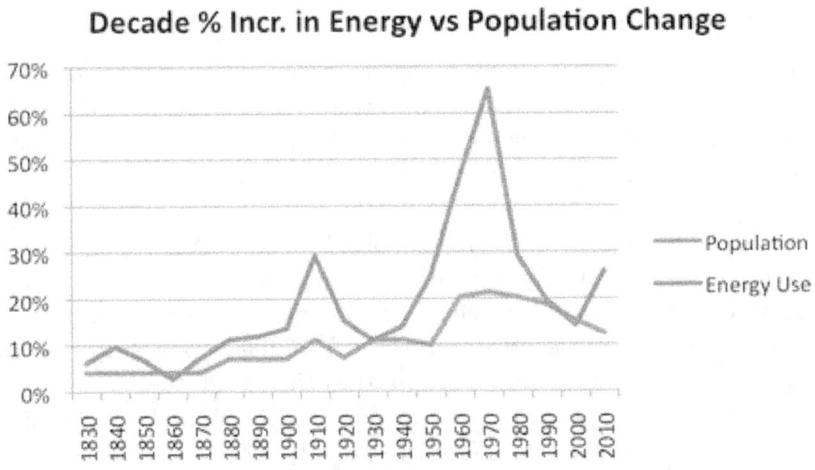

Fig. 8.5. Decade percentage increases in energy use compared to population growth (Tverberg, 2012).

there is no end. Why do we not think "enough" is enough? Because our values are dominated by the whole society, and the mainstream social values are affected by national policy and media advocacy. In many countries, the national policy is encouraging and stimulating domestic consumption to promote economic growth. But the rapid consumption model has led to a series of serious consequences: a surge in demand for resources, especially non-renewable resources, such as fossil energy; huge amounts of wastes, which are difficult to treat, especially hazardous wastes and e-wastes. In terms of media advocacy, it is easy to notice that our daily life is full of different kinds of advertisements urging us to "buy, buy, buy", encouraging us with offers of discounts and "sales". At the same time, it should be noted that more and more people prefer to buy a new article rather than repair the old, since buying a new one may be cheaper. We are used to making money and consumption. The push for economic growth has influenced our values and lifestyle.

We have to rethink our attitude to economic growth and development. Some economists have put forward "degrowth" of an economy, which means we are not only focused on economic developing speed anymore. This new thinking offers a new direction for us. If we compare economic growth with a healthy environment, which one is more important to us? Obviously, the answer is a healthy environment.

Environmental Capital Valuation

In real life, we always encounter many cases that need us to make a choice, for instance, whether a natural item (a tree or a piece of wetland) is worth keeping. We need to give some reasons why we want to keep it, and the common method is to estimate its monetary value.

For example, on the sides of the main street, there are some beautiful old trees, more than a hundred years old. But now the street needs to be widened because of city development planning. Some citizens want to keep these trees, since they have been there for hundreds of years, and some people have grown up together with the trees. That means these trees have a certain emotional value for them. On the other hand, these trees are also beneficial for the city's environment, freshening the air and beautifying the street. Here comes the problem — if we cut them down and widen the street, we could gain, for example, at least US$100,000 worth of benefits directly, based on freeing up this valuable space for road widening, which in turn will attract more investments, such as retail shops, shopping malls and restaurants. But how can we prove the environmental and emotional values in keeping the trees are worth more than US$100,000? Now, some economists have tried to measure the (invisible) value of some natural capital.

However, the problem here is whether we can and should value natural capital? Firstly, we must admit that the value of many natural items cannot be estimated with any certainty, especially their aesthetic value, emotional value, and eco-system services (such as forests absorbing CO_2 and releasing oxygen, thereby freshening air). Secondly, generally speaking, humans will not destroy what they cherish, such as their families, homes, and their own bodies. And all of these too, cannot be valued by money. The common sense perspective is that we should not and cannot lose them. Thirdly, the monetary valuation of natural capital may lead us the wrong way: everything in the world can be measured by money, and can be exchanged by money. Thus, in the context of environmental capital valuation, we should rethink about the invisible value of natural capital.

Biodiversity Conservation — Why and How

Biodiversity conservation is another very important environmental issue. This was earlier discussed in both western and eastern philosophy. Let us now examine this more closely.

Most people think that biodiversity is beneficial, and helps keep the balance of the whole ecosystem since it is a part of the biosphere. Also, most people take the view that it is cruel to kill animals, but it is not wrong to kill and eat animals that are farmed. And it is very strange that most people will feel upset when they hear that someone abused or killed a lovely cat or dog, but no such feeling arises when they hear that a pig was killed. If all life can feel pain and are equal, then why do we treat them differently? Some will say that we are human beings and we need food anyway. Then we should eat vegetables and be vegetarian. Some will say that it has been scientifically proven that some nutrition we can get only from meat, so we "have to" kill the animals. Actually, there is scientific evidence that humans can get all they need from grains, vegetables and fruits. Some scientists say that meat is not good for human health — that taking too much meat may lead to heart disease, high cholesterol, obesity, etc. In ancient India, people believed that meat has "corpse poison", a kind of toxin generated naturally because of fear and horror when a living creature faces death. Thus, there is no conclusion on this issue: whether eating meat is good or bad for human's health on earth. But, at least, we know that it is not necessary for humans to eat meat, because many vegetarians live very well. Then why do we still kill animals for food, every day? The root reason is human's greed. People have become accustomed to eating meat, and cannot say no to the delicious meat. Due to human's unlimited desires, people want to taste various delicious meats, so pangolins, snakes, turtles, bears, and many other wild animals have become food for us.

Hence, there are two main problems of biodiversity conservation: firstly, what should we protect? Should we only protect certain animals or should we protect all lives; another issue is how do we protect biodiversity. For the first question, the answer should be — all lives. All creatures have the right to be alive and to enjoy life. And actually, this goal can be achieved. In fact, most people have compassion, but they just ignore it as they do not give this much thought. When people begin to realise that all lives cherish their life, feel pain and fear the same as humans when faced with slaughter, and their parents will be very sad when they see their children being hurt, our inner compassion and sympathy will be roused. This is the natural instinct of life. Even if now it looks like that some people value money more than everything, including life — when they face the choice of either life or money, I believe most people will choose life.

Then the second question, how do we protect biodiversity? As mentioned above, we need a basis — firstly: consensus of compassion and sympathy for all lives. Without this basis, it is hard to stop the illegal trade of wild animals and the abuse of animals. What is the difference between farming tigers and farming pigs or poultry? To say that laws prohibit the farming of tigers is not an adequate answer since laws are created by humans, and can be changed. Laws can also legalise the farming of tigers. Thus, we need a view that all lives are equal, and the consensus of compassion and sympathy for all lives could be done through education, and here education means education in its broadest sense. Over time, these will be internalised into our values. At the same time, the national policies and regulations should coincide with this, such as economic policies, animal protection laws, and media propaganda. Vegetarianism should be recommended. Maybe at the beginning, it is difficult to be accepted and implemented, but we can accomplish it gradually.

Hence, if we can do as suggested above, then the illegal trade of many animals (including their body parts), such as the pangolins, snakes, marine turtles; tiger parts, rhino horn, ivory, bear paws and bile will be reduced and in time, completely eliminated. Like a public slogan in China: No trading, No killing. No demand, No market. It is possible to respect and protect all lives — this is the real conservation of biodiversity.

Science and Environment — Modern Technology and Ethics

'Science is a systematic enterprise that creates, builds and organises knowledge in the form of testable explanations and predictions about the universe.'[8] 'Technology is the collection of techniques, skills, methods and processes used in the production of goods or services or in the accomplishment of objectives, such as scientific investigation.'[9] Science and technology play a very important role in human society, especially in this modern era. Furthermore, science and technology are essential in dealing with environmental issues.

Science and technology in itself is not good or bad, but it will cause good or bad results depending on the way you use it. Many environmental issues, like pollution, are caused by the abuse of science and technology, or ignorance of the possible consequences of such science and technology. Non-degradable plastics, chemicals causing ozone depletion, pesticides resulting in biological chain damage, and many other similar cases, display the environmental problems caused by science and technology. Rachel Carson's "*Silent Spring*", published in 1962, illustrated the hazards of pesticides for humans and the environment, and for the first time, questioned the quest to "Conquer nature" for so-called society's improvement and development. This book drew worldwide concern for the impact of chemicals on the environment and raised society's awareness. It is necessary to discuss the relationship between science and the environment, and how to use technology to improve the environment.

The development of modern science and technology has resulted in many changes. Now, we can grow vegetables without soil: Soilless cultivation techniques; we can grow crops not based on seasons: Green House cultivation; we can shorten the growth time: Growth-promoting chemicals; we can increase the production: transgenic (GM) technology; we can also enhance food preservation: preservatives. All of these changes challenge the traditional ethical relationship between humans and nature. It seems that human's reliance on nature is getting less and less, which also leads to a reduction in human's concerns towards nature. People rely on technology more than on nature. Thus, we need to rethink this

[8]From Wikipedia, Science. https://en.wikipedia.org/wiki/Science.
[9]From Wikipedia, Technology. https://en.wikipedia.org/wiki/Technology.

change in values. Technology can service our lives. We should adopt a "moderate" attitude towards science and technology, this is recommended in Ancient Confucianism. A moderate attitude means no more and no less, technology should be used wisely and reasonably. We should innovate and develop science and technology based on natural laws, human's basic needs and environmental carrying capacity. Considering the integrity of the environment, the technology should not ignore the laws of nature. As mentioned above, the results of science and technology depend on how we use it. How we use it depends on our motivation. Thus, the correct motivation is the core.

Sustainability

The Brundtland Commission of the United Nations (UN) in March 20,1987 defined "sustainable development" as development that meets the needs of the present without compromising the ability of future generations to meet their own needs" (UNGA, 1987). 'In ecology, sustainability is the capacity to endure; it is how biological systems remain diverse and productive indefinitely' (Lawn, 2011). 'The 2005 World Summit on Social Development identified sustainable development goals, such as economic development, social development and environmental protection' (UNGA, 2005).

In terms of sustainable development, the critical point is "moderate" development. Actually, the opposite of sustainable is unsustainable. An outstanding performance is usually unsustainable — instant success, namely seeking "quick" results, and "more" results. The whole world is seeking speed and quantity today. This will lead to overuse of natural resources and over-production of wastes. Thus, it is critical to reduce and to control the needs of contemporary people. Thus, in environmental management, apart from macroscopic management, like government policies and regulations, self-management is another essential aspect. The first step of self-management is letting people know the impact of their behaviour on the environment, and then letting them understand the possible good and bad results of each type of choice. Only when everyone gets involved in self-environmental management, can sustainability be achieved.

Self-management and Environment

When we mention environmental management, most people will think of governments, enterprises, NGOs, regulations and laws, etc., or environmental protection, pollution control and biodiversity conservation. Of course, all of these are important, but all of these are from an "external" management view. We now have many laws and regulations, many NGOs both national and international, but we still have not solved our pollution problems or the illegal trade in endangered wild animals. Thus, a new perspective of environmental management is needed.

We should understand that the external environment is affected by the internal environment. Our values determine our motivations, and our motivations determine our behaviours. Some of our actions have caused many environmental issues, which means that there is something wrong with the motivations and values behind these actions. Thus, the first step of protecting the environment is to see what is wrong with our motivations and values, namely, we should return to examine our own minds. Thus, self-management is the first key to environmental management.

Human Behaviour Model

Let us have a look at what kind of reason or need will cause corresponding behaviour. There are some theories about the human's behaviour model. In psychology, it believes that all human actions are caused by some motivation. Motivation plays roles in stimulating, facilitating and strengthening human actions, providing incentives. External incentives are only the condition and environment of behaviours while the incentives will work through internal factors. Here we introduce one of the typical theories of human behaviour model: Burrhus Frederic Skinner's Reinforcement Theory.

Skinner proposed an "operant conditioning" theory that people or animals in order to achieve a certain purpose, will take a certain behaviour working on the environment. When the consequences of such behaviour are positive, this behaviour will be repeated in the future; when the consequences are negative, this behaviour will be diminished or disappear. People can use this positive or negative reinforcement approach to affect the consequences of behaviour, thereby correcting their behaviour. This is the Reinforcement Theory, also known as "Behaviour Modification Theory" (Yu, 2001).

Based on this theory, we can find out the reason why people will do something. Firstly, they think or they find that this thing is right (beneficial). Secondly, they have this need. Then, they receive positive results after a period of doing this. Finally, they will continue to do this if all three of the above conditions are met. Once a person decides to do one thing as described above, that person will try his best to perfect the related conditions in this direction, in order to complete it successfully. Thus, human behaviour can be simply classified into two types: initiative behaviour and passive behaviour. Generally, active behaviour always gets better results and will last for a longer time than passive behaviour. When you are willing to do something, you always finish it quickly and well. So if the thinking of harmony between human and nature can be internalised into personal values, then the behaviour and habits of humans can be changed correspondingly. People will save resources and reduce pollution on their own initiative. Then the environmental issues will decrease gradually.

Thus, self-management can l play a big role in environmental management. Now, most people think that the responsibility of protecting the environment lies with governments or companies, and relies on regulations and laws, while ignoring the fact that governments and companies are composed of individuals. When each person can act correctly after rethinking of their values and behaviours, then the environmental issues and many other problems will be solved naturally. We can call this "self-management".

Environmental Regulations and Laws Versus Lifestyle

The emergence and development of modern environmental law was prompted by the emergence of large-scale industrial pollution along with industrial development. From the mid-19th century, a number of capitalist countries began to formulate pollution prevention legislation, such as Britain's "Alkali Industry Law" (1863), and "Rivers Anti-Fouling Law" (1876). The rapid development of environmental law began from the 1950s to the 1960s. At that time, environmental pollution and destruction of the ecological balance became increasingly serious. Environmental Law rapidly became a new and separate field of law.

In environmental management, environmental regulations and laws have played a very important role. But there are some limitations of environmental regulations and laws. Firstly, it must be noted that environmental regulations and laws are only a part of environmental management. Secondly, environmental regulations and laws are usually 'after the event' aiming at solving problems rather than preventing them. Thirdly, another limitation of environmental law is unfairness, between humans and animals. Almost all of the environmental laws are from the perspective of humans and based on human interest, even animal protection laws are aimed at ensuring human's benefit. For instance, killing farmed ducks or pigs is legal, because humans "need" to eat them or to make clothes using their feathers or skins. Killing of wild tigers is illegal now because tigers are protected by law. Now whether tigers should be farmed is a hot topic. But one day, humans may well decide that it is not wrong to farm tigers and kill them for food, medicine, or fur, and then they will legalise tiger farms. Many similar cases in animal protection law show that it is unfair between humans and animals. We protect them for their utility to us rather than for themselves.

Next, there are some gaps between legislation and implementation. Countries with good laws may not have the capacity to implement them, as is the case in the Philippines. Hence, we do need laws to protect the environment, but we cannot be only relying on laws. Environmental law is a supporting role, rather than the protagonist.

From another perspective, environmental regulations and laws only work when there are illegal behaviours while there are many undesirable behaviours that are not against laws. For example, people's lifestyle. Actually, you will find that most environmental issues can be attributed to human lifestyles. And laws cannot stipulate human lifestyle. Advocating solid wood furniture will lead to a demand for more wood, which may lead to excessive deforestation, including the logging of rare trees. Now it is the time that people pursue new technology and the latest designs which speeds up the upgrading of many electronic products, require a lot of rare metals and other raw materials, while producing a large number of e-waste. Global-shopping is also popular — living in Singapore you can enjoy Finnish seafood within a week, which requires advanced preservation technology and fast transport. All of these examples are happening every day, everywhere now. And all of these need a large number of resources and energy and generate a great number of all kinds of wastes. The core of this modern lifestyle is "consumption". We must touch and change the root of this kind of lifestyle, then we can restore and improve our environment gradually. The root cause is our greed and desire. We have to control our greed and desire, which is the core of self-management. Then when we realise this, we can improve our environment step by step.

Case Study

Here we will introduce some good environmental management cases.

1. *Tibetan environment — self-management and management without advanced science*

Firstly, we will discuss the Tibetan environment. As we all know, Tibet is very much a pristine, untouched place, almost without modern industrial and technological invasion, and without ecological destruction. Perhaps in the eyes of many people, Tibet is not a modern or developed place, it is viewed as a backward and poor area and local people must be very unhappy. However, on the contrary, people live peacefully, simply and happily in Tibet. Local people retain their traditional habits and lifestyle. And this is their choice compared to the modern life with science and technology. You cannot easily say which one is better. However, every person who has been there can feel this. Most Tibetans do not pursue advanced technology and the modern life, what they value is the inner joy and happiness. This is because most of them have faith in Buddhism. Thus, most of them know what they should do and what they can do, and they can manage their mind, thoughts and behaviour. Thus, you will see that actually environmental management is not that complex and difficult — it does not need advanced ideas, theories, and technology. In Tibet, because each person manages himself well, there is a sound natural environment, with all kinds of animals living freely. Tibet is regarded as a pure place without troubles and worries.

Another similar place is Bhutan, which developed the concept of Gross National Happiness. Nowadays, people all around the world dream of travelling to Tibet or Bhutan to seek inner peace and escape the hustle and bustle of the city, which means most people have realised that inner joy is more precious than the external enjoyment, and the external environment is related to the internal environment. But note that this does not mean that all poor and backward places have a good environment, and it also does not mean that we have to be poor and backward to achieve a good environment.

2. *Japanese urban environmental management — waste management*

Not all places can be a second Tibet or Bhutan; after all, more than half of the world's population live in cities, and this number will be 66% by 2050 (according to the latest United Nations population report 2014). With the continuous development of urbanisation, the city's environmental management is increasingly important, particularly its waste management.

Japan can be regarded as a mature urban waste management country with a good waste management system.

There are two examples of Japanese urban waste management that are representative of good environmental management. The first is their garbage classification and recycling system. From the 1970s, Japan spent 40 years to gradually refine its method of garbage classification, so that today, it now has a relatively complete and mature system. Japan lacks resources and energy, so all parts of Japan have developed strict waste disposal rules, in order to facilitate garbage disposal and to maximise the use of recycling waste. Domestic waste is generally divided into combustible garbage, non-burnable garbage, resource garbage, hazardous garbage, oversized garbage, and garbage that cannot be recycled. Furthermore, each type of garbage has more detailed classification and description. Apart from rigorous and detailed classification, different wastes have different treatment methods. For example, the different parts of a plastic bottle are divided into different waste categories. In a 27 pages manual printed by Yokohama City Hall, some of the 518 items describe the classification rules. This makes it possible for each family to dispose of its garbage correctly based on this classification. It is also important that in daily life, the Japanese people treat waste management very seriously, and vigorously cooperate with this work. They have formed fixed and good habits of garbage classification. The Japanese have a basic consensus, when you do something, you cannot make trouble for others.

Another example is their "no trash can" principle in a city. When you walk in the streets or the malls of Tokyo, you cannot find a trash can. You have to take your rubbish back home and dispose of them according to the garbage classification system. This principle makes it easy to deal with the different garbage and keep public places clean. It also reduces the cost of maintenance of public facilities. Of course, this cannot be achieved immediately when it was put forward. It needs time to let all people accept this and get used to this.

These two examples from Japan illustrate their successful environmental management, because of self-management working together with national management.

Conclusions and Discussions

In this paper, we have discussed the relationship between human beings and nature from the philosophical perspective.

It is clear that the lifestyle of modern society which leads to high consumption of earth's resources is the main cause for force of environmental degradation. And the real power hidden behind this kind of lifestyle is human's endless greed and desire.

Against this reason, we cannot just rely on advanced technology, ideas or laws to govern the environment. The advanced environmental science and technology, theories, and perfect laws are all just part of environmental management — they play roles of mitigation and are supplementary in solving environmental issues. What we really need is self-management and a change in our inner selves and lifestyles. If we do not save resources do not control our behaviour, no amount of resources is enough. Self-management means one person's behaviour does not adversely impact others, instead, it will benefit others. When we can manage ourselves, then we will have a sound environment, in which animals live freely and happily, plants grow freely and naturally, and finally, humans can live in harmony with nature.

References

Adger, W.N., and Jordan, A. (2009). *Governing Sustainability.* UK: Cambridge University Press.

Baidu Encyclopedia (n.d.). Philosophy. Retrieved from: http://baike.baidu.com/link?url=yto2tDxtfLLjfkXhNNFamOm8aclD5BwD8-biUB3NNddVYwT3hE2l8n_hNS-1BoE4gr2Ixr7iOb2gMGJYjY0mXK.

Bari, J. (1995). Revolutionary Ecology: Biocentrism & Deep Ecology. *Alarm: A Journal of Revolutionary Ecology. Groundhog Day 1995,* **12**(1), 6–7.

Beder, S. (2006). *Environmental Principles and Policies.* Australia: New South Books.

Bell, S., and Morse, s. (2008). *Sustainability Indicators.* London: Earthscan.

Belshaw, C. (2001). *Environmental Philosophy.* Chesham: Acumen.

Bjornstad, D.J., and Kahn, J.R. (1996). *The Contingent Valuation of Environmental Resources.* USA: Edward Elgar Publishing.

Brand, U., Gorg, C., Hirsch, J., and Wissen, M. (2008). *Conflicts in Environmental Regulation and the Internationalisation of the State.* London and New York: Routledge.

Brown, P. (n.d.). *Buddhism and the Ecocrisis: The Role of Buddhism in Enhancing Environmental Philosophy and Psychology in the West Today.* Retrieved from: http://www.buddhanet.net/mag_eco.htm.

Cohen, and S. Marc. (2000). *Aristotle's Metaphysics,* The Stanford Encyclopedia of Philosophy, Edward N. Zalta (ed.), https://plato.stanford.edu/entries/aristotle-metaphysics/.

Des Jardins, J.R. (2000). *Environmental ethics, an introduction of environmental philosophy.* Belmont, CA: Wadsworth Thomson Learning.

Donner, S.D. (2007). Surf or Turf: A Shift from Feed to Food Cultivation Could Reduce Nutrient Flux to the Gulf of Mexico. *Global Environmental Change*, **17**, 105–13.

Environmental Working Group (n.d.). Meat Eater's Guide to Climate Change + Health: Climate and Environmental Impacts. Retrieved from: http://breakingnews.ewg.org/meateatersguide/a-meat-eaters-guide-to-climate-change-health-what-you-eat-matters/climate-and-environmental-impacts/.

Ervin, D.E., Kahn, J.R., and Livingston, M.L. (2004). *Dose Environmental Policy Work?* USA: Palgrave Macmillan.

Fanelli, D. (2007). Meat Is Murder on the Environment, *New Scientist*, 18 Jul. 2007.

Farong, R. (2012). Enlightenment of Taoist thoughts to ecological and environmental protection, An Interview. https://wenku.baidu.com/view/0f9ab16eaf1ffc4ffe47acad.html.

Forestry Commission of Great Britain (n.d.). *Sustainability.* Retrieved on: 2009-03-09.

Global Sherpa (2013). Beijing Air Pollution Exposes China's Health & Environment Risks, Jan 14, 2013. Retrieved from: http://globalsherpa.org/china-air-pollution-health-environment/.

Goodland, R. and Anhang, J. (2009). "Livestock and Climate Change," *World Watch* Nov./Dec. 2009.

Grayling, A.C. (1999). Editor's Introduction, in A.C. Grayling (ed.), *Philosophy 1: A Guide through the Subject*, Vol. 1. UK: Oxford University Press.

Gu C. (2009). *Discussion on Environmental Philosophy in Chinese Traditional Philosophy.* China: Sichuan Institute of Technology. Retrieved from: http://xueshu.baidu.com/s?wd=paperuri%3A%2860714636a933352d1d791106894a8ae8%29&filter=sc_long_sign&tn=SE_xueshusource_2kduw22v&sc_vurl=http%3A%2F%2Fwww.doc88.com%2Fp-1661454494480.html&ie=utf-8&sc_us=1416236095099351123.

Ingram, D., and Parks, J. (2002). *The complete idiot's guide to understanding ethics*, p. 201. USA: Alpha Books.

International Panel for Sustainable Resource Management (2010). "*Assessing the Environmental Impacts of Consumption and Production*". United Nations Environment Programme.

Kantian Ethics (n.d.). Retrieved from: http://www.csus.edu/indiv/g/gaskilld/ethics/Kantian%20Ethics.htm.

Kemp, D.D. (2004).*Exploring Environmental Issues.* London and New York: Routledge.

Kreith, M. (1991), "Water Inputs in California Food Production," Water Education Foundation, 27 Sept. 1991.

Lawn, P. (2011). Is Steady-State Capitalism Viable? A review of the issues and answer in the affirmative. *The Annals of the New York Academy of Science;* 1219. http://onlinelibrary.wiley.com/doi/10.1111/j.1749-6632.2011.05966.x/abstract;jsessionid=E5991F4740A36E09E8494DEA19EB92D6.f03t02?systemMessage=Wiley+Online+Library+%27Journal+Subscribe+%2F+Renew%27.

Leadership, Management (n.d.). David Mcclelland's motivational needs theory. Retrieved from: http://www.businessballs.com/davidmcclelland.htm.

Leopold, A. (1989). *A Sand County almanac, and sketches here and there.* New York: Oxford University Press.

Li, Y. (2015). *Analysis of Confucian Thought of environmental protection.* (Thesis). China: Henan University.

Management (n.d.). McClelland's Theory of Needs. Retrieved from: http://www.netmba.com/mgmt/ob/motivation/mcclelland/.

Mastin, L (2008). "Deontology", in *The Basics of Philosophy.* Retrieved from: http://www.philosophybasics.com/branch_deontology.html.

Oxford English Dictionary Online (2015). Philosophy, definition of "philosophy, n." Retrieved from: http://www.oed.com/view/Entry/142505?rskey=uk0M8u&result=1 (accessed August 05, 2015).

PETA (n.d.). "Vegetarianism and the Environment". Retrieved from: http://www.peta.org/issues/animals-used-for-food/animals-used-food-factsheets/vegetarianism-environment/.

Quinton, A. (1995). "The ethics of philosophical practice". In T. Honderich (ed.), *The Oxford Companion to Philosophy,* p. 666. UK: Oxford University Press.

Robbins, P., Hintz, J., and Moore, S.A. (2010).*Environment and Society.* UK: Wiley.

Schmidtz, D., and Willott, E. (2012). *Environmental ethics: what really matters, what really works,* 2nd ed. New York: Oxford University Press.

Teichmann, J., and Evans, K.C. (1999). *Philosophy: A Beginner's Guide.* UK: Blackwell Publishing.

Thampapillai, D.J., and Sinden, J.A. (2013). *Environmental economics — concepts, methods and policies,* 2en version, pp.190–216.

The Catholic Encyclopedia (n.d.). "Teleology", in *The Catholic Encyclopedia.* Retrieved from: http://www.newadvent.org/cathen/14474a.htm.

Thich, T.Q. (1996). "Buddhism and Environmental Protection". Retrieved from: http://www.budsas.org/ebud/ebdha006.htm.

Treehugger (n.d.) Meat Eater's Guide: Get to Know the Carbon Footprint of Your Diet. Retrieved from: http://www.treehugger.com/green-food/meat-eaters-guide-get-to-know-the-carbon-footprint-of-your-diet-lamb-beef-cheese-are-the-worst.html.

Tverberg, G. (2012). World Energy Consumption Since 1820 in charts. Retrieved from Our Finite World: https://ourfiniteworld.com/2012/03/12/world-energy-consumption-since-1820-in-charts/.

United Nations General Assembly (UNGA) (2005). 2005 World Summit Outcome, Resolution A/60/1, adopted by the General Assembly on 15 September 2005. Retrieved on: 2009-02-17.

United Nations General Assembly (UNGA) (1987). "Our Common Future, Chapter 2: Towards Sustainable Development", in *Our Common Future: Report of the World Commission on Environment and Development.* Retrieved from: http://www.un-documents.net/ocf-02.htm (accessed 1 March 2010).

US EPA (2002). "*Review of Emission Factors and Methodologies to Estimate Ammonia Emissions from Animal Waste Handling,*" State of North Carolina, Department of Environment and Natural Resources, U.S. Environmental Protection Agency, Apr. 2002.

Vasconcelos, V.V., and Martins Jr., P.P. (2011). Teleology and Randomness in the Development of Natural Science Research: systems, ontology and evolution. *Interthesis*, **8**(2), 316–334.

Weinberg, Z. (2011). "What Would Buddha Do? A Buddhist Approach to Environmental Protection". Retrieved from: http://modernserenity.com/2011/08/what-would-buddha-do-a-buddhist-approach-to-environmental-protection/.

Wikipedia (n.d.). "Biocentrism ethics". Retrieved from: https://en.wikiedia.org/wiki/Biocentrism_ (ethics).

Wikipedia (n.d.). *Ethics.* Retrieved from: https://en.wikipedia.org/wiki/Ethics#cite_note-iep.utm.edu-1.

Wikipedia (n.d.). *Environmental philosophy.* Retrieved from: https://en.wikipedia.org/wiki/Environmental_philosophy#cite_note-Belshaw-1.

Wikipedia (n.d.). *Technology.* Retrieved from: https://en.wikipedia.org/wiki/Technology.

Yu, J. (2001). Skinner's reinforcement theory and its application in teaching. *Journal of Qujing Normal University*, **20**(1), 92. Retrieved from: https://wenku.baidu.com/view/d96a54e6aef8941ea76e058e.html.

9

EVALUATING EIA FOLLOW-UPS IN CHINA AND SINGAPORE

DENG MAO and RICK REIDINGER

The process of Environmental Impact Assessments (EIA) has become the mainstream pre-project environmental procedural safeguard internationally. However, there have been insufficient efforts placed on the follow-ups, such as the implementation of mitigation and monitoring works. These are an integral part of the EIA to verify the predictions and ensure that there are sufficient environmental safeguards during the actual project implementation. This study focuses on the EIA follow-up legislative and management systems of China and Singapore. Using the robust system in Hong Kong as a comparison, it aims at identifying shortcomings and recommending improvements in both systems. China lacks a robust management framework and the effectiveness of its EIA process is severely undermined by the absence of qualified/accredited third parties as well as conflicts of interest. In contrast Singapore lacks a mandatory EIA but good environmental performance is ensured by stringent administrative controls by various government agencies. Still, the lack of mandatory EIA, follow-up framework and guidance are inadequacies in Singapore's system. This study proposes further improvements to resolve the shortcomings in each country's system.

Introducing EIA Follow-Up Principles

The Environmental Impact Assessment (EIA) process was first initiated by the United States through its National Environmental Policy Act (NEPA) in 1969, and has since been widely adopted internationally through the United Nations 'Goals and Principles of Environmental Impact Assessment' (UNEP, 1987) and the World Bank's Environmental Assessment Operational Directives (OD) and Performance Standards on Social and Environmental Sustainability.

Conducted at the planning and design stage, the significance of environmental impacts is usually assessed based on various assumptions or conditions on how the works will be carried out. (Hollands and Palframan, 2014) It is equally important that considerations or recommendations at the planning stage provide sufficient guidance through the life cycle of projects, enabling projects to proceed without damaging the environment

(Sadler, 2004). In reality, the insufficient linkage between the EIA and project implementation has been identified as a significant flaw of the EIA process. (Sanchez and Hacking, 2002)

The People's Republic of China has embraced the EIA process and also passed a law mandating the conduct of EIAs, but it is still being plagued by some of the worst environmental damage and pollution issues. Singapore, in contrast, is a small city-state known to be a clean and green 'Garden City' but it does not have an EIA law. Instead, it has developed a more unique environmental assessment system (Lye, 2008). This study aims to examine the differences of EIA follow-up practices between China and Singapore, to identify shortcomings and suggest improvements to both systems.

Good Principles and Evaluation Criteria

Leading academics and EIA practitioners such as Morrison-Saunders and Arts (2004) have discussed substantially the core-components and principles for EIA follow-up. In their opinion, the key purpose of the EIA follow-up is to determine the outcome of the EIA by incorporating a process of monitoring and feedback, enabling continuous learning and improvement of the EIA process by the project proponent, regulator and EIA practitioner.

According to Morrison-Saunders and Arts (2004), the four core-elements of EIA follow-up are: *Monitoring, Evaluation, Management and Communication.* The corresponding best-principles for this EIA follow-up are summarised below in Table 9.1.

In practice, it is a challenging task to obtain details of the EIA follow-up procedures and methodologies, and even harder to evaluate the outcome critically given the lack of publicly available project information in countries of interest. In this study, the effectiveness of EIA follow-up will be reviewed in the context of legislation, management frameworks and

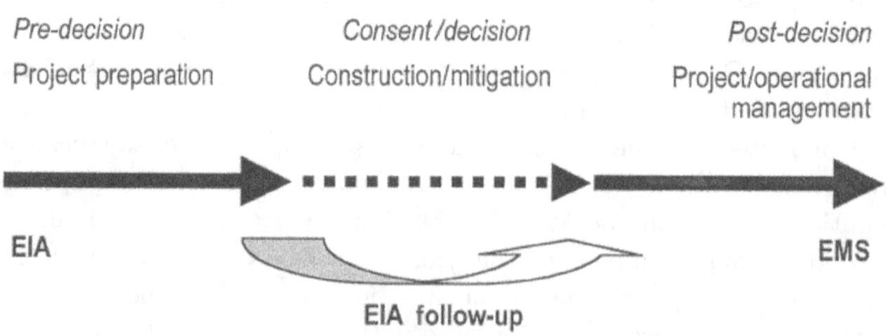

Fig. 9.1. EIA Follow-up's Role in Bridging the Pre- and Post-Decision Gap (Marshall, 2004).

Table 9.1. Best follow-up planning and design principles (Marshall *et al.*, 2005).

No.	Principle	Key concepts
1	Follow-up is essential	EIA is incomplete without follow-up, as without it the consequences of EIA planning and decision will be unknown
2	Transparency and openness	Stakeholders have the right to feedback on EIA follow-up, actions and decisions; active engagement in the follow-up is preferable.
3	EIA to include follow-up commitments	Parties involved in EIA need to have clear commitments to undertake follow-up in the pre-decision stage
4	Community involvement	Communities should be informed of EIA follow-up outcomes, or be engaged directly in the implementation
5	Clear division of roles, tasks and responsibilities;	EIA Follow-up should be defined with clear roles, tasks and responsibilities involving different parties and competent independent practitioners for monitoring, audit and review.
6	'Fit-for-purpose' practicable and feasible;	Follow-up should be tailor-made for the anticipated environmental effects, and designed to fit the project activities, timeline and dynamic context.
7	Clear performance criteria and methodologies	Follow-up should include the setting of clear performance criteria, with well-defined methodologies;

guidance, focusing on several major criteria raised by the above-mentioned international best practice, namely:

- ***Linkage*** between planning and implementation, i.e. how strong is follow-up required and linked back to the EIA;
- ***Organisation*** and managerial efficiency, i.e. whether roles and responsibilities are effectively defined and administered;
- ***Clarity*** in technical scope and performance criteria, i.e. are monitoring/auditing and the response actions properly designed;
- ***Transparency*** in outcomes of the follow-up, i.e. whether the follow-up programme embraces community engagement or public disclosure.

The Exemplary Hong Kong EM&A Model

Requirement on EIA Follow-up and its Linkage

Hong Kong Special Autonomous Region (SAR) has its own legal system on environmental protection that is widely regarded to have the best EIA in

Asia. It is an exemplary model with its follow-up Environmental Monitoring and Audit (EM&A) programme, which is a statutory requirement stated in the Environmental Permit. Under the EM&A, a strong linkage between the content and scope of the EM&A and the EIA Report is ensured by the requirement of completion of the EM&A manual prior to the commencement of construction, based on recommendations raised by the EIA Report and the conditions specified in the Environmental Permit.

Organisational Framework

During the construction phase, the project proponent needs to engage internally an Environmental Team (ET) to supervise and manage the implementation of mitigation measures, to organise environmental monitoring tasks and to prepare routine EM&A reports. Detailed monitoring and auditing are separate and non-overlapping tasks to be conducted by independent third-parties based on their expertise and accreditations specified.

EM&A incorporates an overall professional review by introducing an Independent Environmental Checker (IEC) accreditation. The Environmental Protection Department (EPD) requires the IEC to be a qualified environmental consultant with 7 years of experience in conducting EM&A, who is to conduct an audit on the overall environmental performance. He must be given access to the site and all project related documentations. The IEC shall not be from the same organisation of the ET or involved in the implementation of mitigation or monitoring works.

Overall, the organisation structure of Hong Kong's EM&A programme is presented in Fig. 9.2. It has very clear definition of roles and responsibilities involving project proponent staff and the independent accredited reviewer, ensuring that they work collaboratively without conflict of interests, and build an inclusive EM&A programme.

Clarity of Technical Guidance

Technical aspects of the environmental mitigation and monitoring for EM&A are specified by the HK EIA Ordinance Section 16 — Technical Memorandum. The Technical Memorandum provides principles, procedures and guidelines for the environmental monitoring and audit requirements.

This Technical Memorandum also contains specific criteria covering all major aspects of the environment, including air quality, noise, water, waste management, ecological, fisheries and visual and landscape impacts.

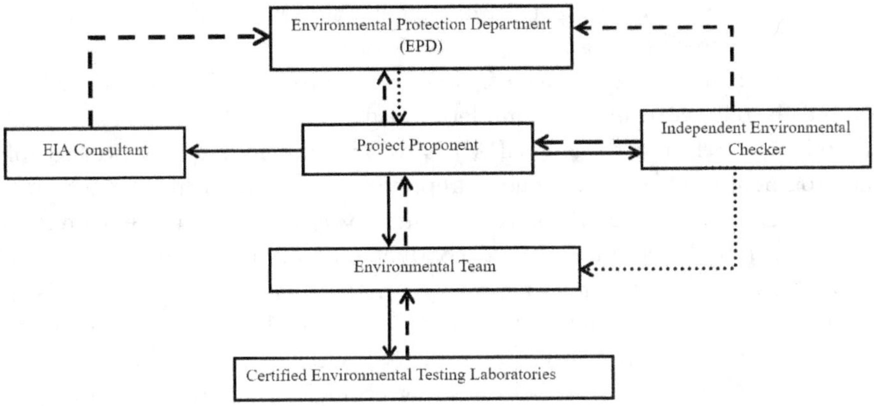

Fig. 9.2. Organisation chart of EIA and EM&A in Hong Kong. Dashed line arrows denoting reporting, dotted-line arrows denoting monitoring/auditing service provision, solid-line arrows denoting service engagement.

The EIA Ordinance recommends that the environmental monitoring results gathered during the EM&A are to be traced back to the EIA predictions for assessing the accuracy of EIA analysis, and for assessing the effectiveness of the mitigation or management system.

Public Disclosure

Under the Hong Kong EIA Ordinance, the environmental performance of the project during its implementation, reported by the EM&A process is to be released for public comment. In practice, the Hong Kong authorities have far exceeded this requirement by establishing an online EM&A database for public access, and holders of Environmental Permits are required to upload their EM&A Reports onto the dedicated project environmental reporting page (Arts *et al.*, 2001).

Evaluating EIA Follow-ups in China and Singapore

Two Vastly Different Systems

In China, EIA has become a mandatory process for major developmental plans and policies as well as for individual industrial plants and construction projects. Presently, EIA in China is governed by the EIA Law implemented on September 1, 2003. The first EIA law in China broadly states that the 'plan-preparation institutions' shall organise 'tracing assessment' and report the results to the 'approval institutions'. Construction projects are required to 'simultaneously implement the environmental protection countermeasures', thus forming the basis of EIA follow-up.

Upon completion of the construction phase and before the project is put into operation, a 'Completion Environmental Protection Audit' (hereinafter referred to as 'CEPA') is to be conducted by the responsible environmental protection agencies approving the EIA, within half a year of project commissioning. Prior to the audit, the developer needs to engage a qualified professional entity to conduct environmental monitoring and submit the CEPA report. In most cases, the qualified professional is an environmental monitoring unit under the provincial or local environmental agency. Certified EIA consultant organisations are sometimes engaged to conduct the monitoring works involving ecological surveys for major EIA projects with an ecosystem impact assessment.

Singapore has taken a completely different path compared to the mainstream practice, as it does not have a formal EIA law or process. While being widely acclaimed as a Garden City for its lush urban greenery and cleanliness, Singapore has sacrificed a huge proportion of its native ecosystem and biodiversity for urbanisation (Lye, 2008). Recognising the sensitive nature of some development projects, government agencies have started to conduct internal environmental assessments in the name of EIA or EIS by engaging experienced consultants, or sometimes, Biodiversity Impact Assessments (BIA). The follow-up requirements are normally incorporated into the contract specifications to be undertaken by the developer as an Environmental Monitoring and Management Plan (EMMP). There is no technical guidance for EMMP works, and the mitigation and monitoring results are mostly documented internally and not disclosed for public comment.

Environmental assessment for industrial projects in Singapore is conducted in the name of a "Pollution Control Study" (PCS), to be reviewed with the plant and building design by the Central Building Plan Department (CBPD), the technical review department of the National Environment Agency (NEA). In practice, the PCS assessment is to show compliance with the Environmental Protection and Management Act (EPMA) and its supplementary guidelines and standards published by NEA, such as Code of Practice on Pollution Control (COPPC). Under each environmental aspects required for the PCS assessment such as air, water, waste and noise, consultants are required to report sources quantitatively, to declare their emissions/discharges and to recommend the respective environmental controls in order to ensure that these emissions are limited to acceptable levels.

China's CEPA System and Follow-up

CEPA System and EIA Linkage

China's Project Completion Environmental Protection Audit (CEPA) system is the main EIA follow-up for industrial and construction projects, and a key step for approval by the environmental agency in China. Monitoring is to be conducted by the subsidiary environmental monitoring units (or stations) under the local environmental agency during the commissioning stage of the industrial project. The overall process constitutes a review of relevant technical documents on mitigation measures and systems, site inspections and monitoring of shortlisted pollutants at source and receptors.

Upon commissioning of the project, the construction company will submit an application for the CEPA to the environmental monitoring unit, which will start drafting an environmental monitoring plan based on the EIA report and relevant design documentation for review and acceptance by the local environmental agency. Upon the agency's approval, a contract is signed between the construction contractor and the environmental monitoring unit to officially commence the monitoring works. The CEPA has reasonable linkage to the EIA due to the plan originating from the EIA. However, it cannot be considered to have a strong linkage as it is only to certify satisfactory completion and is not a concurrent process to provide feedback throughout the construction period.

CEPA Organisation Framework

The monitoring under CEPA process is typically conducted only upon completion of the project construction phase. The results are documented in a required format by the environmental monitoring units appointed, to be submitted to district-level environmental protection agencies for review and clearance. CEPA monitoring and clearance are a prerequisite for the construction phase to be concluded, before the project can be handed over to the owner and operator. Therefore, it is inappropriate for the owner, or even worse, the contractor, to appoint and pay the environmental monitoring unit conducting such works, which would affect the objectivity of the data.

In many provincial or city level projects, district-level environmental protection agencies review the acceptance monitoring reports submitted by their own subsidiary monitoring body. This is clearly an ineffective management arrangement as it is like the 'left hand checking the right hand'. Mr Liu Fang from China's Environmental Monitoring Centre in Beijing, states that this leads to a conflict of interest as the approving party and monitoring party belongs to the same environmental protection agency (Liu, 2011).

Environmental monitoring units are 'naturally' empowered to conduct environmental monitoring based on their double identities — as a branch of the administrative body and as service provider. This may discredit the quality of monitoring outcomes especially in remote administrative regions. In contrast, environmental monitoring and analyses are the job of qualified third-parties in both Hong Kong and Singapore, who compete in the market in order to maintain their accreditation as professionals. The organisational framework of China's EIA Follow-up is presented in Fig. 9.3.

When any violation of environmental standards is found, the current management framework allows the developer to simply pay for re-monitoring, which provides a loophole for them to apply temporary

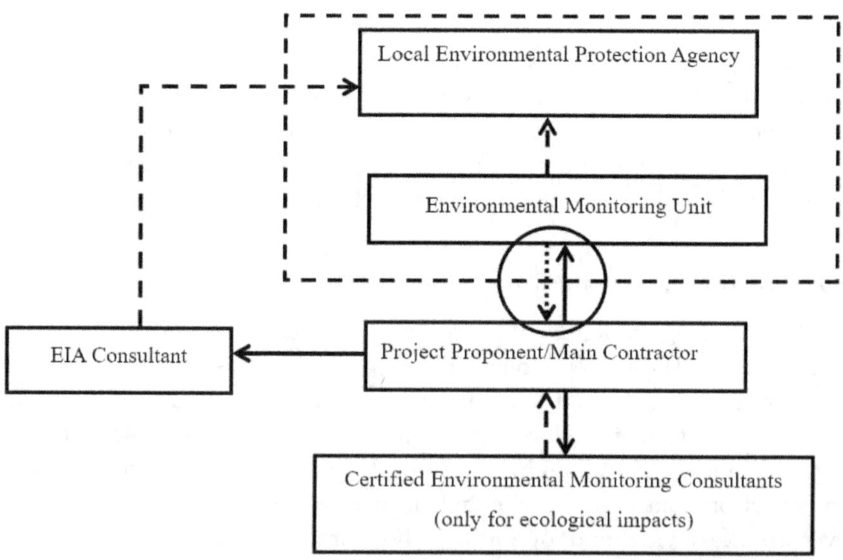

Fig. 9.3. Organisation chart of EIA and CEPA in China. Dashed line arrows denoting reporting, dotted-line arrows denoting monitoring/auditing service provision, solid-line arrows denoting service engagement, and circle denotes major conflict of interest.

measures in order to 'be compliant', instead of implementing permanent mitigation measures to make real improvements.

Technical Clarity of CEPA

Technical requirements on project-based environmental safeguards have been long developed and in certain industrial or environmental aspects, some even pre-date the EIA law itself. The Ministry of Environmental Protection (MEP) has also published a series of national environmental standards, following GB (short form of 'Guo Biao' or National Standard) codes, which includes emission or discharge standards, and ambient environmental standards such as air quality or surface water quality standards.

These guidance documents also specify in detail the baseline monitoring requirements. For example, projects generating maximum air quality impacts higher than 80% of the respective guideline criteria in terms of ground-level pollutant concentration, or impacts of 10% guideline level, reaching a distance of 5 km and beyond based on modelling study are classified as Class I Air Quality EIA. For such projects, the technical guideline has specified the baseline air quality monitoring programme in Table 9.2.

Despite the amount of technical details for environmental monitoring, there are no guidelines on the standard procedures for routine environmental site inspection and audit during the construction phase.

Table 9.2. Baseline air quality monitoring requirement for Class I Air Quality EIA Projects

Monitoring Frequency	2 episodes in summer and winter
Duration	7-days per episode
Parameters	Criteria pollutants being discharged (CO, NOx, SO_2, O_3 and PM10/PM2.5)
Results	Hourly pollutant concentration and 24-hour average concentration in accordance with GB-3095, the National Standard for Air Quality Guidelines and Measurement
Method	Automatic monitoring station certified by the national environmental monitoring administration, or manual reference methods
Locations	Not fewer than 10 locations representing Air Sensitive Receptors (ASR), each in different directions 45 degrees apart, and more monitoring locations in major downwind direction

Outcome and Transparency of CEPA

China has mandated the public disclosure of EIA Reports since its adoption of EIA Law in 2003, which also provided the legal basis for public involvement in the monitoring of project outcomes, represented by the CEPA monitoring process during the commissioning stage of the project. Environmental Protection Acceptance Test Reports are usually published by the project developer on their own web portal or in the respective government web portal after it is approved. On the other hand, there is a significant lack of project-based public disclosure during the construction phase of the project, and after the CEPA Report is approved and the plant enters its normal operating stage.

This may lead to the failure of the EIA's primary intention of public disclosure to incorporate stakeholder feedback into the decision-making process as EIA recommendations are usually neglected by the contractors, and these considerations are lost somewhere down the project organisation chart.

Singapore's Pollution Control Study (PCS) System and Follow-up

PCS Process and its Follow-up

In Singapore, for private industrial projects, a quick environmental permitting process with a much slimmer scope of coverage, namely the Pollution Control Study (PCS), applies under the Environmental Protection and Management Act (EPMA), with the oversight of the National Environment Agency (NEA). It can be considered a limited scope assessment without consideration of the social and economic contexts of the project. The PCS process is normally specified as a condition in the letter issued by NEA on the Industrial Allocation (IA) application, when the developer first applies for the intended development of the land. The PCS is to be completed together with the overall Building Plan submission.

The PCS is designed to assess the operating impacts of the industrial development, such as the emission of air impurities, waste generation and process wastewater discharge, without considering much of its impacts during the construction phase of the project. Based on the author's professional experience, in many cases the PCS is conducted in parallel with the construction works.

The commitment made in the PCS Report will be submitted to NEA as a basis of review and approval, and will form the specifications for future compliance inspection by NEA's inspectorate from the Pollution Control Department. Thus, it can be said that there is a strong linkage between the PCS and environmental management practices during the operation phase. However very rarely are PCS consultants retained by the developer to play a supervisory role in the operation phase, nor does the PCS Report form binding terms of reference for the plant operators. Thus, this linkage and the effective implementation of the commitments contained in the PCS is dependent on the strict administrative controls by the authorities, namely NEA, rather than the merit of the PCS procedures.

Organisation Framework of the PCS Follow-ups

The PCS can be commissioned by either the project owner, developer or sometimes the main contractor of the industrial project, such as the Engineering Procurement Construction (EPC) contractor of the plant. In most cases, the project owner would prefer to put it under the responsibility of the main contractor undertaking building design and submission works, for easier coordination and lower management cost, given that the details of the building plan and pollution control provisions needed in the PCS are provided by Qualified Persons (QP) who are registered architects and professional civil engineers.

The completion of the mitigation design features and pollution control installations is marked by the inspection by NEA's Central Building Plan Department (CBPD) to certify its satisfactory completion as per the Building Plan submissions prior to the granting of permission for operation of the plant. Mitigation measures for non-quantifiable pollutive effects, such as waste management and recycling, or prevention of accidental spillage, will be described as mitigation measures falling under the factory's internal Safety, Health and Environment (SHE) management system under the oversight of the health and safety manager.

In short, the PCS procedures can be described by the organisational chart on Fig. 9.4. It shows that the qualified third-parties are engaged by the contractor or building designer instead of directly by the project owner and proponent, therefore their roles are limited to certifying compliance with local standards stipulated by the agencies, and have practically little influence on the planning and design aspects of the project.

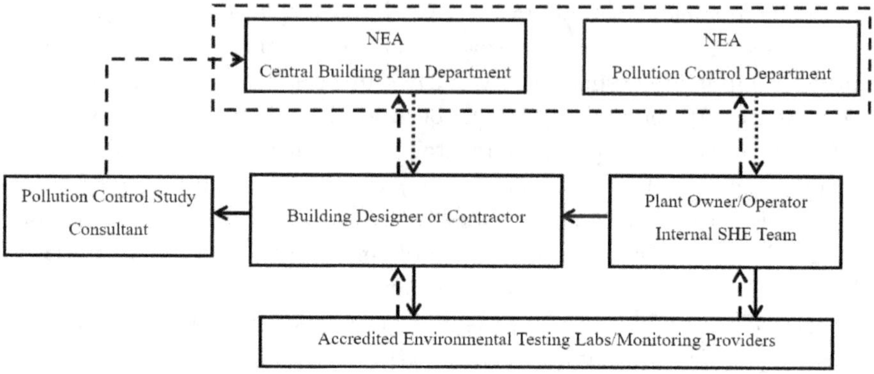

Fig. 9.4. Organisation chart of the Pollution Control Study (PCS) and follow-up in Singapore. Dashed-line arrows denoting reporting, dotted-line arrows denoting monitoring/auditing service provision, solid-line arrows denoting service engagement.

Technical Clarity of PCS Follow-up

The scope of coverage of the PCS process is clearly stated in the Guideline for Pollution Control Study, published by NEA's Central Building Plan Department to supplement Section 36 of the EPMA which requires a study on pollution control to be conducted for industrial developments. The format of the PCS Report must follow the structure in the Guideline to document expected environmental pollution problems based on the process flow of the industrial activities conducted on site. The minimum environmental aspects to be examined following the PCS Guideline structure are listed in Table 9.3.

A PCS requires follow-up monitoring mainly at source/discharge locations during the operation phase to ascertain compliance with the requirements. Typical information that needs to be specified for the monitoring plan of the operation phase includes the type of equipment, test methods, and frequency of monitoring/tests to be conducted. Monitoring and reporting of typical private developments and small industries require intermittent sampling to be conducted by certified testing bodies and documented by the plant operator for routine and ad-hoc inspections by NEA.

The Singapore Accreditation Council (SAC) or NEA certifies qualified third-party testing bodies, usually professional laboratories, to provide such testing services to industrial plants. There is, however, no local technical guidelines developed specifically for the testing method/technique and types of equipment to be installed. Instead, Singapore's statutes

Table 9.3. Typical scope of coverage and monitoring requirements of a PCS.

Environmental Aspects	Mitigation Requirement	Monitoring Requirement
Air	Propose measures to ensure compliance with the emission standards and requirements	Types of equipment/tests on air impurities, frequency of monitoring
Water	Water treatment facilities for processing waste water and trade effluents, to meet discharge standards; Safe storage and containment design for oil and chemical storage area	Water parameters to be monitored, specifying equipment/tests, frequency
Noise	Noise pollution control measures to ensure compliance with standards	Types of noise monitoring, specifying equipment/tests, frequency
Hazardous Chemical	Measures of safe storage and handling to ensure compliance with Code of Practice on Pollution Control (COPPC) Policy and procedures to prevent accidents relating to hazardous chemicals	Monitoring equipment to detect leak of hazardous chemical
Toxic Waste	Measures of safe storage and handling to ensure compliance	System of checks (inspection or audit)
Recycling and Resource Management	Recommend reduce, reuse and recycle of wastes; Recommend energy and water conservation	Nil
Land Contamination	Recommend measures to prevent land contamination	Monitoring if appropriate

and technical guidelines generally refers to international best practices embodied by United States or European Union standards.

Ad-hoc Government or Voluntary EIAs

The review of Singapore's legislative framework has identified the lack of comprehensive management frameworks and guidelines. Ad-hoc EIAs or EIS have been commissioned by relevant agencies such as Public Utilities Board (PUB) and NEA for the planning, siting and development of municipal infrastructure and industrial developments of strategic importance, such as oil and gas, power production, waste incineration, water supply and most recently desalination projects. Some other EIAs have been voluntarily commissioned by multi-national corporations (MNCs), often with stricter internal regulations.

For major industrial development projects that need to undergo an EIA in an informal arrangement with the government as the actual project proponent, the PCS still remains a separate process during the building plan submissions. However, planning considerations, such as project siting and choice of technologies being evaluated in the EIA are beyond NEA's jurisdiction and purview. For example, PUB, the national water resource agency, commissioned an EIA in 2010–2011 for the planning and conceptual design of the second desalination plant for Singapore, located in Tuas South district. The winning project proponent Tuaspring Pte Ltd needed to engage a qualified PCS consultant to conduct the PCS during the detailed engineering design and construction phase. Based on an interview with the qualified PCS consultant undertaking this project, the PCS was not entirely a subset of EIA and actually there were certain variations in the scope of assessment. For example, the Design, Build, Own and Operate (DBOO) contractor provided further engineering design details — in particular, for the auxiliary independent power plant which necessitated a proper air quality impact assessment with air dispersion modelling.

Based on the opinions of qualified EIA and PCS consultant being interviewed, the absence of guidance documents on the management framework of EIA follow-ups is to some extent a hindrance to good environmental management and planning. It is ineffective for professional consultants to advise on the time and resource allocations for environmental safeguards in an informal institutional arrangement. Furthermore, inefficient allocation or misallocation of resources may occur. For example, an EMMP with a focus on water quality such as a coastal development project may rely heavily on *ex-situ* sampling analysis of the receptor water bodies, which can take days to yield test results, but may overlook simple yet effective on-shore audits and inspections which can be more economically feasible and responsive, enabling direct intervention and immediate feedback of poor environmental performance. This misallocation of resources is also due to the fact that testing methods and accreditation of laboratories can be more conveniently spelt out in the contract documents, while there is no certification/accreditation for an independent environmental consultant or auditor in Singapore and no related standard work procedures similar to the Hong Kong qualification of IEC.

Another significant shortcoming is that Singapore's PCS clearly does not require or encourage any form of public disclosure of the details and outcome of the actual impacts of the individual industrial projects. While EIAs are conducted on an ad hoc basis, as earlier mentioned, follow-up processes are practically non-existent; monitoring results are treated as confidential and are nowhere to be found in the public domain.

A New Model in Singapore the Rail Project under LTA

LTA's EIA Follow-up Process

Transport infrastructure projects are known to have environmental impacts due to their scale, and they may traverse land zoned for other purposes such as residential estates or nature reserves. Singapore's Land Transport Authority (LTA) has in recent years implemented many internal EIAs on rail planning and construction projects, overseen by its Safety Division.

Based on interviews with LTA's Safety Division, the need for an EIA is decided in the planning stage by a project steering committee chaired by the Directors of Engineering and of Rail Construction. Members of other relevant divisions are invited to join this steering committee. Most of the EIAs commissioned by LTA are part of the consultancy service package under the feasibility study, including the Cross Island Line (CRL) and Jurong Region Line (JRL) in 2014.

The EIA practice of LTA is reasonable since there is no legal requirement for an independent EIA. Under this arrangement, the costs will be significantly lower than conducting an independent EIA. Time is also saved by placing the EIA under the engineering consultancy services so it can be evaluated simultaneously with the engineering design. The only drawback is that objectivity will be almost naturally lower with no real public disclosure. Overall the nature of such self-commissioned EIAs is more of an attempt at due diligence to incorporate environmental best practices into the planning and construction phase, but it can have no real influence over the higher decisions such as planning, design and choice of technologies.

The only tender that called for a standalone comprehensive EIA that may have the potential to influence the authorities' decision on the design and planning of transport infrastructure was the EIA on the Central Catchment Nature Reserve (CCNR) for the Cross Island Line (CRL). However, this is largely due to the high pressure from civil society led by the

Nature Society Singapore (NSS) which published their own EIA opposing this plan to build the MRT line traversing beneath the MacRitchie Reservoir and adjacent primary forests. Upon its completion, LTA gazetted the EIA Report for the Phase I (soil investigation works) for 4 weeks in February — March 2016 for public comment. However, it is unclear whether this form of independent EIA and public disclosure will become a mainstream practice for future projects.

Environmental management and monitoring has long been a compulsory part of LTA's general specification for all rail and road projects. In the latest issue of the general SHE specifications in April 2015, Clause 1.4 and 1.5 have been added specifically for EIA-applicable rail and road projects, which distinguished the Environmental Management Plan (EMP) from the general construction SHE management practices. Two important new concepts — the Environmental Impact Workshop and Environmental Impact Register were introduced. An Environmental Impact Workshop is to be conducted by the project manager 'within the first 2 months upon the signing of contract' to update and discuss the scope of environmental works to be implemented in response to the EIA's identified impact and mitigation works. The Environmental Impact Register is a formal documentation process for the mitigation measures to be prepared by the contractor and submitted on a regular basis.

LTA's Project SHE Management Framework

Overall, the organisation and managerial aspects of environmental safeguards during the road and rail works fall under the overall managerial framework of the Construction Safety, Health and Environment (SHE) Management System. The very basic SHE management staff to be employed typically include:

- Environmental Control Officer (ECO) registered with the NEA;
- Workplace Safety and Health Officer (WSHO) registered with the Ministry of Manpower (MOM)

In addition to the general SHE personnel above, LTA's SHE management framework requires an 'environmental team' solely for the purpose of environmental control which will work directly under the ECO. The full-time SHE staff and environmental team required for a rail or road construction project are determined based on the contract value, with a minimum of WSHO, ECO and two dedicated environmental workers for projects with a contract value above S$1 million (approximately USD $724,000 with exchange rate of 1 USD = 1.38 SGD).

The LTA also requires external qualified professionals to be engaged by the contractor to undertake more professional mitigation and monitoring requirements for specific environmental aspects of the EMP, such as a certified Acoustic Consultant for noise control, and a Qualified Erosion Control Professional (QECP) for surface discharge control. On the other hand, in LTA's framework there is no requirement for a professional with the similar role of the Independent Environmental Checker (IEC) under Hong Kong's laws, to provide general oversight on the EMP's implementation. Based on the views of LTA personnel interviewed by the first writer, the EIA consultant is typically not held responsible after the EIA stage has been completed, and the LTA's Safety Division is overall in charge of the oversight of the Environmental Management Plan's implementation by the contractor's ECO and environmental team.

In general, the typical organisational framework of the EIA follow-up for LTA's rail and road projects can be represented in the organisational chart in Fig. 9.5. Although such self-commissioned EIAs may not be able to influence planning considerations, they can effectively determine the actual implementation of EMP in the construction phase once the EIA

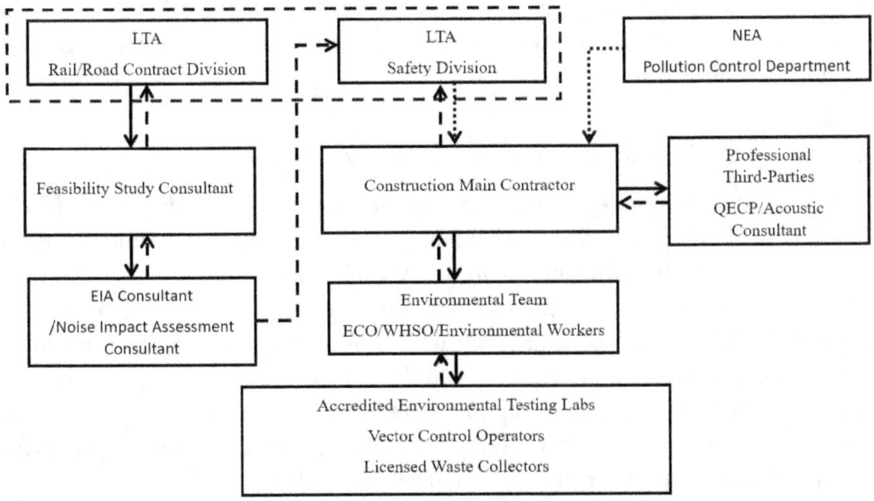

Fig. 9.5. Organisation chart of EIA and EMP for LTA rail & road projects in Singapore. Dashed line arrows denoting reporting, dotted-line arrows denoting monitoring/auditing service provision, solid-line arrows denoting service engagement.

is translated into LTA's contract documents. This system relies mainly on the effective oversight by LTA's internal environmental engineers during the construction phase of the project. In the author's opinion, the most significant improvement of LTA's EIA-EMP framework is indeed the clear requirements specified for professional third-parties to be engaged both in the planning and operation phases of the EMP.

Technical Scope of LTA's EMP

Singapore's LTA has gone further in the SHE management of rail and road works and has published its own series of guidebooks for 'Best Environmental Practices' on Earth Control Measures (ECM), construction waste management, noise control, vector control and water resource management.

In the latest SHE general specifications published in April 2015, the scope of environmental management has been expanded for EIA-applicable construction projects, with its scope and requirements summarised in Table 9.4 below.

It is clear that the higher level of details in the three major aspects, namely vectors, noise and surface effluent discharge from Earth Control Measures (ECM), are largely due to their potential to cause public nuisance and complaints or to violate local environmental standards stringently enforced by NEA with routine inspections and fines. Whereas for other less-focused environmental aspects, the local environmental standards in the construction industry are still unclear, without quantifiable or enforceable standards and limits.

Outcome and Transparency of LTA's EMP

Based on the analysis of the previous sections, EIA-EMP under LTA's latest 2015 SHE framework will likely generate positive outcomes in more stringently regulated and monitored environmental aspects.

Also, LTA's SHE framework requires the main contractor of the project to keep track of and document public feedback/complaints, and take action to address them as a standard procedure. For major projects which exceed SGD \$20 million investment value, the main contractor is required to assign a Public Relations Officer (PRO) directly reporting to the project manager, to take charge of public relations components such as the issue of circulars/notices and Meet the Residents sessions in relation to noise management during the construction phase of the project.

Table 9.4. Mitigation and monitoring requirements of environmental aspects in LTA's EMP.

Environmental Aspects	Number of Sub-Clause	Detailed Mitigation Requirement	Monitoring Requirement
Air	2	Nil (air emission compliance; general housekeeping measures)	Nil
Water	8	Nil (general construction good practice and housekeeping measures)	Nil
Vector	20	Yes (e.g. use of Bacillus thuringiensis isrealensis Bti; zoning method)	Yes (e.g. monitor mosquito population using gravi-traps)
Waste	11	Yes (e.g. bins with air-tight covers of not less than 85 L for organic waste storage)	Nil
Noise	52	Yes (e.g. allowable operating time for piling works; noise barriers or acoustic noise enclosures);	Yes (e.g. baseline noise survey; noise simulation; noise measurement at affected buildings; real-time noise monitoring with alerts)
Earth Control Measures	17	Yes (e.g. surface erosion control measures; sediment control measures; water treatment system	Yes (e.g. monthly ECM inspection; real-time TSS meter, CCTV system and alerts at ECM discharge and downstream drain)

Evaluation and Recommendations

Evaluation Summary of China and Singapore's Systems

A summary of the findings in the above sections, the pros and cons in the EIA follow-up systems of China and Singapore are tabulated below, in Table 9.5.

Table 9.5. Summary of comparison of EIA follow-up systems and identification of key weaknesses.

Evaluation Criteria	China EIA and CEPA	Singapore PCS	Singapore LTA's EIA and EMP
Linkage	Linkage of EIA and implementation phase through review of EIA recommendations and approval conditions in the CEPA in project commissioning. Lack of mandated EIA follow-up monitoring and audit procedures in the construction phase.	Good linkage between PCS and operation- phase environmental monitoring and testing ensured through BP submission and NEA inspection prior to commissioning. PCS follow-up does not cover construction phase of the industrial plant.	Good linkage between EIA and EMP ensured through Contract Specifications of the Construction Project. EMP covers construction and operation phase of rail and road projects.
Organisation	EIA directly commissioned by the project owner or proponent. No requirements on the minimum dedicated environmental staff to be engaged by the contractor. Environmental monitoring and testing conducted by the subsidiary unit of local environmental agency could lead to conflicts of interest and lack of objectivity of results.	Limited scope compliance assessment usually commissioned by the building design or construction contractor. No requirements on the minimum dedicated environmental staff to be engaged by the contractor. Accredited environmental monitoring and testing bodies to conduct EMP testing for contractor/developer.	EIA commissioned normally as part of Engineering Feasibility Study. Clear requirement of minimum full-time environmental staff during the construction phase. Lack of independent consultant to oversee the implementation of EMP. Clear requirement of professional third-parties to be engaged by the contractor on the planning and operation levels.

Clarity	Clear in overall legislative framework of environmental assessment and management. Clearly defined format and technical contents of CEPA upon commissioning. Lack of standard procedures in construction phase and operation phase audits/monitoring. Clear in technical scope of CEPA environmental monitoring/testing and performance criteria at both source and receptors.	Legislative requirements for environmental assessment are scattered under the jurisdiction of different agencies. General environmental monitoring requirements for private industries are in place in the PCS Guideline. Unclear environmental monitoring standards, techniques and performance criteria for construction project apart from high priority aspect, e.g. ECM or trade effluent discharge monitoring.	Environmental assessment and management provisions, including informal EIA are required by LTA's internal SHE framework. Comprehensive mitigation and monitoring requirements applicable to infrastructure development construction projects, such as noise/vibration, vector and silty water/ECM discharge aspects. Unclear environmental monitoring standards, techniques and performance criteria for less-focused areas such as air quality, land contamination and waste management.
Transparency	Transparency and public disclosure is limited to EIA Report and final approved CEPA Reports but significantly lacking in the follow-up process	Very low transparency and no public disclosure requirements for most private projects. NEA publishes regularly confirmed results of environmental offence investigations following public complaint and prosecution.	Low transparency and no public disclosure requirements for most projects. Contractor are required to respond to complaints, and for large scale projects >20 million, to engage stakeholders actively.

Recommendations for China's CEPA System

In the Chinese system, there is a need for the legislative framework of EIAs to be extended to include follow-up efforts during project implementation rather than only checking the completion results. An 'Environmental Protection and Monitoring Act' adding to the existing EIA framework would be greatly beneficial to mandate routine follow-up procedures during the construction and operation phase. This law should:

- Clearly stipulate that in the context of the EIA document, the environmental management and monitoring practice shall be an additional requirement;
- Be read in conjunction with the existing guidelines and standards on environmental monitoring to guide the planning and design of the environmental management and monitoring plan, but focusing more on routine procedures during construction and operation phases;
- Provide standard format and content requirements on regular reports with minimum coverage on important environmental aspects;
- Require the necessary organisation framework to implement routine environmental management and monitoring based on the nature of the project, its scale and complexity.

In addition, the management framework of the EIA follow-up needs significant improvements to encourage higher competencies of the industry, to strengthen the linkage between EIA and follow-up, and to reduce conflicts of interest. In particular, there is a need:

- To develop qualifications for dedicated environmental staff to be engaged by the contractor to supervise the implementation of routine site environmental management works;
- To reduce or remove the role of the local environmental agency in providing monitoring services to the contractor and encourage healthy market competition for qualified environmental monitoring and testing service providers;

Recommendations for Improvements to Singapore's System

Singapore has strong and effective administrative agencies, but improvements can be sought in designing a more systematic framework for project-based environmental management and assurance.

First, it is beneficial to design a national technical guidance document, for instance, a 'Code of Practice for Project Environmental Management Plan', that is universally applicable to development projects of any nature, integrating the existing system of audit, inspections and imposition of penalties by NEA. This document should fulfil the following functions:

- Include aspects such as air quality, land and groundwater contamination with minimum site inspection/auditing requirements,
- Provide a basis for designing monitoring plans with technical requirements such as scoping, parameters, locations, duration/frequencies and monitoring/testing techniques;
- Recommend performance criteria such as warning limits and critical limits, as well as corresponding response actions;
- Allow flexibility to be further substantiated with sector-specific scope of focus, technical standards and performance criteria imposed by other supervisory agencies.

In addition, for projects of larger scale or sensitivity, it is beneficial to standardise the requirement of independent third-party verification similar to the IEC in Hong Kong system. To ensure objectivity of this third-party verification, the certified third-party verifier should not be allowed to undertake any actual monitoring works apart from necessary site inspections.

Summing Up

The review of EIA follow-up efforts in China and Singapore revealed two vastly contrasting project-based environmental management and assessment systems. China has top-down implementation of national laws, statutes, technical guidelines and national standards but poor enforcement. To tackle this issue, a robust management framework needs to be in place ensuring sufficient resources are in place, such as a dedicated environmental working team. Most importantly, independent and objective professional third-party involvement in auditing and certifying good environmental performance is critical, which is apparently lacking in the current system.

In Singapore, no EIA law is in place and environmental assessment and management is broken up into different processes by different relevant agencies. The role and responsibility of the environmental agency, (NEA) for individual industrial projects becomes limited to ascertaining and enforcing compliance. This system has clear limitations as the environmental effects of certain plans or projects are above NEA's sole jurisdiction. Sector-specific environmental management frameworks with *ad-hoc* EIA and EMP requirements such as LTA's EHS framework for rail and road projects are certainly a good response to this. However, it will be beneficial to formalise this to ensure a holistic framework and comprehensive guidance to enable good environmental performance to be incorporated into projects more autonomously and systematically.

References

Arts, J., Caldwell, P., and Morrison-Saunders, A. (2001). Environmental impact assessment follow-up: good practice and future directions — findings from a workshop at the IAIA'00 conference. *Impact Assessment and Project Appraisal*, **19**(3), 175–185.

Au, E., and Hui, S. (2004). Learning by Doing: EIA follow-up in Hong Kong. In: A. Morrison-Saunders, A., and Arts, J. (eds.) *Assessing Impact: Handbook of EIA and SEA Follow-up*. London: Earthscan.

Hollands, R., and Palframan, L. (2014). EIA and EMS integration: not wasting the opportunity. *Impact Assessment and Project Appraisal*, 1–12. Online publication date: 22-Jan-2014.

Land and Transport Authority. (2014). Environmental Impact Assessment in LTA. *LTA Safety News Issue 30*. [Online] Available at: http://www.lta.gov.sg/content/dam/ltaweb/corp/PublicationsResearch/files/ReportNewsletter/SafetyNews/30_Safety_News.pdf [Accessed 12 January 2016].

Liu, F. (2011). Difficult Problems Analysis for Environmental Check and Acceptance Monitoring of Construction Project Completion. *The Administration and Technique of Environmental Monitoring*, **23**(1), 64–66.

Lye, L. H. (2008). A Fine City in a Garden — Environmental Law, Governance and Management in Singapore. *Singapore Journal of Legal Studies*, 68–117.

Marshall, R. (2004). Can industry benefit from participation in EIA follow-up? — the Scottish Power experience. In: Morrison-Saunders, A. and Arts, J. (eds.) *Assessing Impact: Handbook of EIA and SEA Follow-up*, London: Earthscan.

Morrison-Saunders and A., Arts, J. (2004). Introduction to EIA follow-up. In: *Assessing Impact: Handbook of EIA and SEA Follow-up*, James and James, 1–21. London: Earthscan.

National Environment Agency. (2011). Environmental Protection Division Annual Report 2011. Retrieved from http://www.nea.gov.sg/docs/default-source/annual-reports/epd-2011.pdf?sfvrsn=2.

Sadler, B. (2004). On evaluating the success of EIA and SEA. In: A. Morrison-Saunders, A., and Arts, J. (eds.) *Assessing Impact: Handbook of EIA and SEA Follow-up*, London: Earthscan.

Sanchez, L. E., and Hacking, T. (2002). An approach to linking environmental impact assessment and environmental management systems. *Impact Assess Project Appraisal*, **20**(1), 25–38, DOI: 10.3152/147154602781766843.

Singapore Standards Council. (2013). Code of Practice on Pollution Control Incorporating Corrigendum No.1. Singapore Standard SS593: 2013. ICS 13.020.40.

Sinkule, B. J., and Ortolano, L. (1995) *Implementing Environmental Policy in China.* Westport, Conn: Praeger.

United Nations Environmental Program. (1987). Goals and Principles of Environmental Impact Assessment. [Online] Available at: http://www.unep.org/regionalseas/publications/reports/rsrs/pdfs/rsrs122.pdf [Accessed 15 January 2016].

World Bank Group. (1991). Operational Policies 4.01 on Environmental Assessment. [Online] Available at: http://siteresources.worldbank.org/INTFORESTS/Resources/OP401.pdf [Accessed 2 December 2015].

10

LARGER POPULATION FOR SINGAPORE: A UTILITIES PERSPECTIVE

ANG CHUN-WEI ALAN and VICTOR R SAVAGE

In response to the public controversy over the 2013 Population White Paper, this study attempts to evaluate Singapore's electricity, water and waste management infrastructure capability to support a population of 6.9 million by 2030, and 10 million by 2100. An IPAT calculation identified increasing consumption due to increasing affluence of Singaporeans as the primary contributor to Singapore's environmental impact, a situation that can be aggravated by a larger population. An analysis was done based on extrapolating existing utilities consumption data.

Singapore's electricity infrastructure could potentially sustain a population of 10 million; however, the nation's dependence on foreign fuel supply remains a challenge. The Singaporean government is able to diversify Singapore's fuel source, and reduce price volatility and the nation's carbon footprint by sourcing liquefied natural gas (LNG) from global sources, and through continuous investment in clean renewable energy. The closing of the water loop is a game changer; Singapore could attain self-sufficiency in water in the near future through the intensification of desalination projects, and increase in NEWater production. As a populace gains increasing affluence, the amount of waste generated increases. While incineration was able to curb Singapore's landfill needs since 1979, incineration ash has begun exceeding non-incinerable waste as the primary component of landfill waste and the gap continues to widen. There is a crucial need for land-scarce Singapore to reduce not just waste bound for landfills but to also reduce waste generation and increase the recycling rate, to reduce waste bound for incineration.

The advancement of technologies and proliferation of global trade allowed Singapore to grow beyond its natural carrying capacity and will continue to allow it to do so, albeit at the cost of increasing its ecological footprint and resource vulnerability. Increasing geopolitical instabilities and climate change impacts continue to be threats on natural resources and commodity supplies. As Singapore's population is set to grow both in number and level of affluence, there is a key need to address the compound effect on Singapore's utilities beyond supply-side intervention and rhe internalising of resource needs; a paradigm shift in consumer behaviour is required.

Introduction: A Growing Population

Singapore is a city-state with a land area of approximately 716 km^2 (Singstat, 2015). Despite facing physical and natural resources constraints, the former British Crown Colony and Malaysian Federation State gained full independence in 1965, and progressed from a low-income economy country to a high-income country status in 22 years, with a Gross Domestic Product (GDP) per capita of USD$52,052 (WorldBank, 2015). Today, Singapore is ranked the World's 7th and Asia's 2nd most liveable city (Nicole, 2015). Like most industrialised nations, Singapore faces the demographic–economic paradox; an inverse correlation found between a country's wealth and fertility rates, where a higher degree of development and GDP per capita led to fewer child births.

Despite falling and non-replacement fertility rates, Singapore has been registering continuous population growth due to its liberal immigration policy, and the population of non-residents and permanent residents has grown rapidly since 1990 (Yeoh and Lin, 2012). Singapore's current population stands at 5.5 million, consisting of citizens, permanent residents and non-residents with a total fertility rate of 1.25, and population density of 7,697 persons per km^2 (Singstat, 2015), making Singapore the world's 3rd most densely populated country (WorldBank, 2015). In February 2013, the Singapore Government released the Population White Paper (PWP) detailing the need and plan to build a sustainable population for a dynamic Singapore. Three key pillars were identified to realise the plan: strong families, an adequate population to fuel the workforce and the development and maintainance of a high quality living environment. A controversial point mentioned in the Population White Paper is the intention to gradually increase Singapore's population to between 6.5 and 6.9 million by the year 2030 to meet Singapore's future economic and social needs. This will increase Singapore's population density to 9,718 persons per km^2, a 22% increase from today's population figures. The higher population density in land and resource constrained Singapore would create higher societal, resource and infrastructure stress. While the Population White Paper has stoked Neo-Malthusian sentiments among Singaporeans, their primary concerns are from a societal aspect, rather than from an environmental aspect.

Singapore's Carrying Capacity and Ecological Footprint

The World Wild Fund (2015) defines a country's ecological footprint as the sum of all productive lands required to provide for all the resources, infrastructure needs and waste assimilation capacity a country needs. According to Global Footprint Network, an NGO think tank, Singapore's per capita ecological footprint should stand at 1.7 global hectares (gha) in order for Earth to sustain our long-term resource demands. As of 2011,

Singapore's ecological footprint stands at 5.9 gha per capita, the 6th highest ecological footprint per capita in the world. (GFN, 2015) That means Singaporeans are drawing nearly 3.5 folds more resources than Earth can sustain. Advancement in technologies and the proliferation of international trade allows Singapore to support a much higher population, an example of Cornucopian views.

To better understand the contributing factors of our ecological footprint, we can use the I=PAT formula to identify the contribution weightage of three environmental impact factors; population, consumption and carbon emission.

IPAT Equation

Formulated by Commoner, Ehrlich and Holdren in the 1970s (Ehrlich and Holdren, 1971; Commoner, 1972), the I=PAT equation seeks to describe how growing population, affluence and technology contribute towards environmental impacts in a simplified manner; first as a model to quantify contribution to unsustainability, and subsequently as a model to assess the most viable path to sustainability.

The P variable refers to the population of an area. The A variable refers to the affluence level of the population. The T variant refers to the level of technology as an indication of how efficient and environmentally-friendly technologies are in the generation of goods, while the I resultant refers to the cumulative environmental impact of the variables; such as consumption, pollution and waste generation. For the below calculation (Table 10.1), we use GDP per capita to represent the A variable and CO_2 emitted per capita to represent the P variable.

Table 10.1. Population, GDP and CO_2 emission of Singapore.

Year	Population	GDP (USD/Capita)	CO_2 (Metric Ton/Capita)
1960	1,646,400	427.88	0.85
1965	1,886,900	516.29	1.34
1970	2,074,507	925.28	8.78
1975	2,262,600	2,558.66	10.85
1980	2,413,945	5,003.89	13
1985	2,735,957	6,781.82	12.2
1990	3,047,132	12,766.19	15.4
1995	3,524,506	24,937.31	13.4
2000	4,027,887	23,793.04	12.2
2005	4,265,762	29,869.63	7.1
2010	5,076,732	46,569.69	2.66

Table 10.2. Singapore IPAT analysis.

Year	Population	GDP	CO_2	Impact (Number of fold compared to 1960)	Impact (Number of fold compared to Previous Year)
1960	1.00	1.00	1.00	0.00	0.00
1965	1.15	1.21	1.58	1.18	1.18
1970	1.26	2.16	10.33	27.15	11.91
1975	1.37	5.98	12.76	103.90	2.73
1980	1.47	11.69	15.29	261.24	1.50
1985	1.66	15.85	14.35	377.04	0.44
1990	1.85	29.84	18.12	999.45	1.65
1995	2.14	58.28	15.76	1965.87	0.97
2000	2.45	55.61	14.35	1951.59	−0.01
2005	2.59	69.81	8.35	1509.81	−0.23
2010	3.08	108.84	3.13	1049.25	−0.3

Using 1960 as a baseline, we can calculate the environmental impact of development over the years in Singapore, as shown in Table 10.2.

We can see a drastic increase of CO_2 production between 1960 and 1970 due to the industrialisation programmes that seek to stimulate massive growth in Singapore's labour-intensive low-tech manufacturing of goods such as furniture, textiles and toys, in order to reduce reliance on entrepot trade and as a solution to overcome high unemployment. Jurong Town Corporation (JTC) was established in 1968 to spearhead the industrial development of Singapore (HistorySG, 2010).

In 1961, Shell established Singapore's first oil refinery, followed by British Petroleum in 1964, Mobil in 1966 and Esso in 1968 (HistorySG, 2014). In USA, refineries are the 3rd largest source of greenhouse gases after the power plant, and oil and natural gas sectors (EPA, 2014), and this could explain the drastic spike in CO_2 emissions per capita and GDP per capita figures in 1970. With industrialisation on track, coupled with increasing pollution, the Singapore government formed the Anti-Pollution unit in 1970; and in 1972, the Clean Air Act was established to limit air pollutants emitted by factories (Oon and Kartini, 2014). In the 1980s, the government pushed for a focus on high-tech and value-added manufacturing of goods such as computer parts and silicon wafers (EDB, 2015). This could explain the second GDP spike between 1980 and 1995. From 1990 onwards, the amount of CO_2 per capita begins to fall drastically despite increasing affluence and population, largely due to the various environmental laws being enacted. In 1994, NEA implemented the code of practice on pollution control that stipulated the control requirement of pollutants, hazardous substances and toxic industrial wastes. Subsequently, the Environmental Pollution Control Act and Environmental Protection and Management Act were enacted in 1999, and the Energy Conservation Act in 2012 (NEA, 2015). Singapore also switched from fuel oil to the less carbon intensive natural gas as the primary fuel for electricity generation, from 26% of the market share in 2001 to 95% today (EMA, 2015b). The IPAT equation suggests that the greatest environmental impact Singapore faces now is no longer pollution emission, but increasing consumption and waste generated due to the increasing affluence of Singaporeans, a situation that can be aggravated by an increasing population.

Resource Management in Singapore

By adopting the Boserupian stand, the Singaporean government believes that continuous population growth is the key to address its shrinking and aging population, and ensuring continuous economic vitality.

As a small open economy that imports virtually all its energy, Singapore's immediate concern is the volatility of energy prices and supplies that could be affected by regional and international events. The inflation-adjusted crude oil price amounted to US$15.38 per barrel in 1946, before rising to US$51.62 in the wake of 1970s energy crisis, and peaking at US$116.05 in 1980. The 2000s energy crisis that developed in the wake of Middle-Eastern geopolitical tension (Hamilton, 2010) saw oil prices peaking at US$144.38 per barrel. Following a weak global economy and demand, oil prices fell to US$36.69 per barrel in December 2015 (Macrotrends, 2015). This is exceptionally critical as elecritical supply forms the foundation of Singapore's public and utilities infrastructure, and population growth would have a compound demand on Singapore's electricial grid. Singapore has also been diversifying its energy sources and investing in self-generated alternative energy, such as solar power, to reduce the risk of foreign supply interruption (Tan, 2015).

The emphasis on energy and water efficiency have been the forefront of many government policies to reduce demand on these resources. However, the Jevons paradox shows that while technological progress increases the efficient consumption of a resource and produces short-term saving, it could still bring about an increase in consumption in the long-term (Alcott, 2005). Saunders (1992) showed that the Khazzoom-Brookes postulate, a re-look at the Jevons paradox from an energy consumption point of view, is applicable in a range of scenarios under neo-classical growth theory. He brought forward two points: (1) that increased energy efficiency reduces the cost of consumption, thus encouraging increased usage, and (2) that increased energy efficiency leads to overall economic growth, resulting in higher energy demand. The law of deminishing returns could also mean that the allocation of more resources to research and implement every subsequent unit of efficiency improvement will be required. Hence, an increase in resource generation and consumption efficiency should not be seen as the sole solutions to cope with increasing resource demand.

Smeesters *et al.* (2003) identified a Knowledge Gap, Situational Constraints and Unfairness Beliefs as the hindering stones to recycling. The knowledge gap arises due to lack of procedural knowledge (of, for example, the recycling process) and the lack of declarative knowledge (where the individual is unable to obtain information on the end-goals of recycling and how they can make a difference). Situational constraints, such as lack of space, the effort required to sort waste, low frequency or odd schedule of waste collection, proximity of collection points, etc., could interfere with habit development and persistency. Unfairness beliefs arise from the lack of standardisation between different districts and the lack of enforcement on non-compliant individuals. Smeesters *et al.* believe that it is crucial to make recycling knowledge easier to understand rather than overload users with the details. Dawes (1980) states that the consumer will always adopt a defective behaviour that will minimise personal costs out of self-interest, unless the majority cooperate; only then will a cooperative behaviour emerge.

Growing affluence has been proven to negatively impact the environment through a lifestyle of abundance. People began to generate more volume and variation of waste due to the rapid acquisition of lifestyle goods. The compound effect would eventually lead to an increase in demand for electricity and water, as well as an increase in the country's waste management capacity. Despite Singaporean's positive attitude towards environmental conservation (NCCS, 2012), its domestic and overall recycling rate are less than ideal (Boh, 2015). Philomin and Siau (2014) believe that while Singaporeans engage in some form of recycling and conservation, most have yet to understand their role in the bigger picture. Dr Neo, from the National University of Singapore (NUS), believes that Singapore should focus on changing specific behaviours, such as reducing food waste or improving recycling, rather than on consuming less. Currently, there is no legislation in Singapore that mandates waste segregration, and of the waste that are sent for recycling, contaminants and unsuitable materials can make up 20% of the volume (Cheam *et al.*, 2010). A paradiagm shift in how people consume resources and view ecosystem services is fundmentally required.

Powering Our Society

The Energy Market Authority (EMA) has regulated the National Electricity Market of Singapore since the liberalisation of Singapore's electricity and gas markets in 2001. Commercial and industrial consumers with electricity demands of 2 MW[1] and above have the option to choose their electricity retailers. EMA is working towards Full Retail Competition, where all consumers are able to choose between remaining on regulated tariff or to purchase electricity from an electricity retailer at market prices. Electricity tariffs comprise of energy cost, network cost, market support services fees and market administration and power system operation fees. (EMA, 2015) This study will focus on the electricity demand of Singapore. Six primary electricity producers meet Singapore's electricity demand with a combined capacity of 12,191 KW. Private companies and waste-to-energy incineration plants also contribute excess electricity back into the national grid. The average local electricity demand in 2014 was 5300 MW, with a utilisation rate of 43.47% of the electricity producers' maximum capacity (EMA, 2015a). In 2014, Singapore consumed 46,403 GWh[2] of electricity and 16,508 GWh of non-power generating natural gas, of which, the non-domestic sector consumed 85% and 96% respectively (EMA, 2015a).

Current Energy Demand

Between 2005 and 2014, electricity consumption has risen by an average of 2.69% annually, from 35,489 GWh to 46,403 GWh, while per capita electricity consumption has risen by an average of 5.5% annually during the same period, from 6,096 GWh to 6,936 GWh. Domestic per capita electricity consumption makes up 14.95% of Singapore's total energy demand, a decline from 17.18% in 2005, signifying that electricity demand in the non-domestic sector has outpaced domestic demand. Singapore's

[1]Megawatt (MW), 1 Megawatt Hour (MWH) is equal to 1 MW of electricity consume for an hour. 1.628 MWh is equvialent to 1 barrel of oil.
[2]Gigawatt (GW). 1.628 GWh of energy is equivalent to 1,000 barrels of oil.

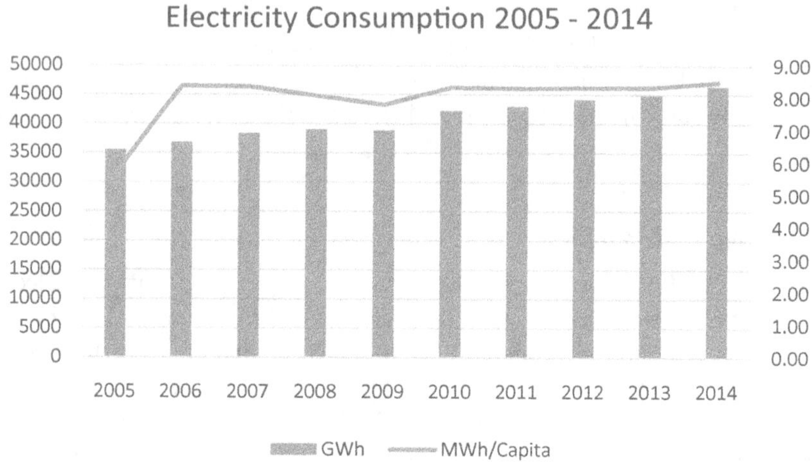

Fig. 10.1. Singapore Electricity Consumption 2005–2014 (EMA, 2015).

direct natural gas consumption has also risen by approximately 49%, from 39,788 TJ[3] in 2009 to 59,427 TJ in 2014, due to increased industrial sector consumption with per capita consumption increasing from 7.98 GJ[4] per capita in 2009 to 10.86 GJ per capita in 2014 (EMA, 2015a). The rises in non-domestic electricity and direct natural gas demands are driven by expanding industrial, utilities and infrastructural development. While the growing population contributes to the rise of electricity consumption, the affluence level of society also contributes to it. Singapore consumed 8,690 kWh of electricity per capita in 2012 — more than the amount consumed per capita in countries like Japan (7,752 KWh), Hong Kong (6,026 KWh) and the UK (5,452 KWh) (WorldBank, 2015). This signifies a potential for the implementation of stronger energy conservation policies and public education.

Meeting Future Energy Demand

From the extrapolated projection, assuming all variables are constant, Singapore would need an additional 26.15% and 82.8% of energy resources to cater to the demands of a 6.9 million and 10 million population, respectively. This demand translates to a projected power generating

[3]Terajoule (TJ). 1 TJ of energy is equivalent to approximately 170 barrels of oil.
[4]Gigajoule (GJ) 5.86 TJ of energy is equivalent to 1 barrel of oil.

6.9 Million and 10 Million Population Projection			
Year	**2014**	**2030**	**2100**
Consumption (GwH)	46,403	58,512	84,800
Population	5,469,724	6,900,000	10,000,000
Consumption (MwH/Capita)	8.48	8.48	8.48
Domestic Household Consumption (GWH)	6,935.8	8,747.54	12,677.6
Natural Gas Consumption (TJ)	59,427.4	74,967.05	108,647.9
Growth % Compared to 2014	-	26.15	82.8

Fig. 10.2. Singapore's Energy Demand Projection.

capacity of 6,679 KW and 9,680 KW, respectively. This means that Singapore's electricity generating capacity of 12,191 KW can sustain a 10 million population, though additional capacity is required to improve power plant availability factor. With the non-domestic sector utilising 85% of Singapore's electricity supply, the compounded effects of growing infrastructure and economic activities due to a larger population could put more pressure on the infrastructure than the growing population itself. The diversification of Singapore's energy sources and a reduction in energy consumption is crucial to lowering Singapore's exposure to supply risk.

Diversifying Energy Sources

Singapore should intensify the adoption of alternative clean energy, such as solar power, to reduce Singapore's reliance on fossil fuel. Plans must be targeted at catering for unexpected peak demands caused by increasing dry weather due to climate change. For example, when water levels at Johor's Linggiu reservoir, Singapore's primary source of imported water, fell to an unprecedented low level, production at NEWater and the desalination plants were ramped up to ensure adequate water supply is available (Kotwani, 2015a). Many Singaporean turn to air-conditioning for relief when affected by transboundary haze and increasingly warmer temperatures (Lai, 2015). These actions will create larger electricity demands as Earth's climate becomes more volatile. As natural gas is Singapore's primary source of fuel for electricity generation, it is crucial for Singapore to diversify its energy sources to mitigate the risks of competitive demand, politics and disruption. In 2013, the International Energy Agency forecasted that ASEAN's energy demand is set to increase by 80% by 2035 — a rise equivalent to the total energy required to power Japan (IEA, 2013). Indonesia's energy regulator will start to import LNG by 2018 to meet growing domestic demand, though they will still honour their long-term

LNG (export) sales contract with Singapore (Wulandari and Mellor, 2013). In 2007, the Indonesian government threatened to terminate the gas export contract if Singapore failed to comply with certain conditions set out in the contract (Krismantari, 2007). The possible negative effects of Singapore lossing Indonesia as its LNG supplier was seen in two instances. In August 2002, disruption of gas supplies from Indonesia deprived 30% of Singapore of electricity for two hours. In June 2004, gas disruption from Indonesia caused businesses to suffer losses up to S$6 million and affected up to 300,000 households. It was estimated that 30% of Singapore was affected then as well (WildSingapore, 2006),

In 2014, Singapore opened its first LNG terminal with an initial capacity of 5 mtpa[5] and a total capacity of 9 mtpa by 2017. Development of a second LNG terminal is in the pipeline (EMA, 2014). The opening of the LNG terminal means that Singapore can import natural gas from global sources by sea, rather than be solely dependent on piped gas from neighbouring sources. Lee (2012) believes that the vast shale gas development, cost of sea freight, and technological advances, coupled with Singapore's strategic sea connectivity could result in making the importation of LNG from global sources a more cost-effective solution than importing via piped sources.

Energy Conservation

The Energy Conservation Act (ECA) is a mandatory energy management requirement enacted in 2013 to mandate large energy users to monitor and report energy consumption and greenhouse gas emissions, and submit energy efficiency improvement plans (NEA, 2013). The Energy Efficiency National Partnership was launched in 2010 to allow companies to access various government incentive schemes such as the Energy Efficiency Assistance Scheme and the Grant for Energy Efficient Technologies, Design for Efficiency Scheme tax rebates, and opportunities to network and gain knowledge with government agencies and industrial partners. As of June 2015, 207 companies have joined EENP (E2, 2015).

The Green Mark Scheme was launched in 2005 to rate buildings' energy and water efficiency, environmental protection, indoor environmental quality and innovative building performance features (BCA, 2015). As of 2008, all existing buildings with GFAs exceeding 2,000 m^2 undergoing major

[5]Metric tonnes per annum (MTPA).

retrofitting are required to comply with the requirement. With the non-domestic sectors utilising 85% of electricity in Singapore, these programmes play a crucial role in mitigating demand as Singapore's economy grows.

NEA launched the Mandatory Energy Labelling Scheme (MELS) and Minimum Energy Performance Standards (MEPS) in 2008 and 2011 respectively, to eliminate household appliances that do not meet the minimum specified energy efficiency levels and empower consumers with the information to make informed purchases through the comparison of energy efficiency performance and life-cycle cost, as indicated on the MELS labels (NEA, 2013b). Energy SAVE, a joint programme between Housing Development Board (HDB), which provides 82% of housing in Singapore, NEA and EMA, seeks to incorporate passive and active energy saving designs into HDB developments. Occupants are encouraged to use energy saving appliances to reduce electrical demand and expenses (HDB, 2015). As Singaporeans becomes more affluent, consumption in basic needs and luxury will rise. Energy intensive devices, such as air-conditioners have become increasingly common. The MELS and MEPS are tools that policy-makers can use to reduce household energy demand as the population grows.

Though Singapore has a tier system charge based on electricity tension levels, its electricity pricing framework is linear (MyPower, 2015). A conservation tax and non-regressive tier billing system can be implemented to discourage excessive electricity consumption, such as the 5-tier system implemented by California (Borenstein, 2008) and Hong Kong's 7-tier system (HKE, 2015).

Renewable and Sustainable Energy

Due to Singapore's geographical landscape, solar power is the only viable sustainable energy to meet Singapore's energy demands through large-scale deployment (NCCS, 2013). Government agencies have conducted test-bedding projects by installing solar panels on rooftops of public housing and water surfaces of reservoirs (CNA, 2015b). The potential is tremendous as HDB housing makes up 82% of Singapore's housing landscape and Singapore's reservoir system occupies over 3,707 hectares (5.2% of Singapore's surface area). With the declining cost of solar panels (Parkinson, 2015), the adoption of solar energy in commercial and industrial buildings can be stimulated with government incentives and schemes, reducing the load on the national grid. Solar panels have been installed on 636 buildings in Singapore with a combined grid-connected capacity of 25.5 MWac[6] (EMA, 2015a). However, urban shading and intermittency resulting from weather conditions as well as the need to ensure adequate energy storage remains a challenge (Luther and Reindl, 2013).

While Singapore is unable to utilise conventional hydroelectric power, there is still an opportunity to harness hydroelectric power through its existing water supply network. Lucid Energy had successfully installed a 1,100 MW micro hydroelectric turbine system on a stretch of water pipes in Portland, Oregon (Gewin, 2015). With Singapore's extensive water supply network and the creation of underground reservoirs, there is an opportunity for continuous clean energy to be produced so long as water is flowing through consumer's taps. The Pluvia system, a smaller micro-turbine system that utilises rainwater runoff from buildings to create energy, is currently being test in Mexico City (Coxworth, 2015).

Adopting New Technology

In 2010, the Singaporean government conducted a two-year pre-feasibility study that concluded that nuclear energy is not ready for deployment in Singapore (Neo, 2015). In 2014, a S$63 million Nuclear Safety Research and Education Programme was launched to train scientists and engineers in radiochemistry, radiobiology and assessment (Kotwani, 2015). A breeder reactor, compared to a conventional reactor, is self-sustaining; it produces more fuel than it spends and theoretically produce little to no waste (Karam, 2006). Through improvements in nuclear and safety technologies and public opinion, an underground or offshore nuclear plant could be feasible in the future to alleviate Singapore's fossil fuel supply risk and reduce the nation's negative environmental impact.

Singapore's Water Challenges

The Public Utilities Board (PUB) is the national water agency of Singapore. PUB's primary roles are the collection of rainwater, production and distribution of potable water, collecting and reclamation of wastewater. Through the Environment and Water Industry Programme Office, S$700 million has been spent annually since 2006, to improve Singapore's capability in water

[6] Megawatt Alternating Current (MWac).

management. Today, Singapore is home to over 200 water related companies and research centers. (EDB, 2015). PUB successfully closed the water loop in 2002 with the introduction of reverse osmosis technology to reclaim waste and seawater, turning them into potable water. Singapore's water demand is fulfilled through the Four National Taps; local catchment water, imported water, desalination and NEWater. Political tension with Malaysia since independence motivated the Singapore government to strive towards water self-sufficiency. PUB plans to meet 55% and 25% of Singapore's water demand through NEWater and desalination, respectively, by 2060 (PUB, 2015).

Under the sole active agreement with Malaysia, Singapore can draw up to 1.1 million m^3 of water daily from the Johore River till 2061 (Chew, 2011). Currently, NEWater and desalination plants are able to meet 55% of Singapore's total water demand. Singapore's seventeen reservoirs occupy 3,707 hectares with a catchment area spanning two-thirds of Singapore (PUB, 2015). Singapore's most urbanised reservoir, located in Marina Bay, is able to meet 10% of Singapore's water demand (Ghangaa, 2015). Purified potable water are stored in fourteen service reservoirs islandwide.

Current Water Demand

Singapore's total domestic potable water consumption has increased from 271.4 million m^3 in 2008 to 291.2 million m^3 in 2014, an approximately 7.3% increase over the 6 years. However, per capita consumption have fallen from 56.08 m^3 to 53.24 m^3, approximately 5.3%, during the same period (PUB, 2015).

Total non-domestic potable water consumption has increased by 12.5% from 191.2 million m^3 in 2008 to 215.1 million m^3 in 2014. Though per capita consumption has fallen from 39.51 m^3 to 39.33 m^3 in 2014 during the same period, a rebound is noted between 2011 and 2014 (PUB, 2015). This suggests a strong growth in non-domestic demand from 2009 onwards that exceeded PUB's demand-side conservation policies. With Singapore's economic-centric policy and the planned expansion of water intensive industries, non-domestic consumption is anticipated to rise further (Shilov, 2014; 2015). Population affluence levels also contribute to an increased consumption of non-domestic potable water due to leisure activities. NEWater consumption has increased by 77.4% from 66 million m^3 to 117.1 million m^3, while industrial water consumption has been

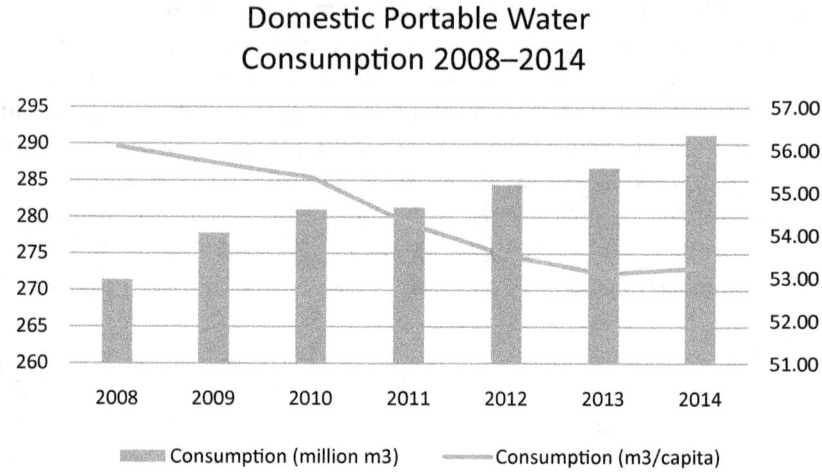

Fig. 10.3. Singapore Domestic Portable Water Consumption 2008–2014 (PUB, 2015).

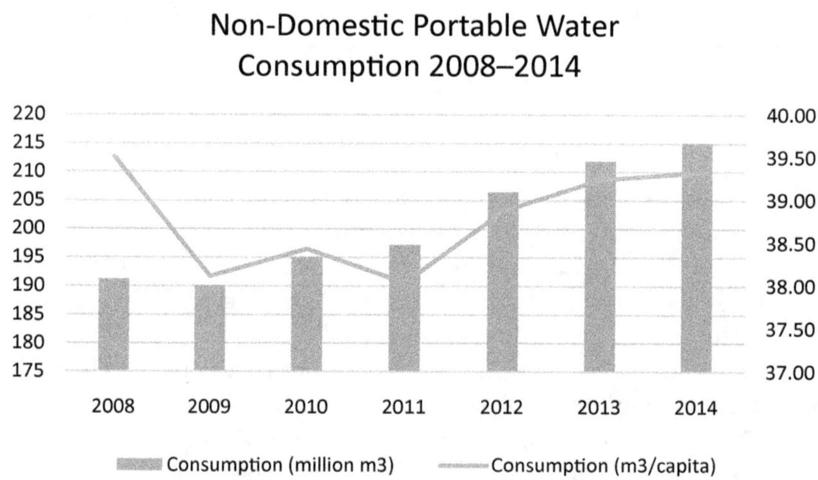

Fig. 10.4. Singapore Non-Domestic Portable Water Consumption 2008–2014 (PUB, 2015).

growing by 16.5% from 23.7 million m^3 to 27.6 million m^3. It is likely that PUB will eventually replace industrial water with NEWater to leverage on economies-of-scale production and its suitability for both human and industrial consumption.

Table 10.3. Singapore's Water Demand Projection.

6.9 Million and 10 Million Population Projection			
Year	2014	2030	2100
Total Water Consumption (million m3)	651	821.31	1190.3
Total Sewerage Treated (million m3)	585.21	738.23	1069.9
Population	5,469,724	6,900,000	10,000,000
Growth % Compared to 2014	—	26.1	82.8

Meeting Future Water Demand

From the extrapolated population projection, assuming all variables are constant, the total annual water consumption is expected to rise to 821.31 million m^3 and 1,190.3 million m^3 under populations of 6.9 million and 10 million, respectively. In 2013, PUB projected total water demand to double to 1,280 million m^3 by 2060 with non-domestic water consumption making up 70% of the demand. This suggests that the PUB is anticipating a stronger non-domestic rather than domestic driven demand.

By 2020, Singapore's daily NEWater and desalination production capacity will grow by 50.5% from today's 989,880 m^3 to 1,490,262 m^3. This annual supply of 543.9 million tonnes of water can meet 83.5% of Singapore's 2014 total water demand of 651 million tonnes. The Deep Tunnel Sewerage System (DTSS) was envisioned in the 1990s to renew and expand Singapore's sewerage system and play a crucial role in ensuring NEWater sustainability, reducing land use and minimising cross contamination risk between water catchments and used water systems (PUB, 2015). An extrapolated water demand projection to 2020 with a population of 6 million would suggest a total water demand of 714.18 million m^3, with 319.4 million m^3 coming from domestic demand. This would theoretically mean that before factoring in maintenance shutdowns and operational water loss, Singapore could produce 76% of its total water demand by 2020, excluding reservoirs and imported water. The annual shortfall of 171.18 million m^3 of water or 469,000 m^3 per day can be potentially fulfilled by the construction of another two plants with capacities similar to Tuaspring desalination plant or Changi NEWater plant. Based on current technology, Singapore would need to build another two and six desalination and NEWater plants to meet the projected needs of a population of 6.9 million and 10 million, respectively.

Adopting of Co-generational Technology

The introduction of a co-generation (concurrently desalinating and power generating) capability at Tuaspring presents a new frontier in utilities efficiency and industrial symbiosis. The combined-cycle power plant will power the facility and channel excess electricity into the national grid. Another example is the Water Reclaimation Plant integrated with a NEWater factory will be constructed in Tuas under DTSS Phase 2. An Integrated Waste Management Facility (IWMF) that could potentially manage up to 50% of Singapore's total waste treatment capacity will be co-located with Tuas WRP to capitalise on the joint used-water and solid waste treatment process and engineering synergies. Food waste and sewage sludge are processed together to produce biogas. The electricity generated from the IWMF can be transfered to Tuas WRP to reduce or eliminate Tuas WRP's electricity demand on the grid. Treated water from Tuas WRP can, in turn, be used for cooling and cleaning purposes at the IWMF. Nuclear desalination, the concept of a co-generational nuclear power and desalination plant, can be a possibility for Singapore to explore in the future. The greatest advantages of nuclear desalination is its ability to hedge against volatile fossil fuel prices affecting water processing costs (WNA, 2015). The International Atomic Energy Agency found that the doubling of oil or natural gas prices will lead to a 70% increase in water costs, while a doubling of the price of uranium would only lead to a 5% increase (Khamis, n.d.).

Conservation

Singapore's Domestic consumption per capita has fallen from 165 liters per day in 2003 to 146 liters per day in 2014. PUB aims to lower the figures to 140 liters by 2030 (PUB, 2015). If PUB is able to reduce domestic per capita consumption from 145.86 liters per day to 140 liters per day based on an estimated population of 10 million, up to 21.4 million m^3 of water can be saved per year. Currently, PUB has an array of programmes poised towards assisting both domestic and non-domestic customers to monitor and optimise their water consumption. In 2009, PUB launched the Water Efficiency Labeling Scheme (WELS), a mandatory water fittings registration programme that mandates a minimum permissble performance for water fittings. The Water Efficiency Management Plan (WEMP) is a mandatory programme launched in 2000 targeted at non-domestic consumers, to assist them in monitoring and improving their water usage. There are also voluntary programmes like Water Efficient Homes (WEH) and Water Efficient Building (WEB) that seek to help residents and managing agents reduce water consumption by encouraging the adoption of water saving habits. The water conservation tax was introduced in 1991 to discourage excessive water consumption. Water consumption for all building types and purposes are levied a tax of 30% of tariff, while domestic households that consume more than 40 m^3 of water per month are levied a tax of 45% of tariff (PUB, 2015). There are opportunities for PUB to mandate higher standards of water efficiency through these programmes, and these could be extended to minor but growing areas of public water areas, like water features and water play areas.

Singapore's water tariff is priced lower than the global average (Smets, n.d.). The water tariff is based on a flat-rate billing for non-domestic consumers and a 2-tier system for domestic consumers. A more aggressive conservation tax and progressive tariff can be implement to discourage excessive water consumption. A quota could be identified based on dwelling type and number of residents with a minimum consumption block for human subsistence not subjected to conservation tax. Subsequent consumption blocks should be subjected to exponential conservation tax and higher tariff to make the payment structure non-regressive. Progressive tariffs have been gaining acceptance in Europe to cope with rising water demand.

Waste Management

The National Environmental Agency's (NEA) mission is to ensure a clean and sustainable environment for Singapore with the long-term vision of making Singapore a Zero Waste Nation. Unlike energy and water needs, waste is not a zero-sum game. Besides catering for the capacity, there is a need to manage the end-products of the waste: incineration ash and non-incinerable waste. Domestic waste is collected on a daily basis by Public Waste Collectors (PWC) appointed on seven-year contacts, segregated over nine geographical zones (NEA, 2015). As the number of HDB households grew from 22,928 units to 968,856 units between 1960 and 2015 (Teo, 2015), the National Recycling Programme (NRP) was implemented in 2001, requiring PWCs to provide recycling bins and a recycling collection service to all households to facilitate recycling. There is no mandatory requirement for households to sort their refuse, and a flat waste collection fee is imposed. Unsorted, non-recyclable, or recyclable waste contaminated by wet waste in Singapore are incinerated at one of the four incineration plants. Non-incinerable waste and incineration ash are disposed of at the Semakau off-shore landfill. Between 1970 and 2014, the daily amount of solid waste generated in Singapore has grown 6.6 folds, from 1,260 tons to 8,338 tons per day.

Current Waste Demand

Singapore generated 4.66 million tons and 7.51 million tons of waste in 2000 and 2014 respectively; an increase of 61.15% (NEA, 2015). The increase is marked by increasing population and rising affluence levels. Between these 15 years, the proportion of waste destined for landfills and incineration have steadily fallen, and by 2004, recycling replaced incineration as the primary waste treatment process, representing steady progress in Singapore's recycling efforts.

Non-domestic waste makes up the bulk of waste generated in Singapore and the figures have been rising gradually from 3.29 million tons (or 5.43 tons per capita) in 2000 to 5.86 million tons (or 11.29 tons per capita) in 2014; an increase of 78.1% in raw numbers (an increase of 108% per capita)

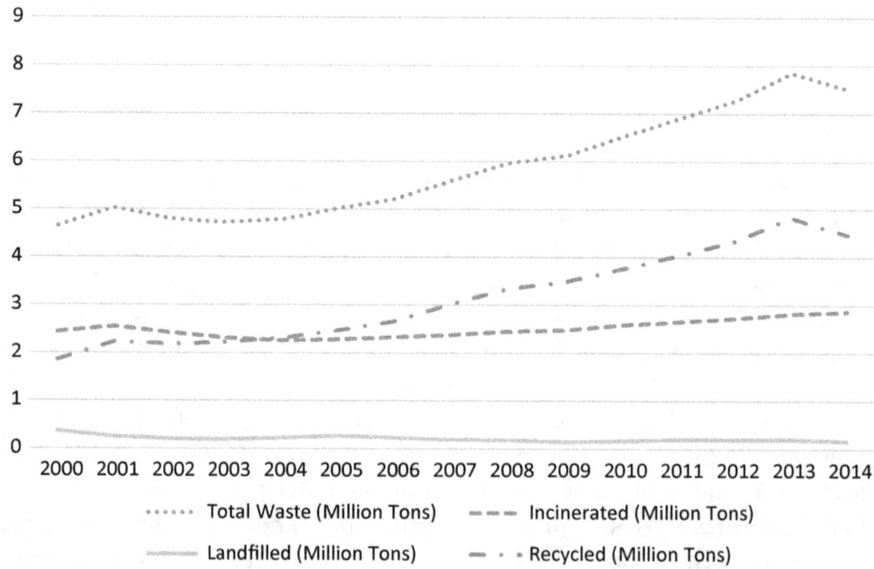

Fig. 10.5. Singapore's Total Waste Generated and Final Treatment Methods 2000–2014 (NEA, 2015).

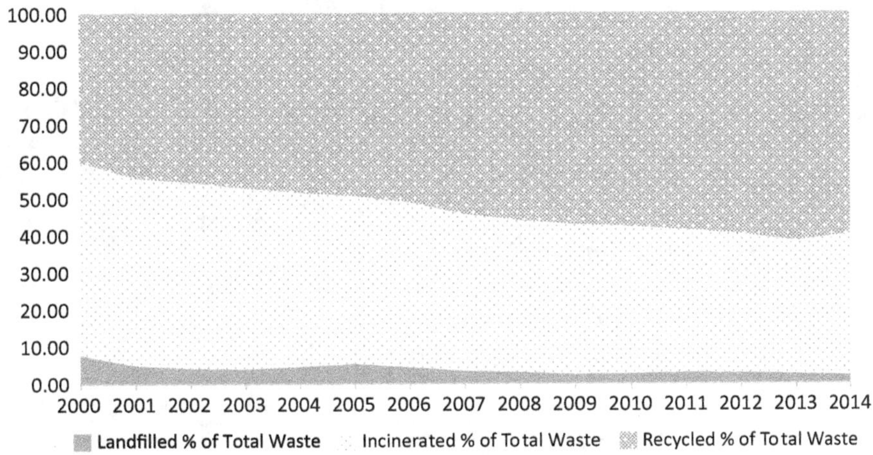

Fig. 10.6. Proportion of Final Waste Treatment Methods (landfill, incineration, recycling) 2000–2014 (NEA, 2015).

(NEA, 2015). This suggests that non-domestic waste growth has outpaced population growth, signifying industrial growth and growing affluence.

While total domestic waste has increased from 1.37 million tons in 2000 to 1.65 million tons in 2014, domestic waste per capita has fallen from 340.13 kg to 301.81 kg in the same period. (NEA, 2015). Domestic waste generation comparison with selected Asian countries shows that Singapore is among the leading domestic waste producers, even exceeding South Korea (Agamuthu and Fauziah, 2013). A domestic recycling rate of 20% still lags behind the overall recycling rate of 60% (Chua, 2014). Stronger NEA policies could encourage consumers to reduce household waste, mitigating the impact of a larger and more affluent population.

Landfill

Singapore's sole landfill is located offshore on Pulau Semakau with a capacity of 63 million m^3 and is expected to serve Singapore till 2045. (NEA, 2015c) Waste disposal demand in landfill have been declining at an annual average rate of 3.94%, from 0.36 million tons in 2000 to 0.17 million tons in 2014. (NEA, 2015) The landfilled per capita volume also declined from 89.38 kg to 31.08 kg in the same period. Historical data shows that incineration and landfill share an inverse relationship; for every spike in incineration demand comes a drop in landfill demand. This reduction meant that Singapore is able to sustain a higher population and waste output in

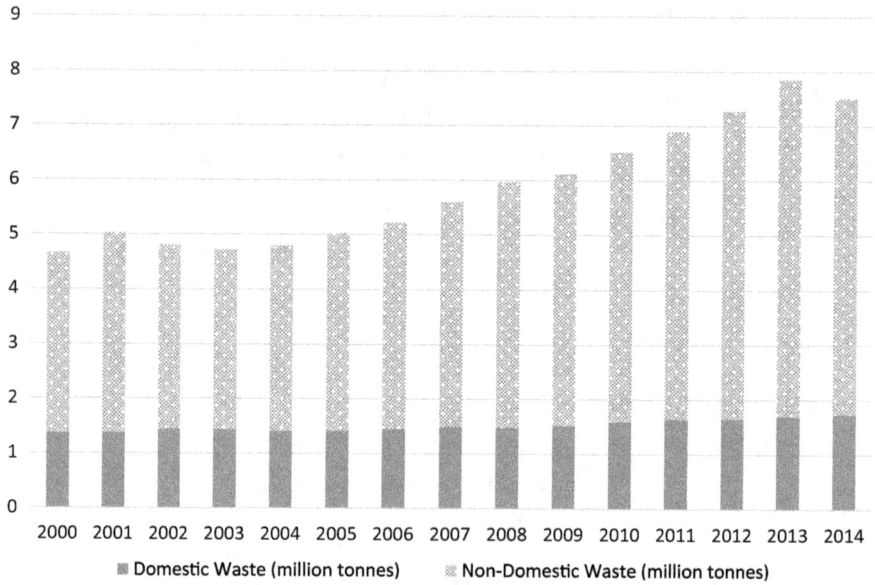

Fig. 10.7. Singapore's Non-Domestic vs. Domestic Waste Volume 2000–2014 (NEA, 2015).

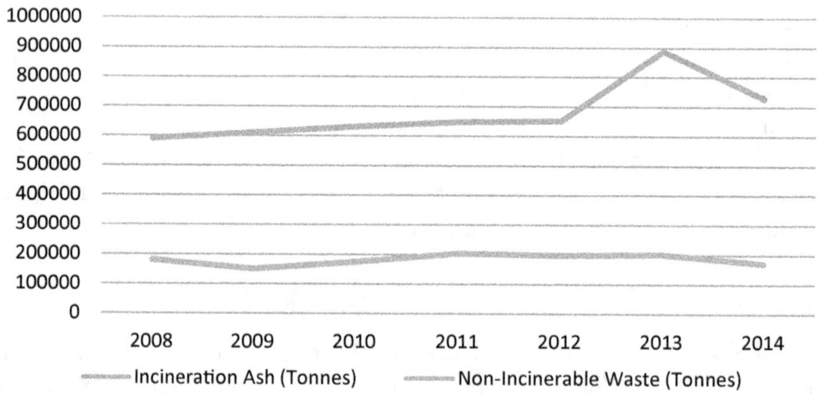

Fig. 10.8. Composition of Singapore's Landfill Waste 2008–2014 (NEA, 2015).

the short-run. Incineration ash that are disposed in landfill are not included in the NEA landfill figures (Fig. 10.8).[7]

[7]It is likely that the incineration ash figures were not included in the landfill waste figures to prevent "double accounting" of waste volume.

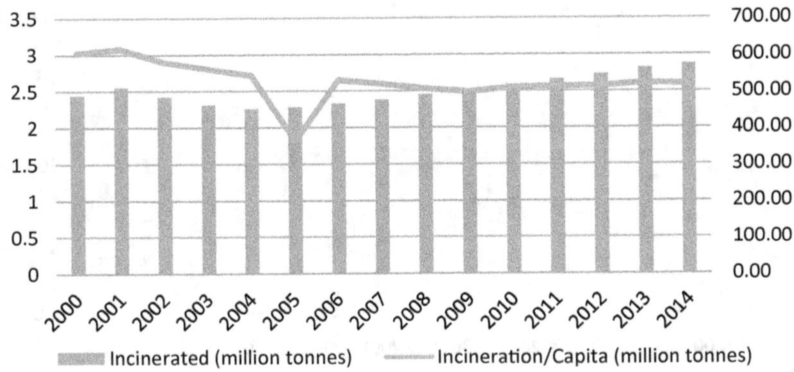

Fig. 10.9. Singapore's Incineration Volume 2000–2014 (NEA, 2015).

In 2014, 728,140 tons or 81% of landfill waste were incineration ash, while 170,983 tons were non-incinerable waste, putting the actual landfill waste at 4.26 folds of the reported landfill figures. In 2008, incineration ash made up 76.6% of landfill waste. Total landfill waste volume has increased by 16.77%, led by the increase in quantity of incineration ash, which offset the reduction in volume of non-incinerable waste. This meant that incineration could never be the long-term solution to the issue arising from Singapore's increasing population. The trend is likely to continue as incineration demand continues to grow alongside population and economic activities amidst a status-quo waste policy.

Incineration

In 2014, 94.4% of Singapore's non-recycled waste was incinerated. Total incinerated waste volume grew at an average of 1.21% annually between 2000 and 2014, from 2.44 million tons to 2.87 million tons. However, the amount of waste incinerated per capita has been falling steadily from 605.78kg per person in 2000 to 519.22 kg per person in 2014. Incineration allows Singapore to reduce 90% of waste volume while recovering energy and scrap metal. In 2002, incineration fees at four plants were raise to S$77 per ton in a bid to encourage waste minimisation and recycling.

Recycling

Between 2000 and 2014, Singapore's recycling volume grew at an annual average rate of 6.66% from 1.86 million tons to 4.47 million tons, an increase of 2.4 fold. The national recycling rate has grown from 39.91% to 59.52%

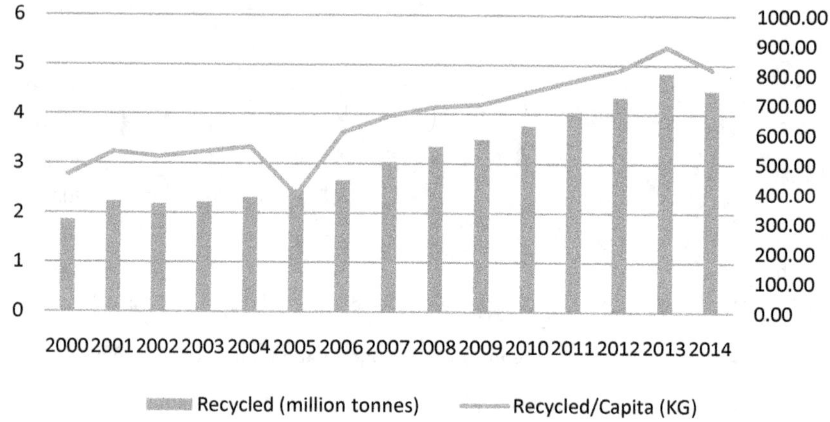

Fig. 10.10. Singapore's Recycling Volume 2000–2014 (NEA, 2015).

Table 10.4. Type and Volume of Waste Recycled and Disposed in Singapore in 2014 (NEA, 2015).

Waste Recycled and Disposed — 2014			
Material	Recycled (Tons)	Landfilled/ Incinerated (Tons)	Recycled Rate
Construction	1,260,000	9,700	99%
Used Slag	361,200	5,100	99%
Scrape	23,100	3,100	88%
Ferrous Metal	1,388,800	57,000	96%
Non-Ferrous Metal	94,700	23,700	80%
Wood	293,500	74,400	80%
Paper/Cardboard	646,500	590,900	52%
Horticultural	163,000	115,300	59%
Glass	15,700	63,800	20%
Ash & Sludge	21,700	126,800	15%
Plastic	80,000	789,000	9%
Food	101,400	687,200	13%
Textiles	16,800	141,800	11%
Others	4,700	355,600	1%
Total	4,471,100	3,043,400	60%

A total of 7,514,500 tons of waste were generated in 2014.

between the same period. Per capita recycling rate grew at an annual average rate of 5.43% from 461.78 kg to 817.23 kg per person, with the bulk coming from non-domestic sectors. (NEA, 2015). However, stagnant per capita recycling over the past eight years suggests that a re-focus of recycling strategy beyond the 3Rs campaign is required.

Table 10.5. Singapore's Waste Management Demand Projection.

6.9 Million and 10 Million Population Projection			
Year	2014	2030	2100
Total Waste (Million Tons)	7.51	9.44	13.68
Incinerated (Million Tons)	2.87	3.58	5.19
Landfilled (Million Tons)	0.17	0.19	0.31*
Recycled (Million Tons)	4.47	5.64	8.17
Population	5,469,724	6,900,000	10,000,000

*Figure is indicative as it is unlikely for landfill to exist in land scarce Singapore in the far future. Through technology and waste management programmes, we will likely see this figure being attributed to recycling or alternative disposal methods.

Meeting Future Demand

From the extrapolated population projection, assuming all variables are constant, the total waste generated is expected to rise to 9.44 million tons and 13.68 million tons annually, under populations of 6.9 million and 10 million respectively. Unlike water and energy, waste management involves managing large amount of incineration ash and non-incinerable waste. There is a need to address its accumulative volume. Approximately 7,863 tons of waste are incinerated daily, exceeding the original capacity of Singapore's four incineration plants. A new Hyflux-Mitsubishi incineration plant with a daily capacity of 3,600 tons will replace the Tuas incineration plant in 2019, increasing Singapore's incineration capacity to 9,500 tons per day (Hyflux, 2013). Assuming no lifestyle change among the populace, this could meet Singapore's projected incineration demand up to 2028 or for a population of 6.7 million. To accommodate a population of 10 million, Singapore would need an additional 3020 tons of incineration capacity per day. Assuming all the landfill waste are incineration ash weighing $1700\,kg/m^3$, 16.7 million m^3 could potentially contain 28.39 million tons of solid waste. Assuming no lifestyle change, Phrase 2 could potentially accommodate Singapore's accumulative landfill demand up to 2041 or for a population of 7.78 million. It is crucial to reduce the amount of waste being landfilled with the long-term goal of closing the waste loop through recycling. The key to an effective waste programme is always to reduce waste at the source, before even resorting to recycling.

Reducing Excessive Waste

In 2007, NEA launched the Singapore Packaging Agreement (SPA), a voluntary programme that encourage industries to reduce their packaging waste. As of December 2015, the 168 signatories have cumulatively reduced 26,000 tons of packaging waste and made savings of S$58 million over an 8-year period (NEA, 2015e). The Bring Your Own Bag Day, started in 2007 as a voluntary programme by the Singapore Environmental Council, has seen relative success (Chiang, 2007). The Fairprice supermarket chain has saved 46.5 million plastic bags between 2007 and 2014 — 9 million plastic bags were saved in 2014 alone (CNA, 2015a). However, the number pales in comparison to the estimated 3 billion plastic bags dispensed in Singapore annually (Reyes, 2013). A nominal grocery rebate or 10-cent penalty chargeable for every plastic bag dispensed is unlikely to deter an affluent population, who would rather opt for the convenience. Unlike Taiwan, Singapore do not have mandatory laws on reducing unnecessary or non-biodegradable waste. It is crucial for manufacturers to minimise the amount of waste that reach consumers, who might not dispose of them properly. Japan and Taiwan vested the responsibility of collecting and recycling products to the respective manufacturers. Recycling fees are payable to the manufacturers to subsidise and promote recycling work (EPA, 2015; JESC, 2012). By vesting ownership to manufacturers, they will be motivated to minimise personal cost by reducing the use of unnecessary material and increasing the amount of recyclables in their products.

In 2014, 687,200 tons of food were disposed, with only 13% being recycled. There is an opporunity for NEA to mandate the recycling of food waste. There is also a need to minimise the cosmetic filtering of unaesthetically pleasing fruits and vegetables. Fairprice launched the Great Taste Less Waste pilot programme which cut and repackaged blemished fruits and vegetables for resale. Positive public responses prompted Fairprice to implement the programme island-wide. Fairprice also launched the Food Waste Index to measure and monitor its annual food wastage in their retail outlets. Fairprice also donates their unsold and near expired canned food to the non-profit food distribution programme Food From The Heart.

Pay-As-You-Throw (PAYT)

While Singapore's water and electricity utility costs are charged in accordance to consumption unit, disposal fees are charged at a flat rate. Reichenbach (2008) highlights the importance of drawing linkage between actual cost of waste disposal and individual waste generation. By paying for the waste actually generated by consumers, they would avoid generating waste. Batllevell and Hanf (2008) reinforce the notion that the perceived fairness would convince people to accept the system if they see their effort being rewarded by having to pay less. PAYT has been implemented in various European countries, Japan, Taiwan, South Korea, etc.. Japan utilises two types of charging systems — a quantity based system and a two-tiered system — where an initial volume is not charged, but excess volume will be charged (Sakai *et al.*, 2008). PAYT can be implemented by selling pre-paid bags of various volumes with individual identification tags that work with an automated system that automatically weighs a bag of waste. The data collected helps the service provider tabulate the monthly bill charged to the household. South Korea's PAYT framework for food wastage has seen a 20% reduction, with restaurants and families preparing less food in a bid to reduce the financial penalty incurred (Hou, 2013). Taiwan also saw a 58% decrease in total waste generation between 1998 and 2011, with domestic waste per capita per day falling from 1.149 kg to 0.673 kg (EPA, 2012).

Optimizing Quality of Waste Stream

Optimising the quality of waste stream is key to high recycling participation. Public waste collectors have highlighted that poor or inappropriate recyclables and food-contaminated recyclables remain a huge challenge. Veolia highlighted that up to 20% of recyclables can be contaminated, while Sembcorp put the figures between 20% to 30%. All contaminated recyclables are eventually incinerated (Cheam *et al.*, 2010; TNP, 2012). A waste management framework without a mandatory waste segregation policy cannot lead to an effective recycling programme.

While NEA, JTC and SMF have been colloborating on various recycling initiatives, an electronic management system could potentially promote industrial symbiosis. An example would be the online waste control and reporting system set up by the Taiwanese government in 2000. Waste disposal statistics from 25,861 waste generators, 4,963 transporters and 865 treatment and disposal facilities, representing 80% of waste generation sources in Taiwan, are required to declare their waste information through this system (Houng and Cheng, 2012). Public housing residents highlight that inconvenience and inaccessibility remain the key obstacles to recycling. Some cited Karung Guni or Rag-and-Bone men, that were common in the 1980s and 1990s, as an affective recycling system (Sile, 2016). Antonides and van Raaij (1998) state that a successful and efficient recycling endeavour would require a separation of waste at source; hence, researchers and policymakers should focus on modifying consumer behaviour to cultivate a desirable disposal habit at the final stage of the consumption cycle. Singapore can emulate the Japanese and Taiwanese governments' policy of waste segregation that can leads to a higher recycling rate due to a higher quality waste stream.

Minimizing Incineration Waste

The implementation of the PAYT framework, reduction of excessive waste, and increase in recycling participation rate in can have a compounded effect on incineration demand. As Incineration Bottom Ash (IBA) makes up 81% of landfill waste, a reduction of incineration demand would in turn reduce landfill demand, contributing to Singapore's goal of being a zero-waste society. After treatment, IBA can be recycled for use in a variety of civil engineering applications (BCA, 2008). In 2014, only 15% of incineration ash and sludge were recycled — representing huge untapped potential. Incineration merely prolongs the lifespan of landfill. An active approach that reduces the need for incineration is the crux to long-term sustainability, rather than a reactive approach of reducing IBA. Ideally, recycling should replace incineration in the long-run to eraducate this "out of sight, out of mind" mentality that undermines waste minimisation and recycling efforts.

Perception and Attitude

MEWR (2013) revealed that 95% of respondents practice some form of recycling, but only 69% believe recycling can help save the environment. Respondents feel that inconvenience and the lack of recyclables deter them from recycling. A knowledge gap was identified when more than half the respondents expressed interest to know more about recycling. Only a quarter of the respondents utilise recycling bins and door-to-door recycling bag collection services. Similar observations were made by an NEA study on the National Recycling Programme (Tan, 2015). Respondents highlighted that, while recycling is the right thing to do, it is not a lifestyle. Shell (2013) found that though Singaporeans believe in the importance of meeting future energy needs, bread and butter issues such as employability and cost of living remain their paramount concern, and less than half are willing to pay more for clean renewable energy.

This suggests that while Singaporeans are environmentally aware, they tend to adopt negative behaviour so as to minimise personal costs and inconvenience, out of self-interest (Dawes, 1980). As a flat rate is levied for waste disposal in Singapore, there is little motivation for ordinary Singaporeans to act on their intrinsic motives and recycle. Jacobs and Bailey (1979) found that combining prizes or monetary rewards with recycling information was effective in modifying recycling behaviour. But De Young (1986) noted that the behaviour only lasted as long as the rewards were in place, and believed that they would not cultivate long-term recycling behaviour as effectively as intrinsic motives and the feeling of having done (altruistic) good. Two strategies can be implemented to facilitate cooperation; an attitudinal strategy, which induces individuals to cooperate for the sake of cooperation, and a structural strategy, which seeks to eliminate the social dilemma by changing the situational environment (Messick and Brewer, 1983). An example is mandatory participation in waste segregation and recycling programmes. It forces individuals to participate in order to avoid undesirable penalties such as fines. Over time, it will invoke a sense of the fufilment of a civic duty, akin to paying taxes and obeying the law, even if the individual's environmental value is non-exsistent (Pieters, 1989). Van Lange *et al.* discovered that people who observe successful defection in peers and neighbours may be tempted to comply due to peer pressure (Van Lange *et al.*, 1997). Once habitualised, the routinisation of sorting and recycling activities will be ingrained into an individual's daily life.

Conclusion

While the Singaporean government is inclined to accept the Boserupian stand on population to stimulate economic growth despite public dissatisfaction towards the increasing population density and societal stress expected to accompany populationg growth, a certain level of emphasis has to be placed on addressing resource vulnerabilities and environmental standards.

Singapore's economic prowess has enabled it to leverage on technology as a resource multiplier and globalisation as a resource avenue diversifier, to grow beyond its natural carrying capacity. Singapore's dependence on resource importation means that its survival is intertwined with its economic status and relative spending power. While both population and affluence have an impact on Singapore's utility demand, higher affluence levels leading to consumerist behaviour can compound the impact on resource utilisation and the nation's ecological footprint. Singapore already has the 6th highest ecological footprint per capita in the world, with its affluence level being the largest contributor to its IPAT equation.

In analysing the Human–Environment equation, there are fundamental facts about Singapore that cannot change — Singapore is a city-state with limited spatial abilities for expansion. This study argues that an increase in population, coupled with increasing affluence, will continue to impact Singapore's Human–Environment equation. Continuous reliance on overseas sources of natural and food resources to support a larger and increasing affluent population will increase Singapore's ecological footprint and resource vulnerability. Increasing threats from geopolitical conflicts, climate change, etc., will impact the supply of commodities, resulting in price increase. While capacity building is the simplest way to address demand in the short-term, there is a urgent need to reduce consumption through a paradigm shift in consumers' behaviour to ensure long-term sustainability. The key to supporting a larger population sustainably calls for the planning of an efficient resource and waste management infrastructure, internalising resource needs, while changing consumer behaviour in the long-run.

References

Agamuthu, P., and Fauziah, S. H. (2013). Sustainable 3R Practice in the Asia and Pacific Regions: The Challenges and Issues. In A. Pariatamby, and M. Tanaka, *Municipal Solid Waste Management in Asia and the Pacific Islands: Challenges and Strategic Solutions* (pp. 15–40). Singapore: Springer.

Alcott, B. (2005). Jevons' Paradox. *Ecological Economics*, **54**, 9–21.

Antonides, G., and Raaij, V. W. (1998). *Consumer Behaviour. A European Perspective.* Chichester: John Wiley & Sons.

Batllevell, M., and Hanf, K. (2008). The fairness of PAYT systems: Some guidelines for decision-makers. *Waste Management,* **28**, 2793–2800.

BCA. (2008). *Sustainable Construction.* Retrieved from A Guide on the Use of Recycled Material.: https://www.bca.gov.sg/SustainableConstruction/others/sc_recycle_final.pdf.

BCA. (2015). *BCA Green Mark Assessment Criteria and Application Forms.* Retrieved from Building Construction Authority: http://www.bca.gov.sg/GreenMark/green_mark_criteria.html.

Boh, S. (2015). *Bigger push to get Singaporeans to recycle.* Retrieved from The Straits Times: http://www.straitstimes.com/singapore/environment/bigger-push-to-get-singaporeans-to-recycle.

Boh, S. (2015a). *NTUC Fairprice launches initiative to discourage food wastage at stores.* Retrieved from The Straits Times: http://www.straitstimes.com/singapore/ntuc-fairprice-launches-initiative-to-discourage-food-wastage-at-stores.

Borenstein, S. (2008). *Equity Effects of Increasing-Block Electricity Pricing.* Retrieved from Center for the Study of Energy Markets: http://www.ucei.berkeley.edu/PDF/csemwp180.pdf.

Cheam, J., Chua, G., and Yong, N. (2010). *Recycling in Singapore: 10 years on — Chute, we got it wrong.* Retrieved from The Straits Times: http://wildsingaporenews.blogspot.sg/2010/12/recycling-in-singapore-10-years-on.html.

Cheam, J. (2013). *Reclamation will add land the size of nine AMK towns.* Retrieved from The Straits Times: http://news.asiaone.com/News/Latest+News/Singapore/Story/A1Story20130202-399527.html.

Chew, V. (2011). *Singapore-Malaysia Water Agreements.* Retrieved from Singapore Infomedia: http://eresources.nlb.gov.sg/infopedia/articles/SIP_1533_2009-06-23.html.

Chiang, G. (2007). *Mixed bag of reactions... But most shoppers respond positively to Bring Your Own Bag Day.* Retrieved from Today: http://www.wildsingapore.com/news/20070304/070419-1.htm.

Chua, G. (2014). *Uphill task to get homes to cut waste.*. Retrieved from The Straits Times: http://news.asiaone.com/news/singapore/uphill-task-get-homes-cut-waste.

Commoner, B. (1972). *The Closing Circle: Nature, Man, and Technology.* London: Jonathan Cape.

CNA. (2015a). *More than 9 million plastic bags saved in 2014 with FairPrice BYOB scheme.* Retrieved from Channel NewsAsia: http://www.channelnewsasia.com/news/singapore/more-than-9-million/1798412.html.

CNA. (2015b). *PUB embarks on study to tap into solar energy via reservoirs.* Retrieved from Channel NewsAsia: http://www.channelnewsasia.com/news/singapore/pub-embarks-on-study-to/2285544.html.

Coxworth, B. (2015). *Rainwater used to generate electricity.* Retrieved from Gizmag: http://www.gizmag.com/pluvia-rainwater-microturbine/31379/.

Dawes, R. M. (1980). Social Dilemmas. *Annual Review of Psychology*, **31**, 169–193.

De Young, R. (1986). Some Psychological Aspects of Recycling. *The Structure of Conservation Satisfactions. Environment and Behavior*, **18**(4), 435–449.

E2. (2015). *Energy Efficiency National Partnership.* Retrieved from E2 Singapore: http://www.e2singapore.gov.sg/Programmes/Energy_Efficiency_National_Partnership.aspx.

EDB. (2015). *Industry Background — Environment and Water.* Retrieved from Economic Development Board: https://www.edb.gov.sg/content/edb/en/industries/industries/environment-and-water.html.

Ehrlich, P. R., and Holdren, J. P. (1971). Impact of population growth. *Science*, **171**, 1212–1217.

EPA. (2012). *1998/Year-Round to 2011/Year-Round Clearance of Municipal Solid Waste by Implementing Agencies — Nationwide.* Retrieved from EPA Environmental Protection Administration: http://edw.epa.gov.tw/eng/reportStatisticTEN.aspx.

EMA. (2014). *Singapore's first LNG terminal launched.* Retrieved from Energy Market Authority: https://www.ema.gov.sg/cmsmedia/Newsletter/2014/04/spotlight-on/singapores-first-lng-terminal-launched.html.

EMA. (2015). Retrieved from Energy Market Authority: https://www.ema.gov.sg/index.aspx.

EMA. (2015a). *Singapore Energy Statistics 2015*. Retrieved from Energy Market Authority: https://www.ema.gov.sg/cmsmedia/Publications_and_Statistics/Publications/SES2015_Final_website_2mb.pdf.

EMA. (2015b). *Piped Natural Gas and Liquefied Natural Gas*. Retrieved from Energy Market Authority: https://www.ema.gov.sg/Piped_Natural_Gas_and_Liquefied_Natural_Gas.aspx.

EPA. (2014). *GHG emissions from large facilities*. Retrieved from Facility Level Information on Greenhouse Gases Tool: http://ghgdata.epa.gov/ghgp/main.do.

EPA. (2015). Retrieved from Taiwan Environmental Protection Administration: https://www.epa.gov/sites/production/files/2014-05/documents/handout-1a-regulations.pdf.

Gewin, V. (2015). *Harnessing hydropower from urban water pipes*. Retrieved from Popular Science: http://www.popsci.com/gregg-semler-turns-tiny-turbines-mighty-generators.

Ghangaa, M. (2015). *Singapore's water supply: Where does it come from?* Retrieved from The Straits Times: http://www.straitstimes.com/singapore/environment/singapores-water-supply-where-does-it-come-from.

GFN. (2015). *Ecological Wealth of Nations*. Retrieved from Global Footprint Network: http://www.footprintnetwork.org/ecological_footprint_nations/ecologi cal_per_capita.html.

Hamilton, J. D. (2010). Historical Oil Shocks. In R. E. Parker, and W. R. M, *Handbook of Major Events in Economic History*. London: Routledge.

HDB. (2015). *Energy Save*. Retrieved from Housing Development Board: http://www.hdb.gov.sg/cs/infoweb/engerysave.

HistorySG. (2010). *Jurong Town Corporation is Established*. Retrieved from HistorySG: http://eresources.nlb.gov.sg/history/events/9206cd21-7c6f-44fa-bd8d-c46e15cbbee3.

HistorySG. (2014). *Esso's Refinery opens at Pulau Ayer Chawan*. Retrieved from HistorySG: http://eresources.nlb.gov.sg/history/events/5e547f82-ef3b-489b-8a2a-79ece9f14cea.

HKE. (2015). *Block Rate Tariff*. Retrieved from Hong Kong Electric: https://www.hkelectric.com/en/customer-services/billing-payment-electricity-tariffs/domestic-tariff.

Hou, L. (2013). *You Waste, You Pay*. Retrieved from Commonwealth Magazine No.519: http://english.cw.com.tw/print.do?action=print&id=14067.

Houng, H., and Cheng, Y. W. (2012). Electronic tracking and management of industrial waste in Taiwan. *Journal of Material Cycles and Waste Management*, **15**(2), 146–153.

Hyflux. (2013). *PUB and Hyflux officially open Singapore's second and largest desalination* . Retrieved from Hyflux: http://hyflux.listedcompany.com/newsroom/20130918_190104_600_6147CE41D66B86FA48257BE900286DFC.1.pdf.

IEA. (2013). *South East Asia Energy Outlook*. Retrieved from International Energy Agency: https://www.iea.org/publications/.

Jacobs, H. E., and Bailey, J. S. (1979). A preliminary analysis of motivational factors in a residential recycling program. *Symposium conducted at the Southeastern Psychological Association meeting.* New Orleans.

JESC. (2012). *Solid Waste Management and Recycling Technology of Japan Towards a Sustainable Society.* Retrieved from Japan Environmental Sanitation Center: http://www.jesco.or.jp.

Karam, P. A. (2006). *How do fast breeder reactors differ from regular nuclear power plants?.* Retrieved from Scientific American: http://www.scientificamerican.com/article/how-do-fast-breeder-react/.

Khamis, I. (n.d.). *Overview of nuclear desalination technologies and costs.* Retrieved from International Atomic Energy Agency: https://www.oecd-nea.org/ndd/workshops/nucogen/presentations/8_Khamis_Overview-nuclear-desalination.pdf.

Kotwani, M. (2015). *Singapore must be prepared to handle nuclear developments.* Retrieved from Channel NewsAsia: http://www.channelnewsasia.com/news/singapore/singapore-must-be/2154300.html.

Kotwani, M. (2015a). *Unprecedented low levels of water at key Johor reservoir.* Retrieved from Channel NewsAsia: http://www.channelnewsasia.com/news/singapore/unprecedented-low-levels/2025108.html.

Krismantari, I. (2007). *Government may terminate Singapore gas contract midyear.* Retrieved from Jakarta Post: http://www.thejakartapost.com/news/2007/05/08/government-may-terminate-singapore-gas-contract-midyear.html.

Lai, L. (2015). *Protection from haze: What docs advise.* Retrieved from The Straits Times: http://www.straitstimes.com/singapore/environment/protection-from-haze-what-docs-advise.

Lee, Y. Y. (2012). *Singapore could get better deal buying gas from US.* Retrieved from Today: http://newshub.nus.edu.sg/news/1206/PDF/GAS-tdy-25jun-p6&p9.pdf.

Luther, J., and Reindl, T. (2013). *Solar Photovoltaic (PV) Roadmap for Singapore.* Retrieved from NCCS: https://www.nccs.gov.sg/sites/nccs/files/Roadmap_Solar_20140729.pdf.

Macrotrends. (2015). *Crude Oil Price History Chart.* Retrieved from Macrotrends: http://www.macrotrends.net/1369/crude-oil-price-history-chart.

MEWR. (2013). *Household Recycling Study 2013.* Retrieved from https://www.mewr.gov.sg/docs/default-source/default-document-library/grab-our-research/mewr_rc_report.pdf.

Messick, D. M., and Brewer, M. B. (1983). Solving social dilemmas: A review. *Review of Personality and Social Psychology,* **4**, 11–44.

MyPower. (2015). Retrieved from My Power: https://www.mypower.com.sg.

NCCS. (2012). *Chapter 6: A Vibrant Ground-up Movement on Climate Change.* Retrieved from National Climate Change Strategy 2012: https://www.nccs.gov.sg/nccs-2012/a-vibrant-ground-up-movement-on-climate-change.html.

NCCS. (2013). *Singapore's Approach to Alternative Energy.* Retrieved from NCCS: https://www.nccs.gov.sg/climate-change-and-singapore/national-circumstances/singapores-approach-alternative-energy.

NEA. (2013). *About Mandatory Energy Management Practices.* Retrieved from National Environmental Agency: http://www.nea.gov.sg/energy-waste/energy-efficiency/industry-sector/mandatory-energy-management-practices.

NEA. (2013a). *Grants and Incentive Schemes.* Retrieved from National Environmental Agency: http://www.nea.gov.sg/grants-awards/grants-and-incentive-schemes.

NEA. (2013b). *Household Sector.* Retrieved from National Environmental Agency: http://www.nea.gov.sg/energy-waste/energy-efficiency/household-sector.

NEA. (2015). Retrieved from National Environmental Agency: http://www.nea.gov.sg/corporate-functions/about-nea/legislation.

NEA. (2015c). *Phase II Semakau Landfill Ready To Meet Singapore's Waste Disposal Needs To 2035 And Beyond.* Retrieved from National Environmental Agency: http://www.nea.gov.sg/corporate-functions/newsroom/news-releases/phase-ii-semakau-landfill-ready-to-meet-singapore-s-waste-disposal-needs-to-2035-and-beyond.

Neo, C. C. (2015). *Move to nuclear power needed if Singapore wants to cut emissions.* Retrieved from Channel NewsAsia: http://www.channelnewsasia.com/news/singapore/move-to-nuclear-power/1960268.html.

Nicole, T. (2015). *Singapore ranks 7th in liveable cities index.* Retrieved from Channel NewsAsia: http://www.channelnewsasia.com/news/business/singapore/singapore-ranks-7th-in/2295506.html.

Oon, S., and B, K. (2014). *The Clean Air Act of 1971.* Retrieved from Singapore Infopedia: http://eresources.nlb.gov.sg/infopedia/articles/SIP_2014-04-07_110024.html.

Parkinson, G. (2015). *Solar PV Costs To Fall Another 25% In Three Years.* Retrieved from Clean Technica: http://cleantechnica.com/2015/05/26/solar-pv-costs-to-fall-another-25-per-cent-in-three-years.

Pieters, R. (1989). *Attitudes and Behaviour in a source-separations program: a garbology approach.* Delft: Eburon.

Philomin, L., and Siau, M. E. (2014). *Sustainable development: Much progress made, much more to be done.* Retrieved from NCCS: https://www.nccs.gov.sg/news/today-sustainable-development-much-progress-made-much-more-be-done.

PUB. (2015). Retrieved from PUB, The National Water Agency.: https://www.pub.gov.sg/ourlibrary/publications/annualreport.

Reichenbach, J. (2008). Status and prospects of pay-as-you-throw in Europe — A review of pilot research and implementation studies. *Waste Management,* **28**(86), 2809–2814.

Sakai, S., Ikematsu, T., Hirai, Y., and Yoshida, H. (2008). Unit-charging programs for municipal solid waste in Japan. *Waste Management,* **28**, 2815–2825.

Saunders, H. D. (1992). The Khazzoom-Brookes postulate and neoclassical growth. *The Energy Journal,* **13**(4), 131–148.

Shell. (2013). *Future Energy Survey.* Retrieved from Shell Singapore: http://www.shell.com.sg/future-energy/foe-survey.html.

Shilov, A. (2014). *GlobalFoundries to upgrade one of its Singapore fabs to 300mm*. Retrieved from Kitguru: http://www.kitguru.net/channel/anton-shilov/globalfoundries-to-upgrade-one-of-singapore-fabs-to-300mm/.

Shilov, A. (2015). *Micron begins to build new $4 billion NAND flash fab in Singapore*. Retrieved from Kit Guru: http://www.kitguru.net/compo nents/memory/anton-shilov/micron-begins-to-build-new-4-billion-nand-flash-memory-fab-in-singapore/.

Sile, A.W (2016). Singapore government focuses on recycling, but residents fail to play their part. Retrieved from CNBC: http://www.cnbc.com/2016/11/14/singapore-government-focuses-on-recycling-but-residents-fail-to-play-their-part.html.

Smeesters, D., Warlop, L., Cornelissen, G., and Abeele, P. V. (Tijdschrift voor Economie en Management Vol. XLVIII, 3, 2003). Consumer Motivation to Recycle When Recycling is Mandatory: Two Exploratory Studies. *Tijdschrift voor Economie en Management Vol. XLVIII, 3*. Luxumbourg: STATEC (National Institute of Statistics and Economic Studies of the Grand Duchy of Luxembourg).

Smets, H. (n.d.). Charging the poor for drinking water: The experience of continent *al European countries concerning the supply of drinking water to poor users*. Retrieved from Water Academy, French: http://www.public policy.ie/wp-content/uploads/Water-for-Poor-People-Lessons-from-France-Belgium.pdf. Singstat. (2015). *Department of Statistics Singapore*. Retrieved from http://www.singstat.gov.sg/statistics/latest-data.

Tan, A. (2015). New plant mixes food waste, used water to produce energy. Retrieved from The Straits Times: http://www.straitstimes.com/singapore/environment/new-plant-mixes-food-waste-used-water-to-produce-energy.

Tan, C. (2015). Promoting Recycling in HDB Households. Retrieved from: http://mem.nus.edu.sg/docs/seminars/MemSem01%20-%20Mr%20Calvin %20Tan%20&%20Mr%20Bu%20Fan%20-%2028%20AUG%202015%20–%20 FLYER.pdf.

Teo, A. (2015). *HDB Statistics*. Retrieved from http://www.teoalida.com/singapore/hdbstatistics/.

TNP. (2012). *Don't go separate wastes*. Retrieved from The New Paper: http://www.asiaone.com/print/News/Latest%2 BNews/Singapore/Story/A1Story20120418-340603.html.

Van Lange, P. A., Otten, W., De Bruin, E. M., and Joireman, J. A. (1997). Development of Prosocial, Individualistic, and Competitive Orientations: Theory and Preliminary Evidence. *Journal of Personality and Social Psychology*, **73**(4), 733–746.

WildSingapore. (2006). Retrieved from Wild Singapore: http://www.wildsinga pore.com/news/20061112/061221-3.htm.

WNA. (2015). *Nuclear Desalination*. Retrieved from World Nuclear Association: http://www.world-nuclear.org/info/Non-Power-Nuclear-Applications/Indus try/Nuclear-Desalination/.

WorldBank. (2015). Retrieved from Country and Lending Groups: http://data.worldbank.org/about/country-and-lending-groups.

Wulandari, F., and Mellor, W. (2013). *Indonesia to Import LNG by 2018 as Demand Rises.* Retrieved from BloombergBusiness: http://www.bloomberg.com/news/articles/2013-03-06/indonesia-to-import-lng-by-2018-as-demand-rises-regulator-says.

Yeoh, B., and Lin, W. Q. (2012). *Rapid Growth in Singapore's Immigrant Population Brings Policy Challenges.* Washington, D.C., USA: Migration Policy Institute.

Part 4

Energy

11

ROLES OF RENEWABLE ENERGY FOR SUSTAINABLE DEVELOPMENT IN INDUSTRIAL PLANTS

TAN CHOON-PUAY and TONG YEN-WAH

It is recognised that energy consumption is one of the key factors required for social, economic and particularly industrial development. An important criteria for development is industrialisation, which is hugely propelled by the consumption of energy as the underlying requirement to achieve greater economic development. For effective and sustainable industrial development, energy production and utilisation must be planned for together with other factors such as technology, human resources, raw materials and finance. The interdependence of energy and the industrial sector is of major concern for governments of developing countries formulating both energy and industrial policies. Industry is a major market for energy, and developments in industry closely affect the energy sector, just as developments in the energy sector affect industry. The size and structure of the industrial sector determines the amount and type or form of energy needed. Similarly, the availability and cost of energy supplies has a major influence on industrial development.

The first generation of energy source comprised of fossil fuels, which not only take many years and multiple natural ecological cycles to replenish, but their consumption also leads to a host of environmental issues that need to be addressed. Many international environmental conventions have recognised these negative impacts on the natural environment and came up with "The Earth Charter" (2012)[1] and "Montreal Protocol" (2011)[2] as some of the early environmental conventions to make known the impacts of environmental degradation that cause climate change, which in turn leads to many other environmental problems. In order to slow down the pace of detrimental impacts and to allow for the sufficient recovery of the environment, mankind's activities need to be made sustainable, in a manner which involves the balancing of social, economic and environmental interests — otherwise more commonly known as the triple bottom line of development. There are three essential aspects of the energy–industry interaction, namely "energy for industry", "industry for energy" and "industrial energy management". "Energy for industry" refers to the development of industrialisation patterns appropriate to and consistent with the local pattern of energy availability; it includes the development or adaptation of energy-efficient or energy appropriate processes and products. There is a strong correlation between the pattern of energy availability,

[1]The Earth Charter Initiative, (2012). *Values and Principles to Foster a Sustainable Future,* <http://www.earthcharterinaction.org/content/pages/What-is-the-Earth-Charter%3F.html> (accessed June 2014).

[2]United Nations Environment Programme, (2011). Montreal Protocol, <http://ozone.unep.org/new_site/en/montreal_protocol.php> (accessed June 2014).

including the type, quality, location and cost, and the corresponding industrial pattern that can be established, such as the sector, size, location, export potential and processes. "Industry" can also refer to a supplier of goods and services for the development of the energy sector — an important industrial sector in its own right — covering a wide range of capital goods and engineering and design services. The timely development of an indigenous energy capital goods sector will enable developing countries to undertake the required energy exploration and production programmes without having to rely on imported technology and equipment. This is an important consideration in relation to the goal of energy self-sufficiency. The conventional sources of renewable energy will be evaluated and measured with case studies that deal with carbon footprint analysis. As there are challenges in harnessing energy out of these renewable sources, there are researches targeted at making utilisation more efficient. This aspect will be covered in the report proper, highlighting the main efficiency-limiting factor(s) and the main technologies adopted for energy conversion.

So far, each and every carbon footprint customisation calculator has its own set of assumptions and conditions, therefore the limitations of the carbon footprint calculator adopted for analysis in this report will also be discussed. As far as the main content of this report is concerned, the carbon footprint calculator selected is just an assessment tool used to enable the quantitative comparison and evaluation of the various renewable energy sources considered in this study.

Introduction

The concept of sustainable development was originally synonymous with that of sustainability and is often still used in that way. Both terms derive from the older forestry term "sustained yield", which in turn is a translation of the German term "nachhaltiger Ertrag" dating from 1713 (Grober, 2000; Finn, 2009). Sustainability science is the study of the concepts of sustainable development and environmental science. There is an additional focus on the present generations' responsibility to improve and maintain the future generations' life by restoring the previous ecosystem and resisting further ecosystem degradation. Sustainable industrialisation, which is characterised by a more balanced and flexible economic and social development with greater international and regional integration and a better ecological environment, tends to close the gap between rapid industrial progression and environmental degradation (Sennholz, 2006). Sustainable industrialisation can be achieved through:

- export-orientation complemented by import-substitution strategy;
- two-stages of industrialisation and modernisation;
- formulation of motivations for development;

- development of human resources; and
- protection of the environment.

Renewable energy is important because of the benefits it provides. The key benefits are:

- Environmental Benefits
  - Renewable energy technologies are clean sources of energy that have a much lower environmental impact than conventional energy technologies.
- Energy for our future generations
  - Renewable energy will never run out. Other sources of energy are finite and will someday be depleted.
- Jobs and the Economy
  - Most renewable energy investments are spent on materials and workmanship to build and maintain the facilities, rather than on costly energy imports. Renewable energy investments are usually spent within the country to create jobs and fuel local economies, rather than going overseas.
- Energy Security
  - Renewable energy sources are not vulnerable to threats by economic fluctuation.

Evolution of Industrialisation

Industrialisation and economic development have historically been associated with man's ability to harness natural energy resources to improve his condition. Based on this definition, two industrial revolutions occurred in the 18th and 19th centuries, where natural resources such as coal (first revolution) and petroleum (second revolution) were widely exploited to produce levels of energy far beyond what could be achieved by human or animal muscle power. More than two centuries of past industrialisation exploited non-renewable energy resources, however, often with undesirable side effects such as pollution and other damages to the natural environment. In the second half of the 20th century, the extraction of energy from nuclear processes grew in popularity, relieving some demands on limited fossil fuel reserves, but at the same time, raising safety and political problems. Meeting the global demand for energy is now the key challenge to

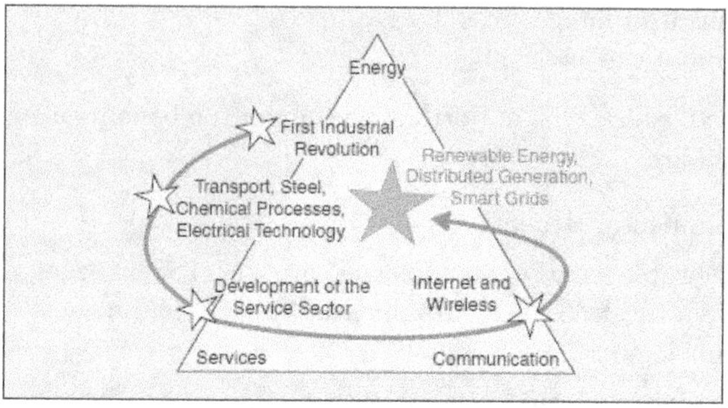

Fig. 11.1. Visual history of industrial revolutions: from energy to services and communication and back again to energy (Liserre *et. al.*, 2010).

sustainable industrialisation. On the other hand, network and wireless communication systems have birthed another modern economic and industrial revolution (Fig. 11.1) (Liserre *et. al.*, 2010).

Importance of Sustainability

Sustainability looks to the future and, in some ways, guarantees an acceptable social and ecological system for future generations. The concepts of sustainability and sustainable development have received much attention among policy-makers and scientists as a result of the existence of limits to growth and the dramatic environmental changes we have witnessed in the last decades (Phillis and Kouikoglou, 2009); more specifically:

- The consequences of a rapidly growing world population and their increasing consumption of finite resource supplies (of which we do not know the precise quantity of remaining resources) — though we know that they will certainly worsen the pressure on ecosystems, and society in general.
- On the other hand, it is clear that environmental problems, such as species extinction, population explosion, global warming, water shortages, exhaustion of fisheries, deforestation, etc., are now global in scale.

Sustainability integrates environmental, economic and societal aspects. It is a concept discussed by politicians, environmentalists, biologists, sociologists, economists, engineers, philosophers, and ordinary citizens who usually assign different meanings to the concepts of sustainability and sustainable development.

Several initiatives, such as the Brundtland Report[3], and Agenda 21 — the 1992 Rio Declaration on Environment and Development[4], urge countries to develop indicators of sustainable development.

Environmental management and corporate social responsibility systems are now adopted by numerous corporations. A large number of leading companies have made a commitment to become sustainable enterprises, and have launched proactive programmes to improve the environmental and social performance of their products, processes, services, and facilities.

Sustainable Development and Renewable Energy in the Various Industrial Sectors

The most exploited renewable energy sources are hydropower, photovoltaic (PV), and wind. Over the years, renewable energies have experienced one of the largest growths in percentage, being comparable with the growth of coal and lignite energy but still below that of natural gas. In 2011, the world's renewable energy production share has been calculated as 21%,[5] with a projection for nearly 25% in 2040.[6] However, 16% is due to hydropower energy production; hence, wind and PV (the most promising renewable sources) energy production is still very modest. Also, these two most promising renewable energies have different requirements. The wind energy production for 2014 was around 370 GW.[7] On the other hand, despite the silicon shortage in recent years, the PV industry is growing at more than 30% per year, and the cost of PV energy will reach the break-even point very soon in many countries. As of 2009, it is evident that wind energy is

---

[3]Brundtland. G.H., (1987). *United Nations Environment Programme — Report of the World Commission on Environment and Development Our Common Future*, <http://conspect.nl/pdf/Our_Common_Future-Brundtland_Report_1987.pdf> (accessed July 2014).

[4]United Nations Conference on Environment & Development 1992, (n.d.). *Agenda 21*, <http://sustainabledevelopment.un.org/content/documents/Agenda21.pdf> (accessed July 2014).

[5]U.S. Energy Information Adminstration, (n.d.). *International Energy Statistics, Electricity, Generation* <http://www.eia.gov/tools/faqs/faq.cfm?id=527&t=4> (accessed March 2015).

[6]U.S. Energy Information Administration, (n.d.). *International Energy Outlook 2013* Chapter 5: Electricity, Figure 83. World net electricity generation by fuel, 2010-40, <http://www.eia.gov/tools/faqs/faq.cfm?id= 527&t= 4> (accessed March 2015).

[7]Global Wind Energy Council, (2015). *Global Wind Statistics 2014*, <http://www.gwec.net/wp-content/uploads/2015/02/GWEC_GlobalWindStats2014_FINAL_10.2.2015.pdf> (accessed March 2015).

becoming well established in world energy production (1%), with forecasts to reach 13%[8] by 2050, while PV energy is experiencing an impressive growth.[9] Other emerging renewable technologies include wave and tidal energy conversion, biomass energy conversion with focus on combined heat and power (CHP), and small-scale hydropower plants (less than 10 MW per site).

Research Problem and Objective of this Paper

Conventional power generating sources and systems are consuming large amounts of fossil fuels which take a very long period of time to form. The natural formation process is not able to regenerate fast enough to keep up with the rate we are extracting these natural resources in the name of development. Looking at the global context, industrialisation will continue to rise and there is a need to find alternative power production methods using renewable energy sources which are readily available, with less detrimental environmental impacts. From recent findings and global conventions, there is a need to change the way the world is using and consuming natural resources. The term "sustainable development" is the way forward for industrialisation in order for future generations to be able to enjoy what is left for them and, at the same time, be able to live in a sustainable manner.

In the context of this paper, it is clear that no single source of renewable energy is the most suitable option for industrial plant use. The selection and adoption of preferred renewable energy sources is location specific, strongly driven policies, subsidies and other economic factors. Renewable energy sources are primarily engaged in power generation industrial facilities. Biomass, solar thermal systems and heat pumps are mostly employed for process heating applications. Conventional geothermal heat is highly location dependent. Transporting heat over long distances is costly, leads to large losses and is feasible in only a few specific conditions. For industrial process heat, the industrial plant must be located very close to the geothermal reservoir. This is unlikely to be possible in any but a few highly specialised applications. Enhanced geothermal systems may make a

[8]International Energy Agency, (2013). *Renewable Energy Outlook — World Energy Outlook 2013* <http://www.worldenergyoutlook.org/media/weowebsite/2013/WEO2013_Ch06_Renewables.pdf> (accessed November 2014).

[9] *"Yearbook 2009: The 2008 world energy review"* (2009). Enerdata, Tech. Rep. Available: <www.enerdata.fr>.

contribution in the long run, subject to the resolution of technology issues. The use of run-of-river hydro for motive power, in kind, has been used for centuries for grinding mills. The use of wind for motive power, for example by driving air compressors, enables the storage of energy in the form of compressed air.

There are many other indicators which can be used to measure the sustainability index for each type of renewable energy, such as, land use, water consumption, social impacts, etc., besides those that are adopted as mentioned in this research's methodology. As this report is not performing any Life Cycle Assessment (LCA) of each renewable energy source, assumptions, boundaries and processes are not formed and analysed in depth.

Various forms of renewable energies are discussed in economical, technological, efficiency, market, application, their individual limitations and carbon emission aspects. In this respect, based on current trends and findings, the information presented is useful to form a reasonable assessment of the future outlook of the role of renewable energy in the different industrial plants. The secondary data presented aids in assessing current and future trends and drawing the conclusion in this report.

Solar Power

Basics

The sun's radiation is used directly to produce electricity in two ways. The primary solar energy technologies for power generation are photovoltaic (PV) and thermal systems. Photovoltaic systems turn sunlight into electricity directly. Solar thermal systems use the sun's heat to heat water, creating steam to turn a turbine and generator (Fig. 11.2). Solar energy technologies offer a clean, renewable, and domestic energy source.[10]

Technologies, Sciences and Application

Concentrating solar power

A new generation of power plants, with concentrating solar power systems, use the sun as a heat source. There are three main types of concentrating solar power systems — parabolic-trough, dish/engine, and power tower.

---

[10] U.S. Department of Energy, Office of Energy Efficiency and Renewable Energy, (n.d.). *Renewable Energy Technology Basics.* <http://energy.gov/eere/energybasics/renewable-energy-technology-basics> (accessed November 2014).

Fig. 11.2. A PV Installation near Hot Springs (courtesy of Shell Solar) (*Source*: Chambers, 2004).

Efficiency, Environmental Impacts and Mitigation Measures

Current photovoltaic technologies and efficiency

The efficiency and cost of the panels have been the two major issues in the development of PV technology, and still are today. In reality, a variety of cells and modules in terms of cost and efficiency are needed to provide an optimal solution for varying power system requirements (Lamont, 2012).

Environmental impacts

Production starts with the mining of the raw materials (i.e., quartz sand for silicon PV; Zn and Cu ores for CdTe PV), and continues with their processing and purification (Fig. 11.3).

Criteria pollutant emissions

Toxic gases and heavy metals can be emitted directly from material processing and PV manufacturing, and indirectly from generating the energy used at both stages (Fig. 11.4).

Heavy metal emissions

Direct emissions

Cadmium is a by-product of zinc and lead, and is collected from emissions and waste streams during the production of these major metals (Fthenakis, 2008).

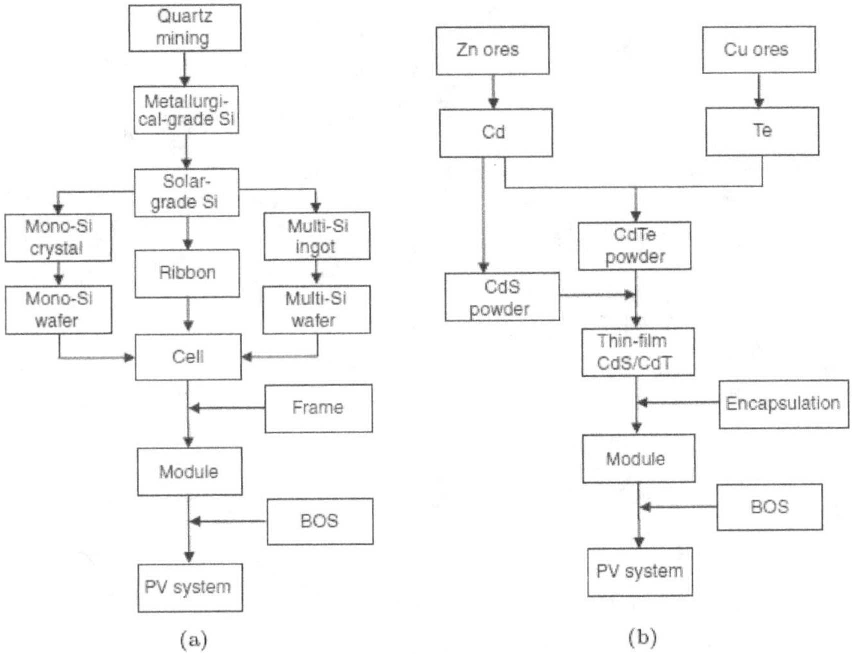

Fig. 11.3. Detailed flow diagram from raw material acquisition to the manufacturing stage of PVs. (a) Silicon PVs and (b) CdTe PVs (frameless). BOS, balance of system; mono-Si, monocrystalline silicon; multi-Si, multicrystalline silicon (*Source*: Fthenakis, 2012).

Indirect emissions

The indirect emissions here are those emissions associated with the production of energy used in mining and industrial processes in the PV life cycle (Table 11.1) (Fthenakis, 2008).

Market and Economic Feasibility

Solar photovoltaic (PV) electricity is on the road to becoming a mainstream energy technology and is currently the fastest-growing renewable energy source in Europe, with a rise in installed capacity of almost 13,000 MW over the past year (see Fig. 11.5), which is the second largest capacity increase in Europe just after gas (Despotou, 2012).

However, to achieve a real paradigm, shift is needed. Policy-makers, regulators, and industry need to work together to drive PV's mass penetration, fostering technological progress, and cost reductions as well

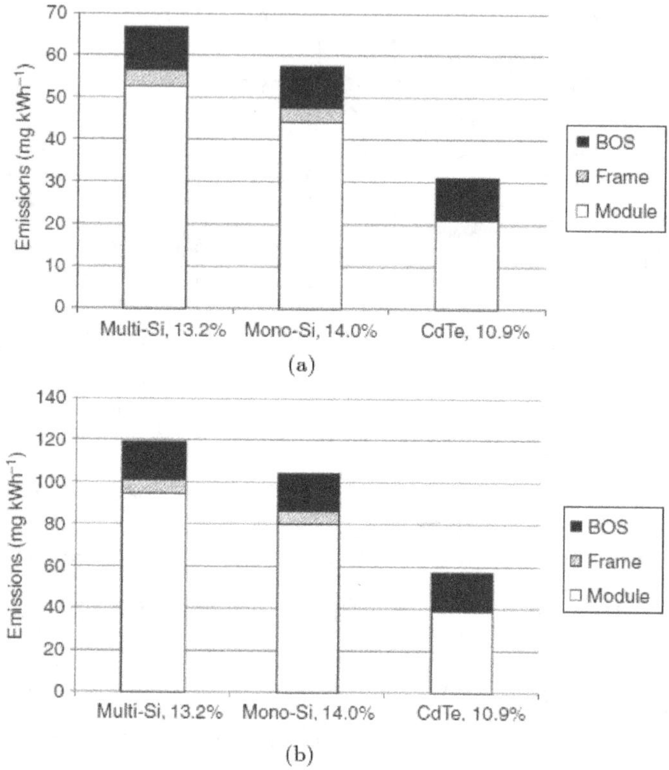

Fig. 11.4. Life-cycle emissions of (a) NO_x and (b) SO_x from silicon and CdTe PV modules, wherein BOS is the balance of system (BOS), that is, module supports, cabling, and power conditioning. The estimates are based on rooftop-mounted installation, Southern European insolation of 1700 kWhm^{-2} yr^{-1}, a performance ratio of 0.75, and a lifetime of 30 years. It is assumed that the electricity supply for all the PV system is from the UCTE grid. Mono-Si, monocrystalline silicon; multi-Si, multicrystalline silicon (*Source*: Fthenakis, 2012).

as creating a predictable regulatory environment that attracts investments (Despotou, 2012).

Disadvantages of Solar Photovoltaic (PV) Panels

Perhaps the biggest disadvantage of photovoltaic (PV) panels is their limited efficiency levels; compared to other renewable energy sources such as solar thermal.

Another disadvantage of PV panels is that they produce direct electric current (DC) which must be converted to alternating current

Table 11.1. Direct, atmospheric Cd emissions during the life cycle of the CdTe PV module (allocation of emissions to coproduction of Zn, Cd, Ge, and In) (*Source*: Fthenakis, 2004).

	Air emissions (g Cd tonne^{-1} Cd^{a})	Allocation (%)	Air emissions (g Cd tonne^{-1} Cd^{b})	mg Cd GWh^{-1b}
Mining of Zn ores	2.7	0.58	0.016	0.02
Zn smelting/refining	40	0.58	0.23	0.3
Cd purification	6	100	6	9.1
CdTe production	6	100	6	9.1
PV manufacturing	3	100	3	4.5
Operation	0.3	100	0.3	0.3
Disposal/recycling	0	100	0	0
Total			15.55	23.3

[a]Tonne of Cd produced.
[b]Energy produced assuming average Southern European insolation (i.e., 1700 kWh m^{-2} yr^{-1}), 9% electrical conversion efficiency, and a 30-year life for the modules.

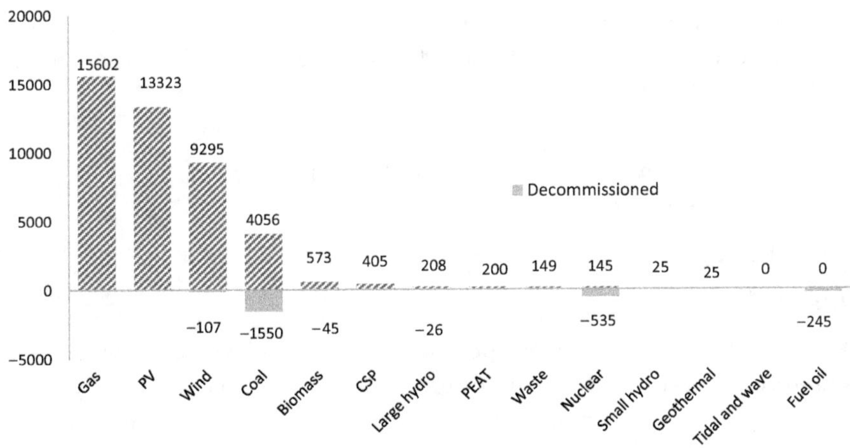

Fig. 11.5. Power generation capacities added in 2010 at EU 27 (*Source*: EPIA Analysis, 2010).[11]

(AC) before the power generated can be consumed (either through transference to the power grid, or directly for own consumption) (Green, 2012).

[11]EPIA (2010) *2015: Global Market Outlook for Photovoltaics until 2015*. Brussels, Belgium: EPIA.

Greenhouse Gas Emissions

Electricity and fuel use during the PV material and module production are the main sources of greenhouse gas (GHG) emissions for PV cycles (Dones *et al.*, 2003; Franklin Associates, 1998).

Case Study: Australian Solar Tower Breaking Records

Australia is home to the world's first commercial scale solar power tower plant, which happens to also be the world's tallest manmade structure and one of the world's largest non-hydroelectric renewable energy generators. The 200 MW plant cost US$431 million to build and began construction in 2003. The tower has a ground level greenhouse area to heat air and force it through a tall, narrow chimney. The flowing air drives turbines to produce power. The tower acts like a magnet in attracting sunrays from the air and from the large body of air underneath the collector system (Fig. 11.6).

The plant is sited near Buronga in Wentworth Shire, Victoria. It produces enough power for about 200,000 homes and offsets the emission of almost 900,000 tons of GHGs annually.

Carbon Footprint Comparison of Solar Energy with Various Fossil Fuels

Wind Energy

Basics

Wind energy, also called wind power, refers to the process by which wind is used to generate mechanical power or electricity. Wind is actually a form

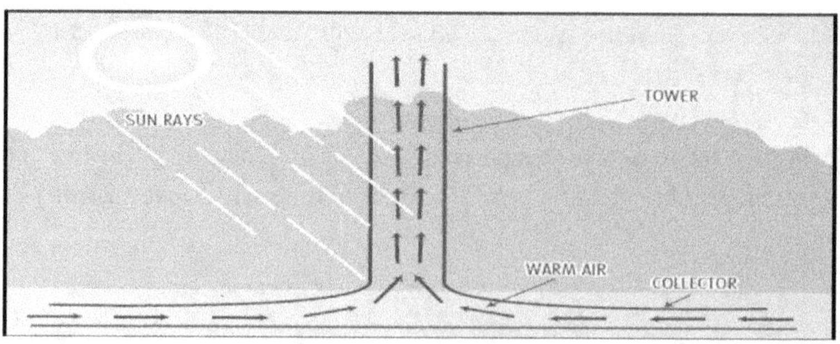

Fig. 11.6. The Solar Power Tower (*Source*: Chambers, 2004).

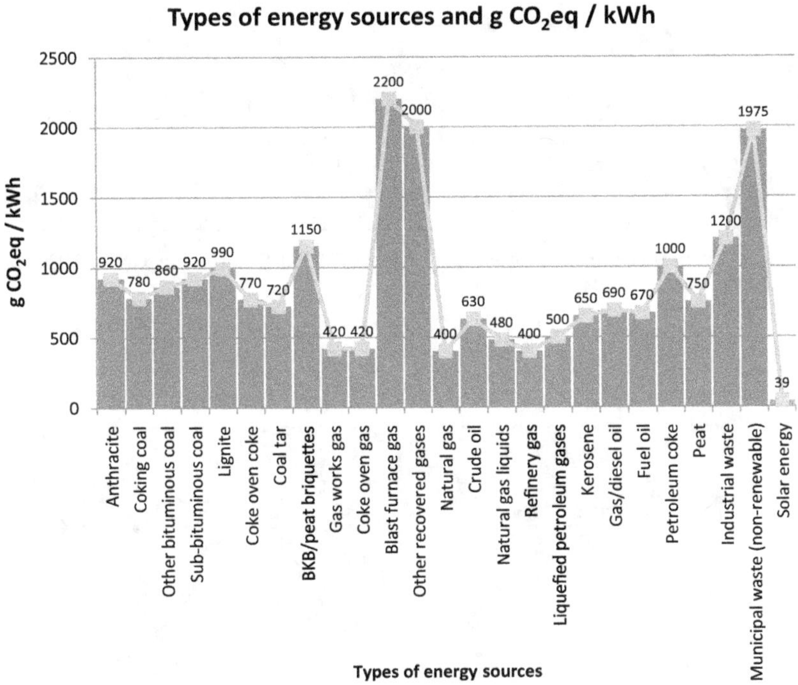

Fig. 11.7. Carbon footprint of solar energy with various fossil fuels (*Source*: 2014 IPCC, Global warming potential of selected electricity sources).

of solar energy because winds are caused by the uneven heating of the atmosphere by the sun, the irregularities of the earth's surface, and rotation of the earth. Wind flow patterns are modified by the earth's terrain, bodies of water, and vegetative cover. This wind flow, or motion energy, is then used by wind turbines to generate electricity.[12]

Technologies, Sciences and Application

Types of turbines

Wind turbines, like aircraft propeller blades, turn in the moving air and power an electric generator that supplies an electric current (Fig. 11.8).

[12]U.S. Department of Energy, Office of Energy Efficiency and Renewable Energy, (n.d.). *Wind Energy Basics.* <http://energy.gov/eere/wind/wind-energy-basics> (accessed November 2014).

Fig. 11.8. Wind Turbines Capture the Power of Wind (photo courtesy of NEG Micon) (*Source*: Chambers, 2004).

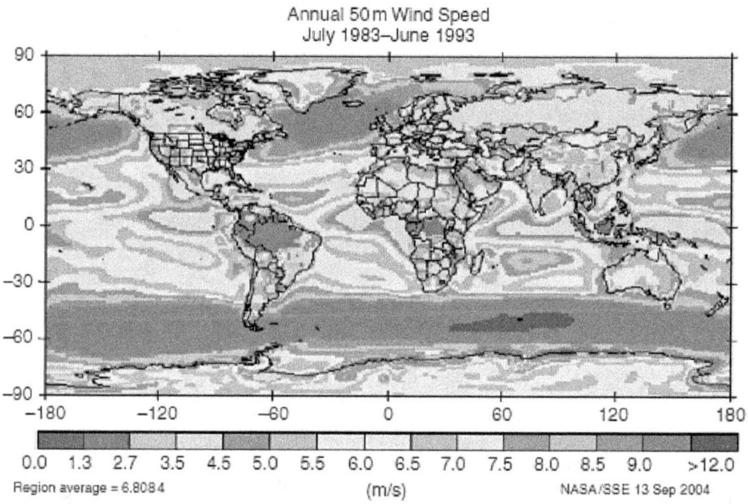

Fig. 11.9. Global wind potential. Reproduced from NASA (www.nasa.gov).

Efficiency, Environmental Impacts and Mitigation Measures

Efficiency

For a given design, the efficiency of a wind rotor depends on the relative speed between the rotor and the incoming wind. Hence, the performance of the rotor is generally expressed in terms of the variations in the power coefficient (C_p) with the tip speed ratio (λ) (Mathew and Philip, 2012). Wind speed is highly unpredictable and volatile and wind potential is not evenly distributed across the globe (Fig. 11.9).

Table 11.2. Main social and environmental benefits of wind power (*Source*: Wizelius, 2012).

Avoid air pollution — reduction of CO_2 emissions
Reduction of water consumption
Fossil fuels saving/substitution
Positive effects on the microclimate of the area
New job positions — employment opportunities
Regional development
Development and support of domestic construction industry and various satellite activities

Environmental benefits of wind energy

Wind energy is one of the cleanest and most environmentally friendly energy sources. It has a long-term positive impact on the environment by reducing the threat posed by climate change. It emits no GHGs, air pollutants or particles that are carcinogenic and that affect human health severely. The development of wind power plants creates employment opportunities and new job positions during equipment construction, installation, and operation of the new power plants (Wizelius, 2012). Table 11.2 highlights the main environmental and social benefits of wind energy.

Avoided air pollution — reduction of CO_2 *emissions*

All electricity generation schemes have a carbon footprint. This means that at some points of their construction and operation, CO_2 and other greenhouse gases are emitted. Wind power is already helping to fight climate change, since each wind-produced kWh avoids a kWh created by the energy mix of coal, oil, and gas — on average $600 - 1{,}000\,g\,CO_2\,kWh_e^{-1}$.[13] Tables 11.3 and 11.4 present the emissions of relevant pollutants produced by various power production technologies including wind power.[14]

Mitigation measures

The most crucial of the impacts belong to the following categories:

- Noise emission level
- Aesthetics and the visual impact

[13]European Wind Energy Association, (2011). <http://www.ewea.org> (accessed November 2014).

[14]Research Center for Energy, Environment and Technology CIEMAT, (2011). <www.ciemat.es> (accessed November 2014).

Table 11.3. Emissions of pollutants per kWh of produced electricity — benefits of wind power versus coal and natural gas.[15]

	Emissions per kWn of produced electricity						Wind power benefits		
	Onshore wind	Offshore wind	Average wind	Hard coal	Lignite	NGCC	vs. Coal	vs. Lignite	vs. NGCC
Carbon dioxide. fossil (g)	8	8	8	836	1060	400	828	1052	392
Methane, fossil (mg)	8	8	8	2554	244	993	2546	236	985
Nitrogen oxides (mg)	31	31	31	1309	1041	353	1278	1010	322
NMVOC (mg)	6	5	6	71	8	129	65	2	123
Participates (mg)	13	18	15	147	711	12	132	696	−3
Sulfur dioxide (mg)	32	31	32	1548	3808	149	1516	3776	117

Table 11.4. Emissions and benefits of pollutants per kWh of electricity produced by wind, nuclear, solar PV, solar thermal, and biomass combined heat and power (CHP) plants.[16]

	Emissions					Wind power benefits			
	Average wind	Nuclear	Solar PV	Solar thermal	Biomass CHP	vs. Nuclear	vs. Solar PV	vs. Solar thermal	vs. Biomass CHP
Carbon dioxide, fossil (g)	8	8	53	9	83	0	45	1	75
Methane, fossil (mg)	8	20	100	18	119	12	92	10	111
Nitrogen oxides NMVOC (mg)	31	32	112	37	814	1	81	6	783
Particulates (mg)	15	17	107	27	144	2	92	12	129
Sulfur dioxide (mg)	32	46	0	31	250	14	−32	−1	218

- Influence on the bird habitat and biodiversity in general
- Impacts on natural reserves, land use, and the landscape.

[15] Research Center for Energy, Environment and Technology CIEMAT, (2011). <www.ciemat.es> (accessed November 2014).

[16] Research Center for Energy, Environment and Technology CIEMAT, (2011). <www.ciemat.es> (accessed November 2014).

The future prospect of environmental impact mitigation is related to the following:

- Proper selection of the wind park siting
- Use of the technology in the equipment and the installations
- Use of IT tools
- R&D for knowledge improvement.

Considering offshore wind parks and the seascape, the correct site selection is even more crucial, since there are many values in the marine environment that need to be taken into account (Kondili and Kaldellis, 2012).

Market and Economic Feasibility

The price of modern turbines around 55–65 m in diameter, for onshore installations, is around US\$1,500 kW^{-1}. Figure 11.10 shows a typical cost breakdown for a complete wind farm (Krohn, 2009).

Long-term plans for wind energy[17] include having wind power supplying 400 GW of power in the EU and 20% of USA's electricity demand

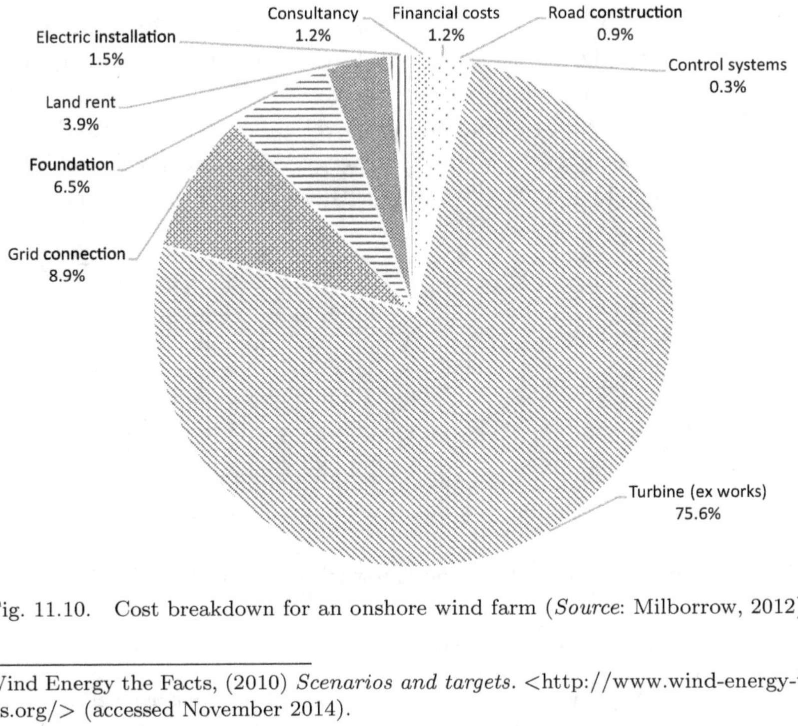

Fig. 11.10. Cost breakdown for an onshore wind farm (*Source*: Milborrow, 2012).

[17]Wind Energy the Facts, (2010) *Scenarios and targets*. <http://www.wind-energy-the-facts.org/> (accessed November 2014).

by 2030.[18] China also plans to achieve a 150 GW installed capacity of wind power by 2020.[19] These plans set the scene for wind power's future prospects and challenge the target of 1,000 GW globally by 2030.

Limitations

Wind energy is not always available, meaning that it is based on non-controllable natural phenomena, which are responsible for the so-called intermittency of wind energy production (Sovacool, 2009). As a result, when considering wind energy, it is imperative to also consider a reserve energy source such as backup conventional power stations, energy storage systems (Chen *et. al.*, 2009), and hybrid units (Kaldellis, 2010), which can be used to complement any energy deficit caused by the intermittency of the primary wind energy system.

Case Study: Brazil's Hydro Crisis Unleashes Wind Demand

Brazil's energy industry is unusual in that it is a large, sprawling country and is powered overwhelmingly through hydroelectric resources. A long-term drought a few years ago, however, punctuated the country's need for a more diversified generation base. As a result, the country is now working to increase the share of natural gas–fired generation in its energy supplies as well as invest in wind energy development. A number of projects, totalling more than 1,000 MW of wind energy, are being developed in Brazil (Figs. 11.11 and 11.12). BP Solar has three solar projects underway in

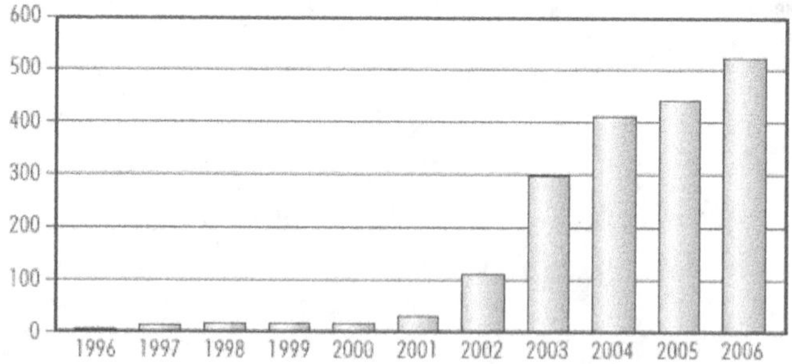

Fig. 11.11. Brazilian Wind Power (*Source*: Chambers, 2004).

[18]American Wind Energy Association (2008). *20% Wind energy by 2030.* <http://www.awea.org/> (accessed November 2014).

[19]Chinese Renewable Energy Industries Association (2010). *China wind energy outlook–2010.* <http://www.greenpeace.org/> (accessed November 2014).

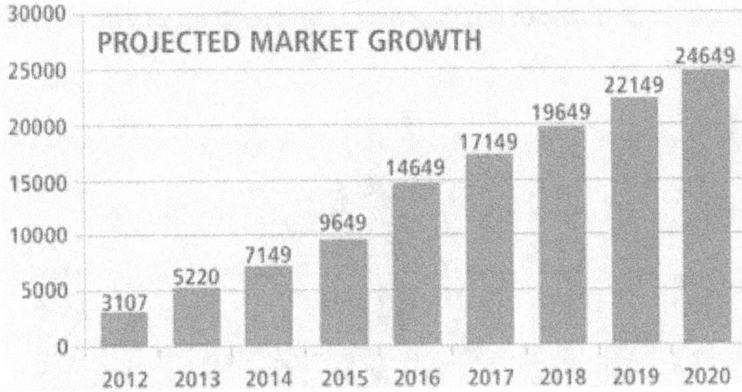

Fig. 11.12. Projected Wind Energy Market in Brazil (*Source*: http://www. brazilwind-power.org/en/).

Brazil. Électricité de France is working to build a US$15 million wind generation plant in Rio de Janeiro state. There is interest in adding offshore wind facilities as well.

Brazil's demand for energy is growing at 6–7% annually, so Brazil needs more capacity, and it does not want to use its domestic coal resources, which are high in ash and sulphur content.

Carbon Footprint Comparison of Wind Energy with Various Fossil Fuels

Bioenergy

Basics

Biomass energy is derived from the energy stored in plants and organic matter. It is used to meet a variety of energy needs including generating electricity, heating homes, fuelling vehicles, and providing process heat for industrial facilities. Biomass also can be converted into transportation fuels such as ethanol, methanol, biodiesel, and additives for reformulated gasoline.[20] Biofuels can either be used in pure form or blended with gasoline (see Table 11.5).

[20]U.S. Department of Energy, (n.d.). *Biomass Basics: The Facts About Bioenergy.* <http://www.energy.gov/sites/prod/files/2014/04/f14/biomass_basics.pdf> (accessed November 2014).

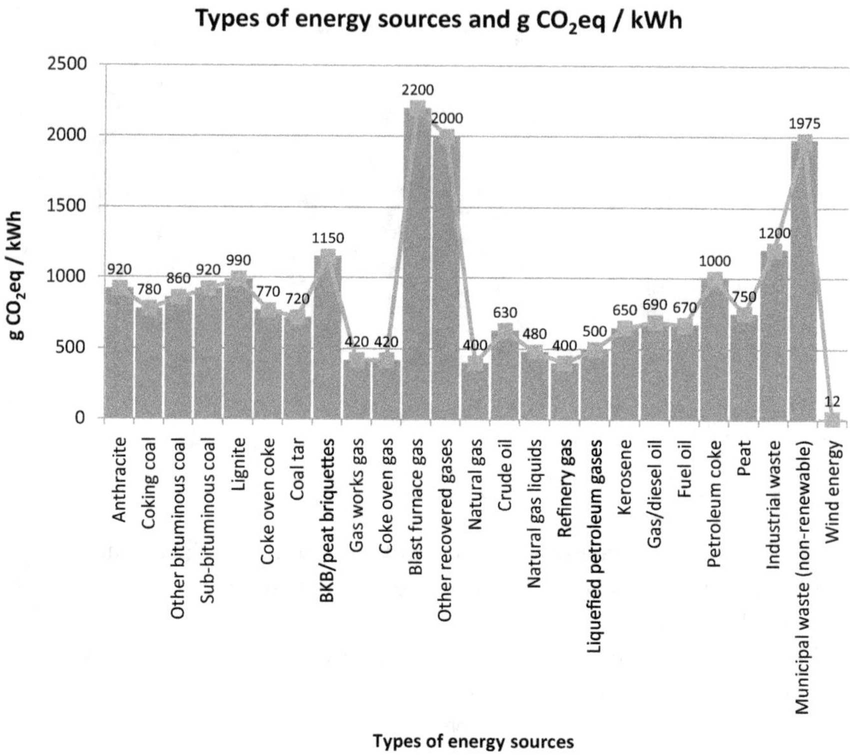

Fig. 11.13. Carbon footprint of wind energy with various fossil fuels (*Source*: 2014 IPCC, Global warming potential of selected electricity sources).

Technologies, Sciences and Application

Technology development

Boilers have been a major part of industrial applications since the industrial revolution in the 1700s and still are. Power and heat can nowadays be distributed more efficiently. Combined heat and power (CHP) is the simultaneous production of heat and power (i.e., electricity). On a large scale, a CHP unit is part of a power station. This is an effective way to improve the efficiency of fuel utilisation from below 50% in a conventional power-only generation plant to 80–90% or even higher in a CHP plant.[21]

[21]U.S. Department of Energy, (n.d.). *Biomass Basics: The Facts About Bioenergy.* <http://www.energy.gov/sites/prod/files/2014/04/f14/biomass_basics.pdf> (accessed November 2014).

Table 11.5. Biomass Use By Type (*Source*: Chambers, 2004).

Fuel	Trillion btus	Fuel	Trillion btus	Fuel	Trillion btus
Wood Energy		*Waste Energy*		*Other Biomass*	
Residential	407	MSW/Landfill Gas	434	Commercial	6
Commercial	43	Commercial	40	Industrial	90
Industrial	1,580	Industrial	104	Electric Power	21
Electric Power	140	Electric Power	290		
Wood Total	2,170	Waste Energy Total	551	*Alcohol Fuek*	
				Transportation	133

Efficiency, Environmental Impacts and Mitigation Measures

Efficiency

Many biomasses contain high moisture content, which will increase heat losses through moisture emitted from the chimney. For very high moisture content fuel, pre-drying of the fuel with waste energy should be considered to improve the efficiency of the process (Malmgren and Riley, 2012).

Environmental impacts

The burning of biomass can make a significant contribution to the government's objectives of CO_2 reduction. It will provide a dispatchable source of renewable energy at a time when the network is becoming more reliant on intermittent wind energy.

Gaseous emissions

The components of gaseous emissions (Malmgren and Riley, 2012) are:

i. Oxides of sulfur (SO_x)
ii. Oxides of nitrogen (NO_x)
iii. Carbon monoxide
iv. Volatile organic compounds
v. Hydrochloric acid
vi. Dioxins and furans

Solid residuals

The components of solid residuals (Malmgren and Riley, 2012) are:

i. Particulates
ii. Heavy metals
iii. Ash[22]

Mitigation

Improving the efficiency and cost effectiveness of the production processes, and broadening the prospective feedstock slate can facilitate market expansion since more products can be produced. Conventional thinking tends to assign liquid biofuels to transport applications, solid biomass to power generation, and biogas to heating applications (Roddy, 2012).

Technology Solutions — New Processes

Anaerobic digestion

Anaerobic digestion (AD) is a microbial process in which complex organic materials are broken down into their simpler chemical components by various enzymes.[23] This occurs in the absence of oxygen and results in the production of biogas and a digestate. Typical feedstock for AD reactors often consist of animal slurries, energy crops, and other agricultural, retail, and industrial wastes (Macias-Corral *et. al.*, 2008).

Advanced biofuel processes

Biomass feedstock consist mainly of cellulose, hemicellulose, and lignin. Of the three, cellulose is the easiest to convert via fermentation into alcohol. The term 'advanced biodiesel processes' usually refers to BtL or biomass-to-liquids (Roddy, 2012).

Algae

The quest for alternative feedstock has led many people to algae. For those who consider that land availability may at some point in the future become

[22]Pitman R., (n.d.). Wood ash use in forestry — A review of the Environmental Impacts, no year stated but references from 2002. <http://www.forestry.gov.uk/pdf/Use_of_ash_in_forestry.pdf/$file/Use_of_ash_in_forestry.pdf> (accessed November 2014).
[23]Boyle G. (2004). *Renewable Energy*. Oxford, UK: Oxford University Press in association with the Open University.

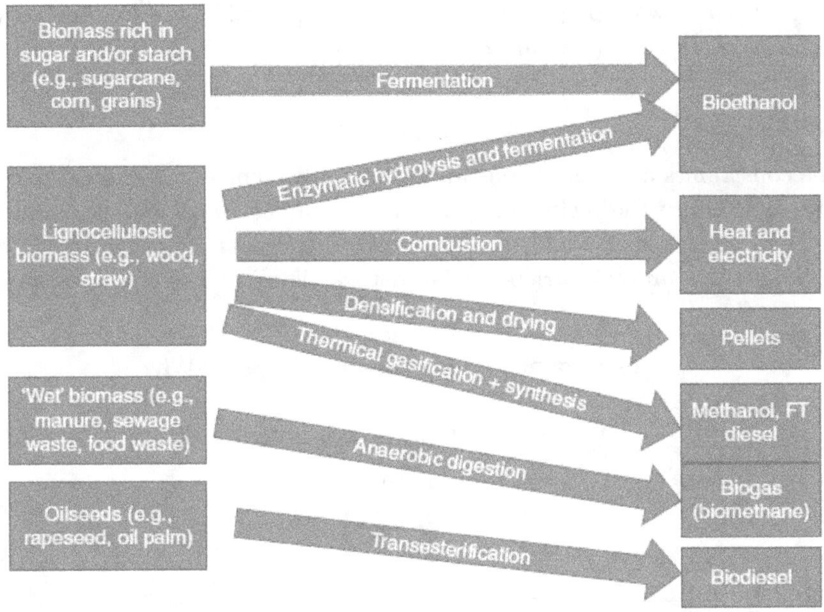

Fig. 11.14. Means of conversion from biomass to biofuels (*Source*: Hammarlund *et. al.*, 2010).

the limiting factor, looking at feedstock that grow in water holds some attractions — especially since the Earth's water surface is significantly greater than its land surface. There are two broad areas of interest: macroalgae (usually known as seaweed) and microalgae (often found in the form of pond scum) (Roddy, 2012).

Market and Economic Feasibility

Bioenergy markets are still a relatively new phenomenon. An overview of different biomass resources and how they may be utilised for energy purposes can be found in Fig. 11.14 (Hammarlund *et. al.*, 2010).

Worldwide, biomass is the fourth largest energy resource after coal, oil, and natural gas. It is used for heating (such as wood stoves in homes, and for process heat in bioprocessing industries), cooking (especially in many parts of the developing world), transportation (in the form of fuel, such as ethanol) and increasingly, for electric power production. There are estimates of about 35,000 MW of installed capacity using biomass worldwide. Most of this capacity is in the pulp and paper industry in combined heat and power systems. Worldwide biopower generation is expected to grow to more than 30,000 MW by 2020 (Baskar, 2012).

Limitations

The constraints and limitations initially tended to be of a technical nature: how to scale up the technology; how to widen the feedstock slate; how to develop engines that could handle larger proportions of biofuel in the fuel blend; how to improve agricultural feedstock yields, and so on. Then serious

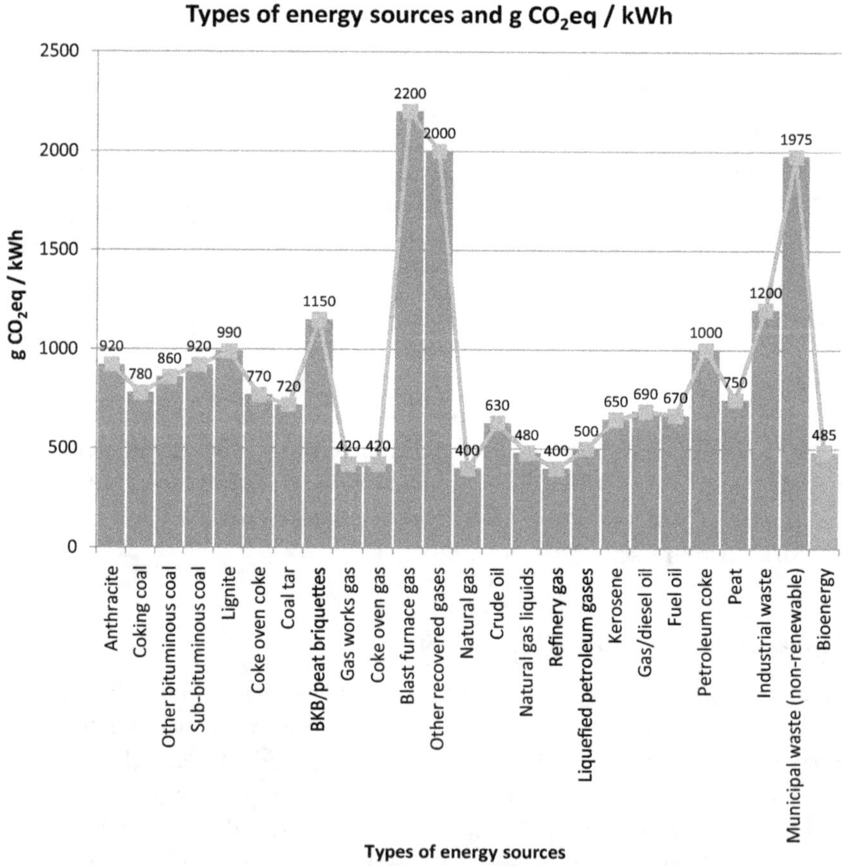

Fig. 11.15. Carbon footprint of bioenergy with various fossil fuels (*Source*: 2014 IPCC, Global warming potential of selected electricity sources).

questions were asked about the sustainability of biofuels, with an array of arguments reaching both negative and positive conclusions (Roddy, 2009); (Searchinger *et al.*, 2008).

Case study: Biomass Use In Bangladesh

On many levels, Bangladesh is a country that is ideally suited for the development of small-scale biomass energy systems. Because the economy is largely dependent on agriculture, the residues needed for such projects are available. Approximately 75% of the 130 million people in the country live in rural areas, and for all practical purposes they are not able to benefit from the national electricity transmission grid. The country is relatively poor. The main need for energy in rural Bangladesh is for cooking, although biomass is also used as housing material and animal feed. A limited amount of biomass is used as feedstock for recycled paper and in pulp mills.

The main sources of biomass in Bangladesh include rice husks, jute stalks, sugarcane stalks, and peanut shells. A thriving business in Bangladesh is biomass briquetting or densification. Briquetting processes require heat and pressure to produce fuel pellets from rice husks and wood chips. Another example of biomass use in Bangladesh is biodigesters. Unlike briquetting, biodigesters have had a mixed record of success. The principal barriers to further commercialisation of the technology are high capital costs and lack of financing options. Despite economic and availability hurdles, it is expected that biomass will continue to play a key role in supplying the energy needs of Bangladesh (Islam & Mondal, 2013).

Carbon Footprint Comparison of Bioenergy with Various Fossil Fuels

Geothermal Energy

Basics

Miles beneath the earth's surface lies one of the world's largest energy resources — geothermal energy. Geothermal energy is the heat contained below the earth's crust. This heat, brought to the surface as steam or hot water, is created when water flows through heated, permeable rock. It is

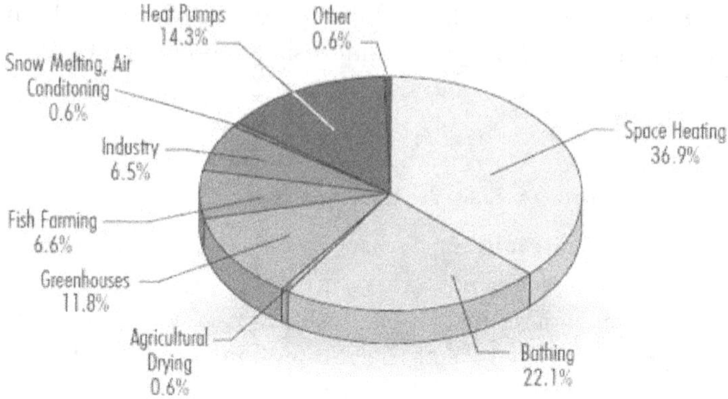

Fig. 11.16. Geothermal Heat, Direct Use (*Source*: Chambers, 2004).

used directly for space heating in homes and buildings or converted to electricity[24] (Fig. 11.16).

Geothermal resources come in five forms:

- hydrothermal fluids
- hot dry rock
- geopressured brines
- magma
- ambient ground heat

Of these five, only hydrothermal fluids have been developed commercially for power generation. Three technologies can be used to convert hydrothermal fluids to electricity. They are conventional steam turbines, flash steam technology and binary-cycle technology. The type of conversion used depends on whether the fluid is steam or water, and its temperature (Fig. 11.17).

Power production

Geothermal electricity is clean, reliable, and cost effective. Geothermal resources represent an abundant, secure source of energy. Today's hydrothermal power plants with modern emissions controls have minimal

[24]Geothermal Energy Association, (n.d.). *Geoenergy Basics*. <http://geo-energy.org/Basics.aspx> (accessed November 2014).

Fig. 11.17. Hot Springs Steam (*Source*: Chambers, 2004).

impact on the environment. The plants release little or no CO_2. Geothermal power plants are very reliable when compared to conventional power plants (Chambers, 2004).

Technologies, Sciences and Application

Geothermal resources are distributed throughout the planet. Even though most geothermal systems and the greatest concentration of geothermal energy are associated with Earth's plate boundaries, geothermal energy may be found in most countries. It is highly concentrated in volcanic regions, but may also be found as warm groundwater in sedimentary formations worldwide. In many cases, geothermal energy is found in populated, or easily accessible, areas. Moreover, geothermal activity is also found at great depths on the ocean floor, in mountainous regions, and under glaciers and ice caps (Axelsson, 2012).

Applications

Geothermal energy was, for the first time in the 20th century, harnessed on a large scale for space heating, electricity generation, and industry. Electric power was first generated from Larderello, Italy, in 1904, and commercial-scale electricity generation began in Larderello in 1913. The first large-scale municipal geothermal district heating service began in Iceland in 1930 (Fridleifsson, 2001).

Table 11.6. Range of efficiencies for various geothermal plant types (*Source*: DiPippo, 2012).

Plant type	Utilization efficiency (%)	Thermal efficiency (%)
Dry-steam	45–55	Not applicable
Single-flash	25–35	Not applicable
Double-flash	35—45	Not applicable
Basic subcritical binary	15–45	5–15
Supercritical binary	16–50	5–15
Binary with recuperator	18–55	14–18
Binary with mixture WF	20–55	15–20

Today, geothermal energy is primarily utilised in three technology categories:

- heating and cooling buildings via geothermal heat pumps that utilise shallow sources;
- heating structures with direct-use applications; and
- generating electricity through indirect use (Davidsdottir, 2012).

Efficiency, Environmental Impacts and Mitigation Measures

Plant performance assessment

Table 11.6 gives typical ranges of the efficiency values for various types of geothermal power plants. The values for all types of binary plants are strongly dependent on the temperature of the geofluid supplied to the plant; generally, the higher the temperature, the higher the efficiency (DiPippo, 2012).

Environmental Considerations

Geothermal energy is considered a renewable and 'green' energy resource; however, there are several environmental impacts that must be considered and are usually mitigated. These are the emission of harmful gases, noise pollution, water use and quality, land use, and impact on natural phenomena, wildlife, and vegetation. Nevertheless, the use of geothermal energy for direct-use is reliable, is renewable, has minimum to no air emission, has minimum environmental impacts, is combustion free, and is a domestic fuel source (Kagel *et al.*, 2005).

Market and Economic Feasibility

Microeconomic analysis

Cost comparison between geothermal projects in different countries and companies is highly sensitive to various factors. The main factors determining the investment costs are the following:

1. Resource conditions
    a. Temperature of reservoir
    b. Depth of reservoir
    c. Drilling conditions
    d. Permeability/fluid yield
2. Geographical conditions
    a. Location of reservoir
        i. Topography, access
        ii. Distance to market
    b. Availability of technical services, workforce, material, equipment, and contractors
    c. Investment environment, taxes, government support, and so on.
3. Global conditions
    a. Development of commodities price, for example, oil and steel
    b. Availability of manufacturers and lead time. Referring to the investment costs mentioned earlier, a typical split of a US$4 million MW^{-1} geothermal project is:

- Exploration: US$0.5 million MW^{-1}
- Reservoir development: US$1.0 million MW^{-1}
- Construction: US$2.5 million MW^{-1}
- Total investment cost: US$4.0 million MW^{-1}.

In addition to this, the cost of developing geothermal projects has been increasing due to more stringent environmental requirements.

Macroeconomic analysis

When undertaking a geothermal project, several key factors need to be in place. Among the most important are access to the geothermal resource, knowledge of how to harness it, capital, market, and the price of energy (Kristjánsdóttir and Margeirsson, 2012).

Limitations

Materials used in high-temperature geothermal wells and equipment in direct contact with geothermal fluid can be subjected to corrosion; this results in high costs associated with the materials, labour, and production efficiency of wells. These issues can be avoided by correct material selection, good engineering design, and proper corrosion and scaling control methods (Karlsdóttir, 2012).

Case Study: Tomato Drying in Greece

A small tunnel dryer has been constructed in northern Greece for the dehydration of tomatoes (Andritsos *et al.*, 2003). The drying tunnel is a 14 m long × 1 m wide × 2 m high construction of polyurethane aluminium panels (Fig. 11.18). Geothermal water at the temperature of 59°C and flow rate of $25\,m^3\,h^{-1}$ is fed into a finned-tube coil air/water heat exchanger to heat the drying air with a capacity of 1.26 MJ, producing air temperature of 55°C. The drying racks have 25 trays, each holding about 7 kg of raw tomatoes. One rack is introduced into the head of the tunnel every 45 min. It then takes about 30 h for the rack to reach the end of the tunnel, reducing the moisture content from about 150% to 10% by dry weight; thus, 100 kg of raw product produces about 9.5 kg of dried product. The long drying time appears to preserve the colour and aroma of the tomatoes as compared to sun drying. Only three persons operate the plant. Up to 5.5 tons have been produced in a year.

Carbon Footprint Comparison of Geothermal Energy with Various Fossil Fuels

Hydro Power

Basics

Hydropower converts the energy in flowing water into electricity. The quantity of electricity generated is determined by the volume of water flow and the amount of head. The greater the flow and head, the more electricity produced.

Hydropower plants provide inexpensive electricity and produce no pollution. Unlike other energy sources such as fossil fuels, water is not destroyed during the production of electricity — it can be reused for other

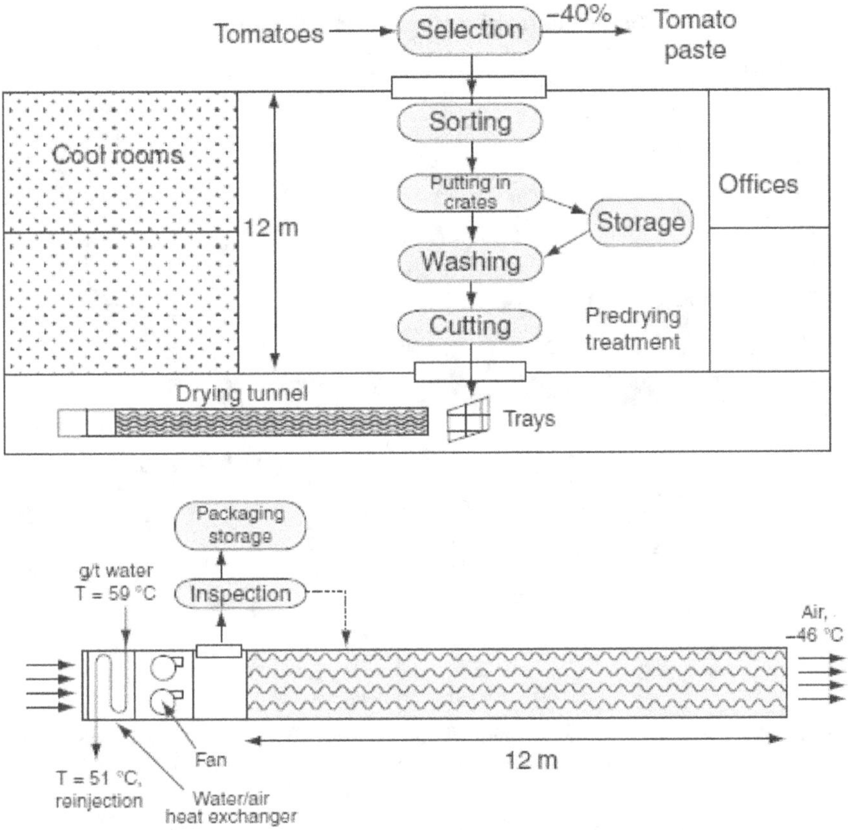

Fig. 11.18. Schematic diagram of the geothermal tomato dryer system. (*Source*: Lund, 2012).

purposes. The best sites for hydroelectric plants are swift-flowing rivers or streams, mountainous regions, and areas with heavy rainfall.[25]

Technologies, Sciences and Application

Types of turbines

In water turbines, the kinetic energy of flowing water is converted into mechanical rotary motion. Theoretical power is determined by the available head and the mass flow rate. To calculate the available power, head losses

[25]U.S. Department of Energy, (n.d.). *Hydropower Basics.* <http://energy.gov/eere/water/hydropower-basics> (accessed November 2014).

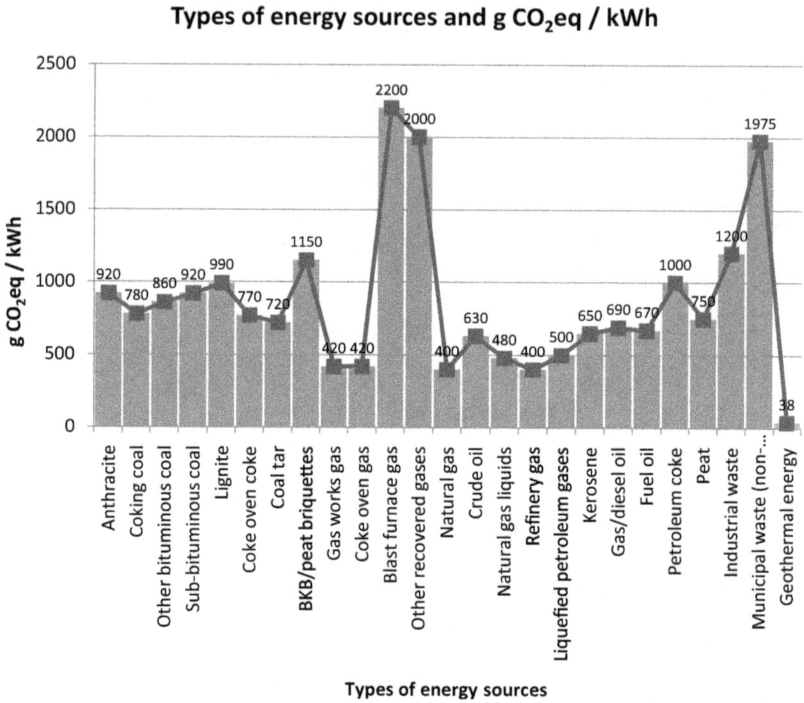

Fig. 11.19. Carbon footprint of geothermal energy with various fossil fuels (*Source*: 2014 IPCC, Global warming potential of selected electricity sources).

due to friction of flow in conduits and the conversion efficiency of machines employed must also be considered. The formula, thus, is the following:

$$P = H_n Q_{\rho g} E_s \text{P(in watts)}$$

where P is the output power in watts; $H_n$the net head = gross head – losses (m); Q the flow in $m^3\,s^{-1}$; g the specific gravity = $9.81\,ms^{-2}$; ρ the specific mass of the water; and E_s the overall efficiency.

Types of dams

The primary purpose of a dam may be defined as to provide for the safe detention and storage of water. There is no nominal structural design life for dams, if the effects of reservoir siltation or similar time-dependent limitations on their operational utility are disregarded. As a corollary to this, every dam must represent a design solution specific to its site circumstances. The design therefore also represents an optimum balance of local technical and economic considerations at the time of construction (Lejeune and Hui, 2012).

Efficiency, Environmental Impacts and Mitigation Measures

Efficiency

Efficiencies of the components of a hydroelectric system are measured as the ratio of energy output to input or to total potential energy at the site. No component is perfect, because its functioning involves lost energy (conversion to heat). The efficiency of a plant or system is the product of the efficiencies of its several components; thus,

$$E_s = E_c E_t E_g E_u E_l E_d \dots \tag{1}$$

where $\boldsymbol{E_s}$ is the over-all system efficiency made up of the product of the several efficiencies of the conduits; $\mathbf{E_t}$ is the efficiency of the turbines, including the scroll case and the draft tube; $\boldsymbol{E_g}$ is the efficiency of the generators, including the exciter; $\boldsymbol{E_u}$ is the efficiency of the step-up transformers; $\boldsymbol{E_l}$ is the efficiency of the transmission lines; $\boldsymbol{E_d}$ is the efficiency of the step-down transformers; and $\boldsymbol{E_c}$ is the efficiency of the canal, the tunnel, the penstocks, and the tailrace (Lejeune and Hui, 2012).

Environmental, social issues and mitigation

Current hydropower technology, while essentially emission-free, can have undesirable environmental effects, such as fish injury and mortality from passage through turbines, as well as detrimental effects on the quality of downstream water. A variety of mitigation techniques are in use now, and environmentally friendly turbines are under development.

Market and Economic Feasibility

The sources of hydropower generation are widely spread around the world. Potential exists in about 150 countries, and about 70% of the economically feasible potential sites remain to be developed. They are mostly in developing countries (Lejeune and Hui, 2012).

With more than a century of operating experience and construction know-how, well-advanced technology and modern power plants providing the most efficient energy conversion process (>90%), hydropower has proven to be a dependable power source.

The production of peak-load energy from hydropower is another economic benefit. It allows for the best use of other less flexible electricity-generating sources to produce the base-load power, notably wind and solar power. Its fast response time enables it to meet the sudden fluctuations in demand in the supply electric grids.

Hydropower plants have the lowest operating costs and the longest plant life compared to other large-scale generating options. Typically a hydro plant in service for 40–50 years can have its operating life doubled.

Hydropower, as an energy supply, also provides unique benefits to an electrical system. First, when stored in large quantities in the reservoir behind a dam, it is immediately available for use when required. Second, the energy source can be rapidly adjusted to meet demand. These benefits are part of a large family of benefits, known as ancillary services.

Limitations and Restrictions

- Hydropower plants require suitable topographic situations and should be adjusted to atmospheric and environmental conditions.
- Construction period could be lengthened without appropriate technological know-how and/or expertise.
- Adverse environmental impacts and archeological problems.
- There are a lot of stakeholders and institutes with diverse interests involved in water issues, such that decision-making sometimes becomes either exhausting or impossible.
- The organisation that is chiefly responsible for developing hydropower projects sometimes does not have as much authorisation as necessary.
- The existence of complexities and ambiguities in water laws and identifying stakeholders needs.
- Providing the required investments are sometimes difficult; large hydropower projects have high (construction) costs and thus need sizeable investments.
- The existence of ambiguities in economic assessment of these projects.

Case Study: The Three Gorges Project in China (Suo, Niu and Xie, 2012)

The Three Gorges Project (TGP) is a multipurpose development project with great comprehensive benefits — mainly in flood control, power

Table 11.7. Key features of the Three Gorges Project.

Normal pool level	175 m (156 m at the initial stage)
Limit level far flood control	145 m (135 mat the initial stage)
Low level in dry season	155 m (140 mat the initial stage)
Pool level with 1% flood	166.9 m
Design peak flow (0.1% flood)	98800 m^3s^{-1}
Design flood level (0.1% flood)	175 m
Check peak flow (0.01 % flood plus 10%)	124000 m^3s^{-1}
Check flood level	180.4 m
Total storage capacity (below the normal pool level)	39.3 billion m^3
Storage capacity for flood control (145–175 m)	22.15 billion m^3
Regulating storage (155–175 m)	16.5 billion m^3
Crest elevation	185 m
Installed gross capacity/guaranteed output	22.4 GW/4.99 GW
Average annual output	90.0 billion kWh
Single unit capacity/number of units	700 MW/32 units

generation, navigation, and supplying water to the downstream during the dry season.[26,27]

Benefits of the Three Gorges Project, China

Flood control

The primary purpose of building the TGP is to protect the middle and downstream of the Yangtze River from floods, as well as to improve the sustainable development of the middle and downstream area along the Yangtze River.

Power generation

The total installed capacity amounted to 22.4 GW, with the corresponding power generation of 90.0 billion kWh, which is equivalent to building a super coal mine with an annual production of 50 million tons of coal or a super oil field with an annual production of 25 million tons of oil. In other words, its power generation capacity ranks the largest in the world.

[26]Changjiang Institute of Survey, Planning, Design and Research, (1997). *Yangtze Three Gorges Project Technical Series (in Chinese)*. Wuhan, China: Hubei Science and Technology Press.

[27]Niu X. and Wang X. (2004). Design and study on the layout of the Three Gorges Project of the Yangtze River. *International Conference on Dam Engineering*. Nanjing, China.

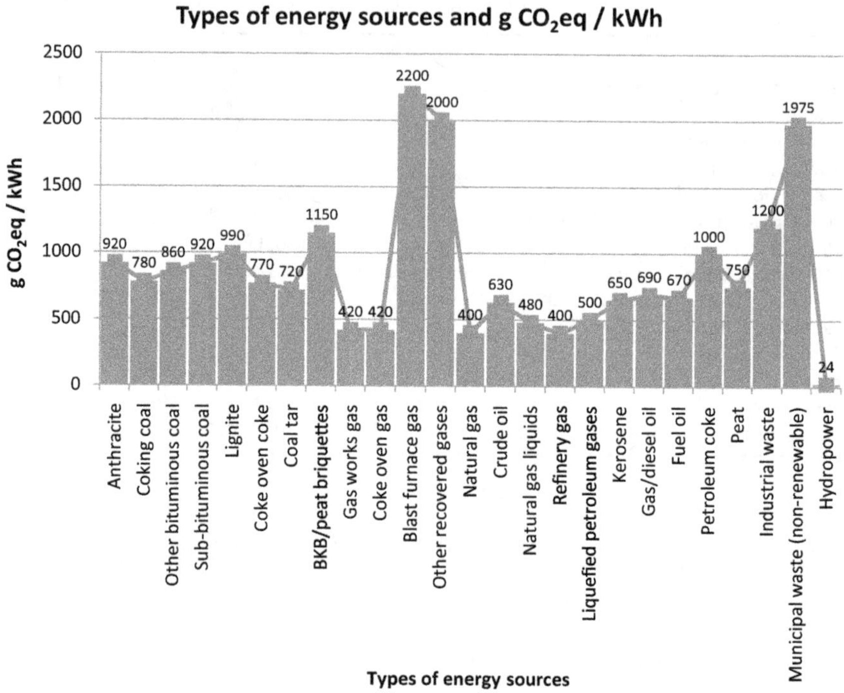

Fig. 11.20. Carbon footprint of hydropower with various fossil fuels (*Source*: 2014 IPCC, Global warming potential of selected electricity sources).

Navigation

The river channel of the Yangtze River has been always called the golden channel because it is a traffic artery connecting the coastal area in the southeast with the hinterland in the southwest, forming a complete inland navigation system.

Carbon footprint comparison of hydropower energy with various fossil fuels

Discussion and Conclusion

Discussion

Industrialisation is a key criterion for development, and renewable energy sources are the focus of recent industrialisation strategies. Studies have shown that industrialisation has progressed at the expense of the environment and its natural resources, which (in economic context) are often termed as "natural capital". By looking at the situation of developed countries, it is obvious that uncontrolled extraction and utilisation of natural resources, such that wastage is generated, will soon lead to resource depletion. According to environmental economists, this relation can be explained appropriately through the first law of thermodynamics, in which matter can neither be created nor destroyed. A direct interpretation of this law for economic activity at a given time: {Sum of matter contained in Source Materials that enter the economy} must equal: {The sum of matter retained within the economy + The sum of matter deposited in sinks}. This became the basis for the Materials Balance Approach to accounting for the role of nature (Thampapillai, 2012).

The depletion of natural resources also leads to a host of negative environmental issues, such as climate change, which is caused by pollution. This report sought to evaluate the various forms and sources of renewable energy; the technological techniques to harness them in an efficient manner suitable for application; and their economic assessment with market feasibility and outlook — each with its own set of limitations — against case studies with carbon footprint analyses for comparison as a measuring tool for assessing their respective environmental impacts.

Industrialisation is deemed necessary for economic progress, hence every country is trying to boost their economic standing by achieving a high level of industrialisation. The pioneer developed countries achieved their current status with irreversible environmental degradation without consideration for appropriate mitigation measures and control. As such, in order for the future generation to be able to live on a safe Earth, today's generation needs to carry out industrialisation in a sustainable manner so that future generations are able to live as comfortably as, if not better off than, their forefathers. Many environmental conventions are being drafted to ensure a sustainable future for everyone.

Since natural capital is not replenished as quickly as it is being consumed, the way forward is to rely on renewable energy to achieve energy security and a stable economy. Renewable energy technological advancement needs to address the exponential demand on energy usage and consumption. One increasingly popular approach in attempting to reflect the environmental significance of energy technology, is to use the resultant carbon emissions to account for the technology's environmental impact based on current methods and approaches. Carbon accounting and analysis covers all GHG emissions and translates them into equivalent carbon dioxide terms. However, while comparisons of the carbon and GHG emissions are useful and central in terms of global climate change, there are also clearly other impacts to consider in order to give a complete picture — e.g. acid emissions and radioactive emissions. According to environmental economists though, the natural environment cannot be valuated once it is used up because the natural recovery process takes ages. Thus in the context of carbon accounting, the limitation is that natural capital is often undervaluated as current technological advancement is still unable to ensure the total and full recovery of environmental assets.

One of the challenges in making renewable energy adoption attractive is to design creative renewable support schemes that are effective and cost-efficient, but also take into consideration existing and planned infrastructure in order to minimise adverse effects.

Conclusion

The concept behind industrialisation has changed over the years with lots of lessons learnt. First world countries, mostly Annex I parties in the Kyoto Protocol (KP) in the United Nations Framework Convention on Climate Change (UNFCCC), have gone through a scenario where lots of environmental attributes have been compromised in the process of industrialisation and development. Recent conventions are beginning to place emphasis on arising environmental issues leading to global warming and the resulting climate change.

There is a rising importance placed on industrial development in a sustainable manner, influenced to a great extent by energy consumption. Carbon emission is the attribute which has direct impact on the environment, and is rising in importance in global conventions — especially those related to climate change scenarios. As such, this aspect associated with the adoption of renewable energy for industrial plants will be discussed and analysed in the context of this report.

From various literatures, bioenergy is ranked highest in carbon emissions due to the by-products released in the combustion process. Its bioenergy feedstock, which is known as natural capital in economics terms, is often undervalued and unfairly accounted in various aspects. On the other hand, wind energy is found to be the "cleanest" source of renewable energy (Fig. 11.21); however, the harnessing process and plant construction often proves to be costlier than other sources. Further emphasis on other aspects can be conducted taking into account notable and significant indicators using life cycle assessment (LCA) approach guidelines for a more holistic

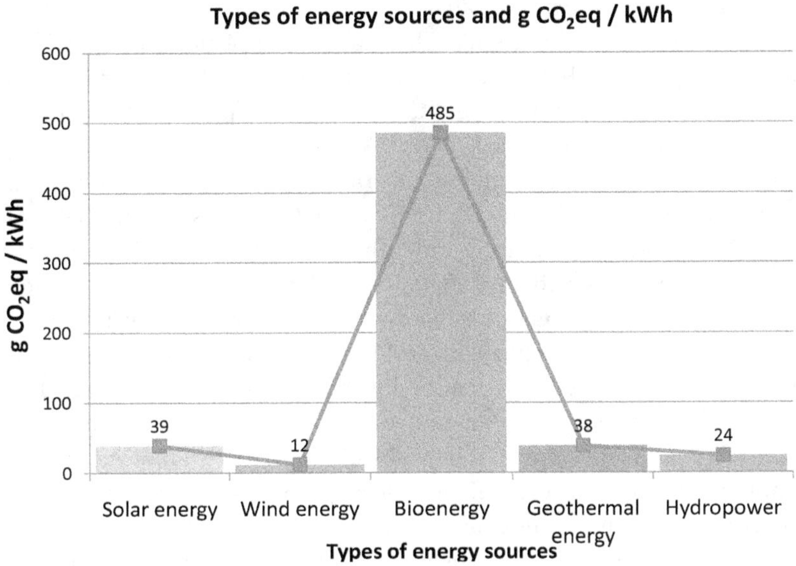

Fig. 11.21. Comparison of carbon footprint of renewable energy sources.

weightage for appropriate and suitable choices to be adopted on different pristine or existing environmental conditions. As life cycle assessment studies are usually only as good as how accurate the assumptions and setting of processes and boundaries are, different sources may give different conclusive analysis.

Some literature reports proposed indicators such as price of electricity generation, availability and technological limitations, efficiency of energy generation, land use and social impacts to rank the different renewable energies (See Table A.4 in the Appendix). For PV systems, the selection of cell material is critical for efficiency, and different composite materials may give different results. The challenge with wind energy is site selection and their visual impact on their surroundings. Bioenergy involves optimising the combustion process for different types of feed stock. Geothermal energy sites have impact on natural phenomena, wildlife, and vegetation. Hydropower energy sites pose maintenance difficulties, corrosion issues and impacts on marine biology.

Across the globe, financial support mechanisms are clearly seen as a key tool to encourage the deployment of renewable energy. However, the cost of such schemes must ultimately be borne by society at large, and there will be tensions, especially in economically straitened times, over how much we are willing to pay to achieve our renewable energy ambitions. Maximising efficiency in support schemes will therefore be a priority for politicians, but this will need to be balanced against the needs of investors in terms of certainty and level of reward as, ultimately, a support scheme that does not work for them will be ineffective.

Renewable energy will play a central role in the world's decarbonised energy future. A study by EURELECTRIC suggests that 38% of the European Union's electricity (and approximately 50% of its generating capacity) will be renewable by 2050. Judging from current trends and technological knowledge, renewable energy projects are mostly much costlier than conventional energy projects and require policy formation and subsidy to move forward, especially when many stakeholders are involved; thus, the policy implementation mechanism needs to address these trying situations. It is highly unlikely that public subsidies for renewable energy will be sustainable in the long term, given the large volumes that would need to be supported. The renewable energy industry will therefore need to prepare for a transition into a world without subsidy, in which renewable energy projects are able to stand on their merits alone (Porter, 2012).

The future for renewable energy is optimistic. Many industrialised countries have recognised the importance of sustainable development. Although sustainability may still remain a far-fetching concept in many developing countries, as they would rather propel economic advancement at the expense of environmental preservation, recent global conventions have begun addressing environmental issues arising from industrialisation, and are adopting measures and policies to mitigate pollution.

References

American Wind Energy Association (2008). 20% Wind energy by 2030. Retrieved from: http://www.awea.org/ (accessed November 2014).

Aydin, N.Y., Kentel, E., and Duzgun, S. (2010). GIS-based environmental assessment of wind energy systems for spatial planning: A case study from Western Turkey. *Renewable and Sustainable Energy Reviews*, **14**, 364–373.

Baker, T.L. (1985).*A Field Guide to American Windmills*, 1st ed. Norman, OK: University of Oklahoma Press.

Barbounis, T.G., and Theocharis, J.B. (2007). A locally recurrent fuzzy neural network with application to the wind speed prediction using spatial correlation. *Neurocomputing*, **70**, 1525–1542.

Baskar, C., Baskar, S., and Dhillon R. S. (2012). *Biomass conversion: the interface of biotechnology, chemistry and materials science.* Berlin: Springer-Verlag.

Blankenship, R.E., Tiede, D.M., Barber, J., Brudvig, G.W., Fleming, G., Ghirardi, M., Gunner, M.R., Junge, W., Kramer, D.M., Melis, A., Moore, T.A., Moser, C.C., Nocera, D.G., Nozik, A.J., Ort, D.R., Parson, W.W., Prince, R.C., and Sayre, R.T. (2011). Comparing Photosynthetic and Photovoltaic Efficiencies and Recognizing the Potential for Improvement. *Science*, **332**, pp. 805–809, DOI: 10.1126/science.1200165.

Blankership, R.E., Tiede, D.M., Barber, J., Brudvig, G.W., Fleming, G., Ghirardi, M., Gunner, M.R., Junge, W., Kramer, D.M., Melis, A., Moore, T.A., Moser, C.C., Nocera, D.G., Nozik, A.J., Ort, D.R., Parson, W.W., Prince, R.C., and Sayre, R.T., Sennholz H.F. (2006). *The third industrial revolution*. Retrieved from: http://www.mises.org/story/2105 (accessed December 2014).

Brundtland Commission (1987). *Report of the World Commission on Environment and Development*. United Nations.

Brundtland. G.H. (1987). *Report of the World Commission on Environment and Development: Our Common Future*. Retrieved from: http://conspect.nl/pdf/Our_Common_Future-Brundtland_Report_1987.pdf (accessed July 2014).

Burger, J., and Gochfeld, M. (2012). A conceptual framework evaluating ecological footprints and monitoring renewable energy: Wind, solar, hydro, and geothermal. *Energy and Power Engineering*, **4**(4), 303–314.

California Energy Commission (2010). Comparative costs of California central station electricity generation. *CEC-200-2009-07. SF*. California Energy Commission, Sacramento, California.

Chambers, A. (2004). *Renewable Energy in Nontechnical Language*. Tulsa, Okla.: PennWell Corp.

Changjiang Institute of Survey, Planning, Design and Research, (1997). *Yangtze Three Gorges Project Technical Series (in Chinese)*. Wuhan, China: Hubei Science and Technology Press.

Chen, H., Cong, T.N., Yang, W., Tan, C., Li, Y., and Ding, Y. (2009). Progress in electrical energy storage system: A critical review. *Progress in Natural Science*, **19**, 291–312.Chinese Renewable Energy Industries Association, (2010). China wind energy outlook — 2010. Retrieved from Greenpeace International: http://www.greenpeace.org/ (accessed November 2014).

Commission of the European Communities, (2008). Energy sources, production costs and performance of technologies for power generation, heating and transport. *SEC (2008) 2872*. Commission of the European Communities, Brussels. Retrieved from: http://ec.europa.eu/energy/strategies/2008 (accessed November 2014).

Despotou, E. (2012). Vision for Photovoltaics in the Future, In *Comprehensive Renewable Energy*, by Sayigh, A. (ed.), pp. 179–198. Oxford: Elsevier.

Dinh, D.D. (n.d.). Sustainable Industrialization — A New Strategic Trend In Asian Developing Economies, Retrieved from: http://www.uvic.ca/research/centres/capi/assets/docs/Dinh_Sustainable_Industrialization.pdf (accessed June 2014).

Dodo, T.J. (2012). Personal Communication. Lee Kuan Yew School of Public Policy, National University of Singapore.

Dones, R., Bauer, C., Bolliger, R., Burger, B., Faist, E.M., Frischknecht, R., Heck, T., Jungbluth, N., and Röder, A. (2003). Sachbilanzen von Energiesystemen. Final report ecoinvent 2000, vol. 6. Swiss Centre for LCI, PSL, Dubendorf, CH.

Enerdata (2009). *"Yearbook 2009: The 2008 world energy review,"* (2009). Tech. Rep. Retrieved from: www.enerdata.fr> (accessed November 2014).

Energy Information Administration (2009). Annual energy outlook. US Department of Energy. Retrieved from: http://www.eia.doe.gov (accessed December 2014).

Energy Information Administration (2011). International energy statistics. Retrieved from: http://www.eia.gov (accessed November 2014).

EnviroMission Limited (2014). *Clean Green Renewable Energy*. Retrieved from: http://www.enviromission.com.au/EVM/content/technology_technologyover.html (accessed October 2014).

Esty, D.C., and Cornelius, P. (2002). *Environmental performance measurement: The global report 2001–2002*. New York: Oxford University Press.

European Photovoltaic Industry Association (2010). *Global Market Outlook for Photovoltaics until 2015*. Brussels, Belgium: EPIA.

European Wind Energy Association (n.d.). Retrieved from: http://www.ewea.org (accessed November 2014).

Evans, A., Strezov, V., and Evans, T.J. (2009). Assessment of sustainability indicators for renewable energy technologies, *Renewable and Sustainable Energy Reviews*, **13**(5), 1082–1088.

Finn, D. (2009). Our Uncertain Future: Can Good Planning Create Sustainable Communities?, dissertation, University of Illinois at Urbana-Champaign.

Forster, P., Ramaswamy, V., Artaxo, P., Berntsen, T., Betts, R., Fahey, D.W., Haywood, J., Lean, J., Lowe, D.C., Myhre, G., Nganga, J., Prinn, R., Raga, G., Schulz, M., and Dorland, R.V. (2007). Changes in atmospheric constituents and in radiative forcing. In: *Climate Change 2007: The Physical Science Basis. Contributions of Working Group I to the Fourth Annual Assessment Report of the Intergovernmental Panel on Climate Change*. Cambridge UK and New York: Cambridge University Press.Franklin Associates (1998). *USA LCI Database Documentation*. Prairie Village, KS: Franklin Associates.

Fthenakis, V. (2003). Overview of potential hazards. In: Markvart, T. and Gastaner, L. (eds.) *Practical Handbook of Photovoltaics: Fundamentals and Applications*, pp. 857–868. Kidlington, UK: Elsevier.

Fthenakis, V.M., (2004). Life cycle impact analysis of cadmium in CdTe PV production. *Renewable and Sustainable Energy Reviews*, **8**(4): 303–334. DOI: 10.1016/j.rser.2003.12.001.

Fthenakis, V.M., and Kim, H.C. (2012). Environmental Impacts of Photovoltaic Life Cycles, In *Comprehensive Renewable Energy*, by Ali Sayigh (ed.), pp. 143–159. Oxford: Elsevier.

Fthenakis, V.M., Kim, H.C., and Alsema, E. (2008). Emissions from photovoltaic life cycles. *Environmental Science & Technology*, **42**, 2168–2174. DOI: 10.1021/es071763q.

Fthenakis, V.M., Kim, H.C., Colli, A., and Kirchsteiger, C. (2006). Evaluation of risks in the life cycle of photovoltaics in a comparative context. In: *21st European Photovoltaic Solar Energy Conference*, pp. 3194–3201. Dresden, Germany.

Geothermal Energy Association, (n.d.). Geoenergy Basics. Retrieved from: http://geo-energy.org/Basics.aspx (accessed November 2014).

Global Wind Energy Council, (2011). Global wind report 2010. Retrieved from: http://www.gwec.net (accessed November 2014).

Green, D. (2012). Pros and Cons of Photovoltaic (PV) panels — solar energy. Retrieved from Green Energy Saving Tips: http://www.greenenergysavingtips.com/pros-and-cons-of-photovoltaic-pv-panels-solar-energy/ (accessed August 2014).

Grober, U. (2007). *Deep roots — A conceptual history of "sustainable development"* (Nachhaltigkeit), [in] Wissenschaftszentrum Berlin für Sozialforschung, pp. 8–9.

Haimes, Y.Y. (1998). *Risk Modeling, Assessment, and Management.* Hoboken, NJ: John Wiley & Sons.

Hirschberg, S., Spikerman, G., and Dones, R. (1998). *Severe Accidents in the Energy Sector*, 1st ed. Switzerland: Paul Scherrer Institute.

Ibrahim, H., Ghandour, M., Dimitrova, M., Ilinca, A., and Perron, J. (2011). Integration of wind energy into electricity systems: Technical challenges and actual solutions. *Energy Procedia*, **6**: 815–824. Information Resources Management Association, (2014). *Sustainable practices: Concepts, methodologies, tools, and applications.* Hershey, Pennsylvania: IGI Global.

International Electrotechnical Commission, (n.d.). Retrieved from: http://www.iec.ch/about/brochures/pdf/technology/renewable_4energies_2.pdf (accessed December 2014).

Islam, M., and Mondal, T. (2013). Potentiality of Biomass Energy for Electricity Generation in Bangladesh. *Asian Journal of Applied Science and Engineering,* **2**(2), 103–110.

Jones, S. (2008). *Standard Loss Definitions for Wind Resource/Energy Assessments Global Energy Concepts (DNV) Paper*, AWEA. Retrieved from: www.talentfactory.dk (accessed December 2014.).

Kaldellis, J.K. (2012). Wind Energy — Introduction, In *Comprehensive Renewable Energy*, by Ali Sayigh (ed.) , pp. 1–10. Oxford: Elsevier.

Kaldellis, J.K. (2010). *Stand-Alone and Hybrid Wind Energy Systems: Technology, Energy Storage and Applications*, 1st ed. Cambridge, UK: Woodhead Publishing.

Kaldellis, J.K., and Zafirakis, D. (2011). The wind energy (r)evolution: A short review of a long history. *Renewable Energy*, **36**, 1887–1901.

Kondili, E., and Kaldellis, J.K. (2012). Environmental-Social Benefits/Impacts of Wind Power, In *Comprehensive Renewable Energy*, by Ali Sayigh (ed.), pp. 503–539. Oxford: Elsevier.

Krohn, S. (2009). *The Economics of Wind Energy.* European Wind Energy Association, Brussels. Retrieved from: http://www.ewea.org (accessed November 2014).

Lamont, L.A. (2012). History of Photovoltaics, In *Comprehensive Renewable Energy*, by Ali Sayigh (ed.), pp. 31–45. Oxford: Elsevier.

Lejeune, A. (2012). Hydro Power — Introduction, In *Comprehensive Renewable Energy*, by Ali Sayigh (ed.), pp. 1–14. Oxford: Elsevier.

Lejeune, A., and Hui, S.L. (2012). Hydro Power: A Multi Benefit Solution for Renewable Energy, In *Comprehensive Renewable Energy*, by Ali Sayigh (ed.), pp. 15–47. Oxford: Elsevier.

Liserre, M., Sauter, T., and Hung, J.Y. (2010). Future Energy Systems: Integrating Renewable Energy Sources into the Smart Power Grid Through Industrial Electronics. *Industrial Electronics Magazine, IEEE*, **4**(1), 18. doi: 10.1109/MIE.2010.935861. Retrieved from: http://ieeexplore.ieee.org/stamp/stamp.jsp?tp=&arnumber=5439057&isnumber=5439039 (accessed December 2013).

Malmgren, A., and Riley, G. (2012). Biomass Power Generation, In *Comprehensive Renewable Energy*, by Ali Sayigh (ed.), pp. 27–53. Oxford: Elsevier.

Mathew, S. (2006). *Wind Energy: Fundamentals, Resource Analysis and Economics*, 1st ed. Berlin, Heidelberg, Germany: Springer.

Mathew, S., and Philip, G.S. (2012). Wind Turbines: Evolution, Basic Principles, and Classifications, In *Comprehensive Renewable Energy*, by Ali Sayigh (ed.), pp. 93–111. Oxford: Elsevier.

Milborrow, D. (2012). Wind Energy Economics, In *Comprehensive Renewable Energy*, by Ali Sayigh (ed.), pp. 469–501. Oxford: Elsevier.

Milborrow, D.J. (2010). What a difference a year can make. *Windpower Monthly*, **26**(1), 41–47.

Niu, X., and Wang, X. (2004). Design and study on the layout of the Three Gorges Project of the Yangtze River. *International Conference on Dam Engineering*. Nanjing, China.

Ofgem, (2009). Project discovery. Energy market scenarios. Retrieved from Office of Gas and Electricity Markets: http://www.ofgem.gov.uk (accessed December 2014).

Olla, P. (2012). Global Sustainable Development and Renewable Energy Systems (pp. 1–354). Hershey, PA: IGI Global. doi:10.4018/978-1-4666-1625-7.

POSTnote, (2006). Carbon footprint of electricity generation. No. 268: 1–4. UK: Parliamentary Office of Science and Technology.

Pasqualetti, M.J., Righter, R., and Gipe, P. (2004). History of wind energy. In *Encyclopedia of Energy*, by Cleveland C. (ed.), pp. 419–433. Amsterdam, The Netherlands: Elsevier.

Phillis, Y.A., Kouikoglou V.S., and Zhu, X. (2009). Fuzzy Assessment of Material Recyclability and Its Applications. *Journal of Intelligent and Robotic Systems*, **55**(1), 21–38. doi:10.1007/s10846-008-9290-x.

Porter, D. (2012). Renewable Energy Policy and Incentives, In *Comprehensive Renewable Energy*, by Ali Sayigh (ed.), pp. 1–4. Oxford: Elsevier.

Pouliezos, A. (2012). Wind Turbine Control Systems and Power Electronics, In *Comprehensive Renewable Energy*, by Ali Sayigh (ed.), pp. 329–370. Oxford: Elsevier.

Rosenbloom, E. (2006). A Problem With Wind Power. Retrieved from: http://www.aweo.org/problemwithwind.html (accessed November 2014).

Saidur, R., Rahim, N.A., Islam, M.R., and Solangi, K.H. (2011). Environmental impact of wind energy. *Renewable and Sustainable Energy Reviews*, **15**(5), 2423–2430.

Sayigh A. (2012). *Comprehensive renewable energy*. Boston, Mass; Amsterdam: Elsevier.

Sorensen, B. (2004). *Renewable Energy — Its physics, engineering, use, environmental impacts, economy and planning aspects*, 3rd ed. Denmark: Academic Press.

Sovacool, B.K. (2009). The intermittency of wind, solar and renewable electricity generators: Technical barrier or rhetorical excuse?, *Utilities Policy*, **17**, 288–296.

Suo, L., Niu, X., and Xie, H. (2012). The Three Gorges Project in China, In *Comprehensive Renewable Energy*, by Ali Sayigh (ed.), pp. 179–226. Oxford: Elsevier.

The Earth Charter Initiative (2012). *Values and Principles to Foster a Sustainable Future*. Retrieved from: http://www.earthcharterinaction.org/content/pages/What-is-the-Earth-Charter%3F.html (accessed June 2014).

United Nations Conference on Environment & Development (1992). *Agenda 21*. Retrieved from: http://sustainabledevelopment.un.org/content/documents/Agenda21.pdf (accessed July 2014).

United Nations Environment Programme, (2011). Montreal Protocol. Retrieved from: http://ozone.unep.org/new_site/en/montreal_protocol.php (accessed June 2014).

UNFCCC (2010). Report of the conference of the parties on its fifteenth session, held in Copenhagen from 7 to 19 December 2009 — Addendum — Part Two: Action taken by the conference of the parties at its fifteenth session. United Nations Framework Convention on Climate Change, 30 March 2010, Copenhagen.

U.S. Department of Energy (n.d.). Biomass Basics: The Facts About Bioenergy. Retrieved from: http://www.energy.gov/sites/prod/files/2014/04/f14/biomass_basics.pdf (accessed November 2014).

U.S. Department of Energy (n.d.). Hydropower Basics. Retrieved from: http://energy.gov/eere/water/hydropower-basics (accessed November 2014).

U.S. Department of Energy, Office of Energy Efficiency and Renewable Energy (n.d.). Renewable Energy Technology Basics. Retrieved from: http://energy.gov/eere/energybasics/renewable-energy-technology-basics (accessed November 2014).

Wikipedia (n.d.). Best Research-Cell Efficiencies. Retrieved from: http://en.wikipedia.org/wiki/File:PVeff(rev100414).png (accessed April 2014).

Wind Energy the Facts, (2010). Scenarios and targets. Retrieved from: http://www.wind-energy-the-facts.org/ (accessed November 2014).

Wizelius, T. (2012). Design and Implementation of a Wind Power Project, In *Comprehensive Renewable Energy*, by Ali Sayigh (ed.), pp. 391–430. Oxford: Elsevier.

Appendices

Table A.1. Millennium Development Goals (MDGs).

189 United Nations (UN) member states and at least 23 international organizations have agreed to achieve by 2015. The following eight development goals have been adopted to improve the social and economic conditions in the world's poorest countries, encompassing universally accepted human values and rights:

1. Eradicate extreme poverty and hunger
2. Achieve universal primary education
3. Promote gender equality and empower women
4. Reduce child mortality
5. Improve maternal health
6. Combat HIV/AIDS, malaria, and other diseases
7. Ensure environmental stability
8. Develop a global partnership for development.

(Source: United Nations Millennium Declaration — 8 September 2000)

Table A.2. Top 10 Permitting Issues of Wind Siting (Chambers, 2004).

1.	Land use
2.	Noise
3.	Birds and other biological resources
4.	Visual resources
5.	Soil erosion and water quality
6.	Public health and safety
7.	Cultural and paleontological resources
8.	Socioeconomics, public services, and infrastructure
9.	Solid and hazardous wastes
10.	Air quality and climate

Source: NVVCC

Table A.3. Largest hydroelectric dams and plants in operation (Lejeune, 2012).

Dam	Country	Year of completion	Capacity (MW)	Max annual production (TWh)
Three Gorges	China	2009	17 600[a]	>100
Italpú	Brazil/ Paraguay	1984–2003	14 000	90
Guri (Simón Bolivar)	Venezuela	1986	10 200	46
Tucurui	Brazil	1984	8 370	21
Grand Coulee	USA	1942/1980	6 809	22.6
Sayano-Shusthenskaya	Russia	1985/1989	6 400	26.8
Krasnoyarskaya	Russia	1972	6 000	20.4
Robert-Bourassa	Canada	1981	5 616	
Churchill Falls	Canada	1971	5 429	35
Bratskaya	Russia	1967	4 500	22.6
Ust-llim skaya	Russia	1980	4 320	21.7
Yaciretá	Argentina/ Paraguay	1998	4 050	19.2
Longtan	China	2009	3 500[b]	18.7
Tarbela	Pakistan	1976	3 478	13
Ertan	China	1999	3 300	17.0
IIha Solteira	Brazil	1974	3 200	
Xingó	Brazil	1994/1997	3 162	
Gezhouba	China	1988	3 115	17.0
Nurek	Tajikistan	1979/1988	3 000	11.2

[a]22 500 when complete.
[b]6300 when complete.

Table A.4. Sustainability rankings (Evans *et al.*, 2009).

	Photovoltaics	Wind	Hydro	Geothermal
Price	4	3	1	2
CO_{2-e} Emissions	3	1	2	4
Availability and limitations	4	2	1	3
Efficiency	4	2	1	3
Land use	1	3	4	2
Water consumption	2	1	3	4
Social impacts	2	1	4	3
Total	20	13	16	21

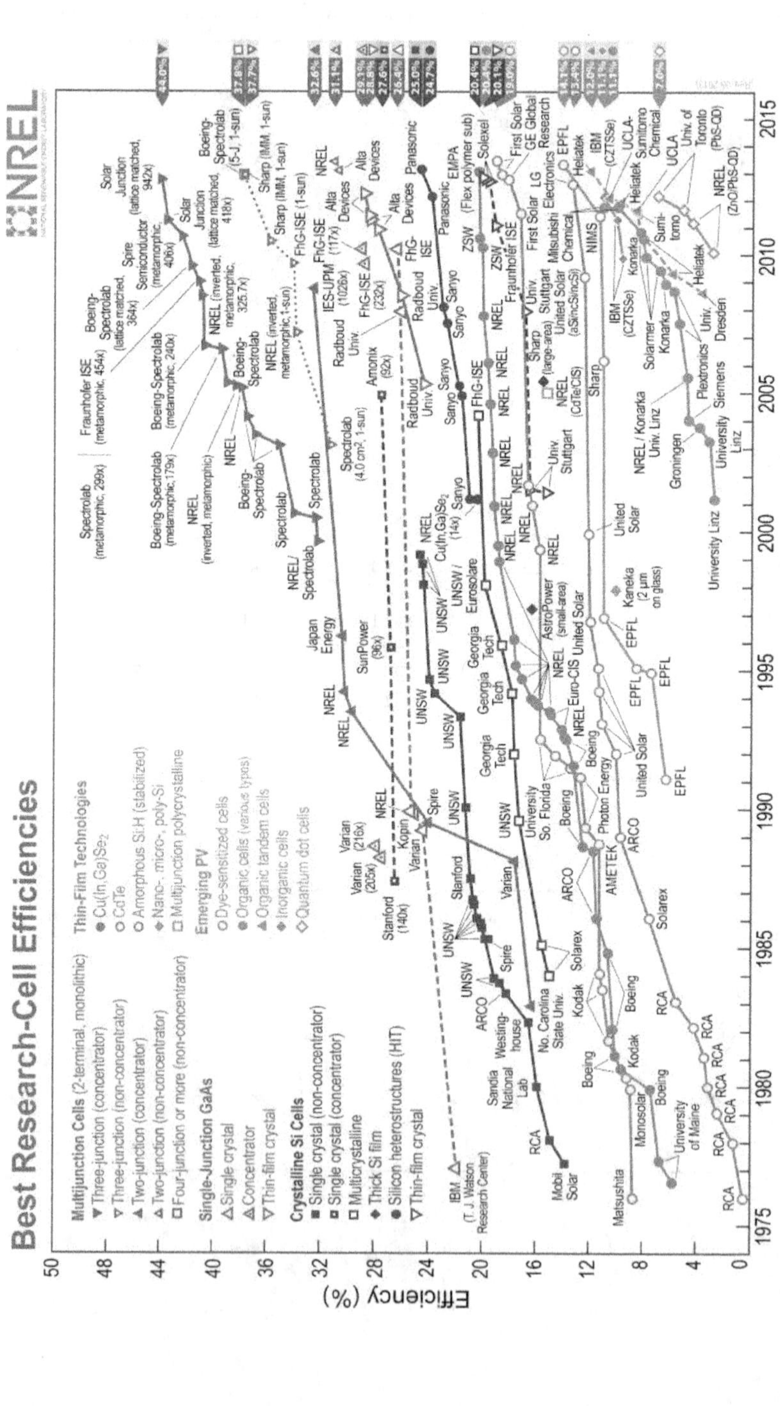

Figure A.1. Best research-cell efficiencies as a function of time for various technologies. (*Source:* NREL website (www.nrel.gov) and Wikipedia (http://en.wikipedia.org/wiki/Solar_cell_efficiency)).

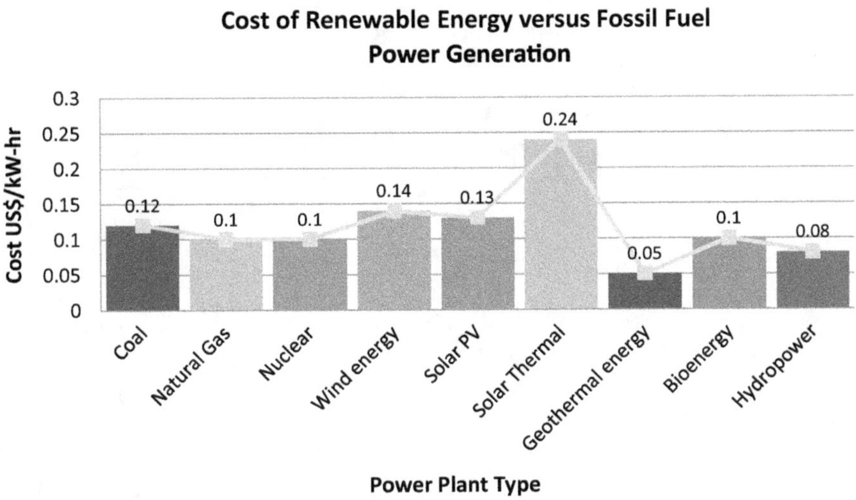

Figure A.2. Cost of Renewable Energy versus Fossil Fuel Power Generation (U.S. D.O.E. Annual Energy Outlook, 2014).

12

A COMPARATIVE ANALYSIS OF THE REGULATORY FRAMEWORK BETWEEN THE UNITED STATES OF AMERICA AND THE EUROPEAN UNION ON THE USE OF 'FRACKING' IN UNCONVENTIONAL HYDROCARBONS DEVELOPMENT

CARMELITA LEOW and NAMRATA CHINDARKAR

The development of 'fracking' has made previously uncommercial unconventional reservoirs commercially viable. The success of unconventional drilling in the United States of America has been hailed as a model for replication in other parts of the world. This method of drilling is not without environmental problems and industry development has progressed rapidly, with legislation struggling to keep up. Current legislation fails to address the social and environmental issues fully and frameworks for approaching the process are flawed. In this paper, we aim to do a comparative analysis of the regulatory frameworks of USA and EU, with opposing regulatory approaches on the management of unconventional drilling and production, showing the benefits and flaws of each system and the need for a reconsideration of a more dynamic method for crafting regulatory frameworks.

Introduction

Crude oil has been the favoured hydrocarbon for exploitation and production, due to its versatile nature and high marketability. Not only is it a fuel source, it is also the raw material used in the production of many chemicals, products and plastics (Swanson *et al.*, 1993). However, due to the decreasing supply of crude oil reserves and growing demand for fossil fuels, interest in capitalizing on unconventional hydrocarbons have been growing.

One of the "uncommercial reserves" that have now become viable is that of tight oil and tight gas; collectively grouped and termed as unconventional hydrocarbons. These reserves were once deemed as uncommercial due to the large number of wells and costly technology required to harvest hydrocarbons, and the low flow rates per well compared with conventional wells and hydrocarbon plays. However, with improvements in technology and increasing prices of oil, there has been an increased interest in unconventional hydrocarbons, their reservoirs and its exploitation (Stevens, 2012; Haddadian and Shadhidehpour, 2015).

The topic of exploiting unconventional hydrocarbons is highly contentious. Faced with the inevitability of more countries looking for alternatives to traditional crude oil, there is an urgent need to be able to regulate the industry for unconventional drilling practices. This chapter is a comparative analysis of the regulatory systems that govern unconventional drilling. By exploring the challenges of regulating the environment, and the different frameworks available for unconventional drilling, it aims to propose an updated framework. Through the comparison and discussion of two case studies, with opposing regulatory approaches on managing unconventional drilling and production, and the analysis of the results, recommendations can be made for countries that are looking to set up similar regulatory frameworks.

The Growth of Unconventional Drilling

The United States is the most successful example of unconventional drilling. Their application of technological innovations and successful drilling program has led to an oil and gas renaissance and great optimism, paving the way for other countries with similar geological structures, to consider their unconventional reserves with renewed interest in developing an exploration and drilling program (Mason *et al.*, 2015).

A 2013 study by the U.S. Energy Information Administration (EIA), assessed 137 shale formations outside of the United States, for a global estimate as to the amount of recoverable unconventional resources available for possible exploitation. Figure 12.1 shows the region and basins assessed as well as the distribution of shale resources globally.

With increased knowledge and better information being published with regard to unconventional drilling and the geology of successful unconventional plays, many countries other than the U.S., such as Algeria, Argentina, Australia, China, India, Mexico, Poland, Romania, Russia, Saudi Arabia, Turkey, Ukraine, and the United Kingdom, have drilled stratigraphic wells in their shale formations to help clarify the geological properties and productive potential of their shale resource (EIA, 2013).

Environmental Concerns for Fracking

The development of unconventional reservoirs and fields are often larger in scale and more intense than that of their conventional hydrocarbon

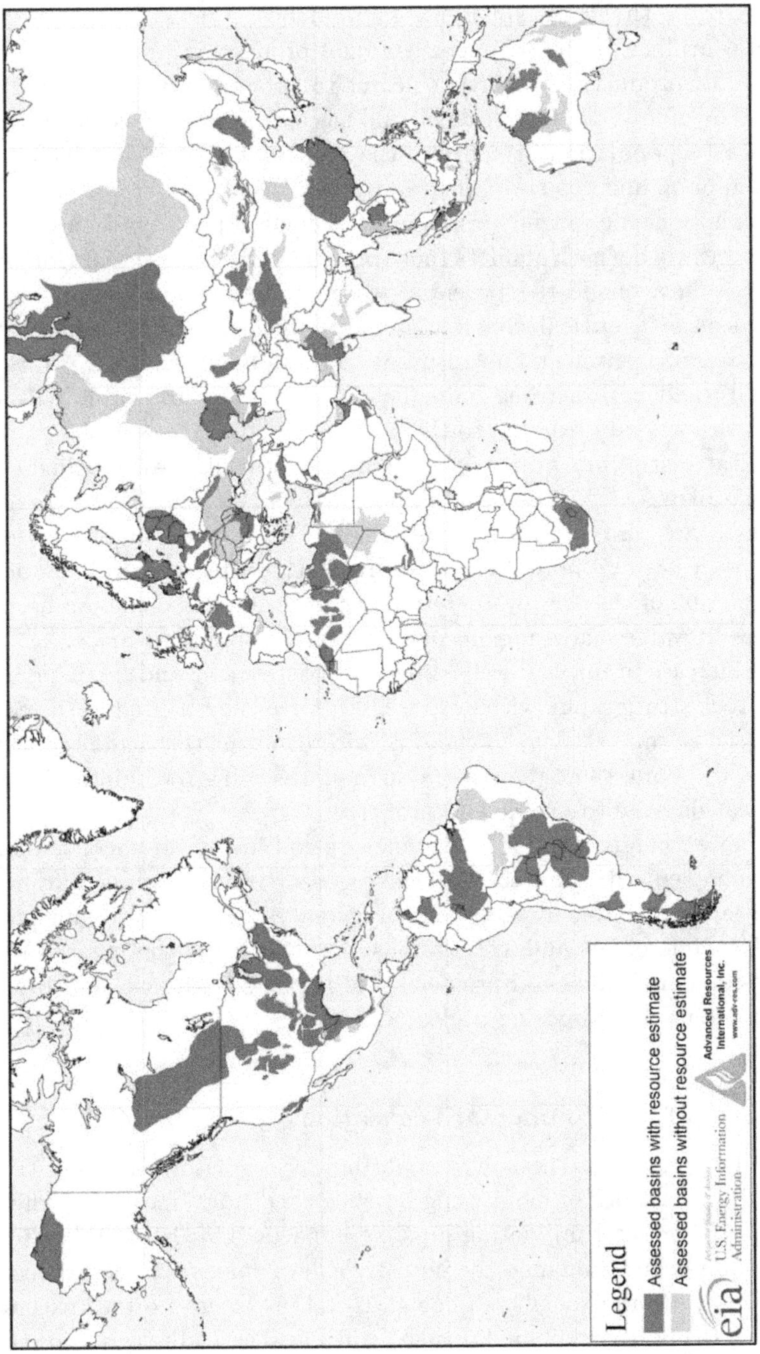

Fig. 12.1. Global distribution of Shale basins and formations (EIA, 2013).

counterpart, with more apparent environmental risks. Moreover, the horizontal drilling nature of unconventional production also increases the likelihood of production occurring nearer to urbanised areas and regions that may be unused to large-scale oil and gas extraction. The increase risks of impact to population strengthen the call for better enforcement and regulation of oil and gas activities.

Fracking chemicals and proppants, pumped extensively beneath the surface, at times beneath aquifers that are tapped into as reservoirs for cities and states, have made the potential water pollution and contamination due to seepage a critical concern. Inadequate measures to regulate the transportation, treatment and disposal of these highly pollutive fracking fluids will result in disastrous consequences to the contaminated areas and communities who are exposed to them.

Habitat disruption from the construction of drill sites, pipelines and processing plants as well as air pollution by methane and volatile organic compounds are also some concerns raised. Moreover, as unconventional drilling processes are more water-intensive both during the drilling and at the completion of the well, water security and freshwater depletion become hot issues in water scarce regions and/or densely populated areas.

The increase in injector wells, injecting waste water and methane back into the rock stratigraphy with deep wells, have also been cited as the cause of increased seismic activity in the U.S. with concerns that as the industry continues to boom, these man-made earthquakes will grow in intensity and cause great damage to people and property.

The recent adaptation of the technology and history of unconventional drilling, coupled with limited knowledge, research and documentation on possible environmental impacts and occurrences has made it difficult to draw clear references and correlations between unconventional drilling activities and environmental problems and impacts in the vicinity, leading to much debate and opposing discourses.

Challenges of Environmental Legislation

As such, in scenarios as these, it is important for regulators and industry to work out a framework of monitoring baseline conditions, measuring effects, minimising risks, and mitigating possible impacts. This balance has to be achieved to safeguard against the industry's development at the expanse of the environment and social well-being, as well as to ensure that excessive concerns do not lead to an impasse, whereby the industry's growth is unfairly stifled.

However, in the consideration of legislation for environmental problems, we are often faced with a conundrum. The relationship between environmental goals and industrial competitiveness has always been thought of as being on opposite ends of the spectrum, whereby there has to be a trade-off between social benefits and private costs. The dilemma, therefore, is how to balance society's desire for environmental protection with the economic burden on industry.

The emergence of a new dynamic paradigm that takes into account environmental sustainability and innovation raises the possibility in which environmental regulations, when properly designed, has the ability to trigger innovation and higher efficiency that will bring about a win-win situation for both industry and the environment (Porter and Linde, 1995). However, three problems still remain.

The first problem is that, despite the change in paradigm, problems with definition and lack of consensus remain. At international conventions, we have seen a move from the economic and social issues to one that embraces the environment, especially that of climate change (Weiss, 1992). While none will dispute the importance of environmental considerations at all scales when planning for development and growth, there is no clear consensus amongst the different groups when it comes to the definition of what constitutes sustainability, the measurement of risk, and the setting up of clear environmental safeguards and parameters (Porter and Linde, 1995).

The second problem is the inability to accurately account for nature in economics. Our current worldview, social and institutional structures are so steeped in economics that the failure to account for environmental costs and benefits in traditional economics has had large repercussions in how we organise resources and exploit them. Many disagree on the method of quantifying nature and assigning a numerical cost to it. The intangible benefits of a pristine environment on an individual, community and nation is difficult to put to quantify and qualify.

The third problem is the transboundary nature of environmental problems. The open nature of environmental issues and resultant impacts makes it difficult to regulate and set boundaries for the start and end of an issue, as well as to assign blame and responsibility when environmental problems only come to light years after its initial inception. In our context, this transboundary nature leads to a contestation of regulatory laws on different scales — for example, the emission of methane in the flaring of shale oil wells — when the impacts that are created in a local state have the potential to have impacts across the state line, on the national level or even internationally (Srebotnjak and Rotkin-Ellman, 2014; EPA, 2011a; EPA, 2015).

Guiding Principles For Environmental Legislation

In recognition of the problems stated above, it is necessary for regulatory frameworks and laws to be guided by certain principles. Binding precautionary and preventive principles must exist to justify decisions designed to protect the public when causation is uncertain (Sheng *et al.*, 2015). These principles, imbued into legally binding directives, direct societal and industrial actions, and ensure that catastrophic or irreversible consequences of human hazards can be prevented.

Practiced by both, the way in which USA and the EU imbue these principles into their structures also show the importance of environmental issues in relation to economics.

The precautionary principle is part of the EU's Consolidated Treaties, affecting the functioning of the EU regarding the protection of the *environment* and *public health*, as well as the *prudent and rational utilization of natural resources*. Article 191, paragraph 2 of the Lisbon Treaty states that:

> *Union policy on the environment shall aim at a high level of protection taking into account the diversity of situations in the various regions of the Union. It shall be based on the precautionary principle and on the principles that preventive action should be taken, that environmental damage should as a priority be rectified at source and that the polluter should pay.*

The US, on the other hand, does not have a constitutional precautionary principle. Instead, the principles are shown in the history of federal case laws where it states in the case of Environment Protection Agency's (EPA) regulations and laws, the agency and associated regulations have a "precautionary and preventive orientation" (Sheng *et al.*, 2015).

No matter how each party decides on implementing the precautionary principle, the question still remains: How conservative should regulators be? The results are often based on the decision-maker's outlook and pessimism (Sheng *et al.*, 2015). Therefore, there is a need for a basket of factors to be considered in forming regulatory frameworks for it to be the most balanced. In the next section, we will look at factors spreading over several categories that affect unconventional drilling and development.

An Updated Framework

Konschnik and Boling (2014), in their study of the shale gas industry, proposed a smart regulation framework (CO/RE) that is an adaptation of adaptive governance. They focused on characterising risk, optimising mitigation strategies, regulation, and enforcement, in order to design tailored governance strategies. Although their framework is useful in targeting specific risks, the narrow scope fails to take into account the linkages across disciplines. It also does not give a big picture view of the environmental issues plaguing the industry and fails to suggest a measurement of success for enforcement. (Rijke *et al.*, 2012; Konschnik and Boling, 2014).

Three core components have been identified in formulating a framework:

1. The list of factors for consideration,
2. Legislation and its formulation of directives for the industry and
3. The ability for successful enforcement of those directives.

Beyond these three elements, the most important measure of a successful framework is the dynamism of it, to change and shift according to the needs and context of a particular industry, country or population.

It is no longer sufficient for a framework to be merely adaptive, as change in adaptive governance seems to lack urgency and can take a very long time to come about. A framework that is dynamic captures the need for urgency, whereby change has to be fast and purposeful. This purposeful change can only occur if there is a means for a methodology that can identify the gaps and problems in the current set-up of the industry and strategies to solve them.

Environmental factors take up a large portion of the factors for consideration. Coupled with the difficulties in trying to accommodate and account for environmental considerations when formulating legislation and regulations, we will focus on the comparative analysis of the environmental legislation portion of the suggested framework.

Acknowledging the importance of legislation having the ability to take into account a variety of multi-disciplinary factors, as well as the ability to have an enforcement agency to ensure that legislation is adhered to, this study is unable to give sufficient attention to all factors. Due to space constraints, the focus is to attempt to fill the gap in current literature by providing a method of comparing and evaluating current legislation.

Whereas previous studies have focused on individual case studies, or a general exposition of the methods in which legislation formulation and mitigation strategies are adopted in a particular context, there is yet to be a method to compare the regulations of two differing legislation styles, such as the stances adopted by the USA and EU.

By classifying the current legislation into environmental categories of water, land and air, as well as showing the targeted stage in which the legislation is applied to; be it the pre-drill, drilling or completion stage of unconventional development, a comparative analysis of the laws of the USA and EU, showing the strengths and weaknesses of legislature that have been passed and/or put in place, will be carried out.

For the USA, we will be looking at the federal laws administered by the Environmental Protection Agency (EPA); and in the case of the EU, the directives studied will come from the report done by the European Commission set up in 2014 for the study on the use of fracking for unconventional hydrocarbons in the EU.

All legislation and guidelines that apply to applications and submission of designs and plans before the construction of infrastructure and supporting facilities of drilling platforms, will be considered to be under the pre-drill stage. Any legislation that applies to operations from the spud of the well to the commercial flow of hydrocarbons will be considered under the drilling stage, with the production stage applying once a well is linked up to a collection facility and is contributing to the amount of hydrocarbons produced. Completion and abandonment is considered to have occurred when a well which ceases production is appropriately cased, plugged and abandoned, with the facilities and platform to be decommissioned.

The comparison of extent and scope of the legislations will be based on these parameters: is it a general piece of legislation regulating all industries, a piece of legislation regulating the mining and exploration industry, a piece of legislation affecting the drilling for oil and gas, a piece of regulation that applies to a particular process of drilling and production, and/or a piece of regulation that is applicable to fracked wells and its subsequent processes.

This study also assumes that both regions have available to them:

1. Sufficient geological prospectivity for the commercial exploitation and development of unconventional hydrocarbon
2. A pool of skilled workforce that has the sufficient expertise
3. The technology for fracking

Comparative Analysis of USA and EU's Legislation

Table 12.1. Comparative Analysis of the legislation between USA and EU.

		Legal document	Legislation elaboration	Stage of Fracking Development: Pre-drill	Drilling	Production	Completion/ Abandonment	Other factors considered
General Legislation	USA	National Environmental Policy Act of 1969	Requires the Federal government to consider the environmental impacts of any proposed actions as well as reasonable alternatives to those actions. This is translated to Environmental Assessments, Environmental Impact Statements and Categorical Exclusion Reviews for projects and developments.	NA	NA	NA	NA	
		Natural Gas Act	Allows for gas pipelines to be constructed by obtaining the Federal Energy Regulatory Commission's (FERC) approval, which supersedes state and local laws.			X		
		Presidential Executive Order 13605	"Supporfting Safe and Responsible Development of Unconventional Domestic Natural Gas Resources" to coordinate efforts of federal agencies in the development of natural gas resources and associated infrastructure.	X	X	X	X	

(Continued)

Table 12.1. (*Continued*)

Legal document	Legislation elaboration	Stage of Fracking Development				Other factors considered
		Pre-drill	Drilling	Production	Completion/ Abandonment	
Vol. 80 Federal Register 16130 (Elaboration of Title 43, CFR Part 3160)	Bureau of Land Management revised its rule relating to fracking that's applicable to federal and Indian lands, effective 24 June 2015.					
	Final ruling ensures that wells are properly constructed to protect water supplies, public disclosure of the chemicals use in fracking liquids, and tighter regulations for details of operator planning, submission of design plans, monitoring services and rigid requirements for management of operations.	X	X	X	X	
Environmental Impact Assessment Directive — 85/37/EC	Projects listed in Annex I (inclusive of hydrocarbons extraction) are considered to have significant effects on the environment and require an EIA with consultation to be carried out between the environmental authorities, public and developer.	NA	NA	NA	NA	

EU	Management of waste from extractive industry — 2006/21/EC, 2004/35/EC	Covers the procedures for management of wastes for extractive industries onshore, in planning for the prevention or minimisation, treatment, recovery and disposal of waste, to ensure minimal waste generation and harmfulness, taking into account available technology.			X	X	Social, Health
	Regulation on the registration, evaluation and authorisation of chemicals (REACH) — 1907/2006/EC	Regulation applies to workplaces and environmental legislation, for registration, regulation and monitoring of hazardous chemicals; their uses, management and mitigation strategies.	X	X	X	X	Social, Health
	Directive on environmental liability — 2004/35/EC, 2006/21/EC, 2009/31/EC, 2013/30	Establishment of a common framework for liability based on the "polluter pays" principle, with a view to prevent and remedy environmental damage, direct or indirect, as long as a causal link between the damage and activity can be made.	X	X	X	X	

(Continued)

Table 12.1. (*Continued*)

		Legal document	Legislation elaboration	Stage of Fracking Development				Other factors considered
				Pre-drill	Drilling	Production	Completion/ Abandonment	
Competition for Water Supply and Water Stress	USA	No Federal regulations	Individual states to manage water resources.	NA	NA	NA	NA	
	EU	EU Water Directive — 2000/60/EC	Individuals and businesses' use of water have to be sustainable, to ensure the protection and improvement of the quality of water resources.	X	X	X		
Groundwater Contamination	USA	Safe Drinking Water Act 2005, Part C	An Underground Injection Control (UIC) permit is required for wastewater disposal through deep well injections when injected fluids contain diesel fuels.				X	
		"right to know" disclosure program — associated with the 1986 Superfund law and Resource Conservation and Recovery Act	Oil and gas activities allowed to keep their patented formula of chemicals used in fracking undisclosed as "trade secrets".		X	X		Economic, Social, Health

		PENDING: EPA is working on subjecting fracking chemicals to toxic substance reporting requirements.					
	PENDING REGULATION	**In 2009, congress urged EPA to conduct a study on the relationship between fracking and drinking water to understand potential contamination risks. Final report is expected to be completed in 2016.**		X	X	X	Social, Health
EU	EU Protection of Groundwater Directive — 2006/118/EC, 1980/68/EEC	Directive on the protection of groundwater against pollution and deterioration by establishing specific measures such as the assessment of groundwater chemical status, identification and reversal of upward trends and control of hazardous substances.	X	X	X	X	

(*Continued*)

Table 12.1. (*Continued*)

		Legal document	Legislation elaboration	Stage of Fracking Development				Other factors considered
				Pre-drill	Drilling	Production	Completion/ Abandonment	
		EU Drinking Water Directive – 1998/83/EC	Establishing standards and preventive parameters with regard to activities within the vicinity of watersheds intended for human consumption, with monitoring, compliance and management.	X	X	X	X	Social, Health
Wastewater Management	USA	Title 40, Code of Federal Regulations Part 435 (Proposed Rule, 2015)	Recommending pre-treatment standards for wastewater discharge, preventing them from being discharged into publicly owned treatment works. Done to protect human health, environment and the operational integrity of publicly owned treatment works.			X	X	Social, Health, Economic
		Title 40, Code of Federal Regulations Part 122-125, 403	Direct discharges from unconventional oil and gas extraction are subject to National Pollutant Discharge Elimination System (NPDES) permitting and regulations, with indirect discharges to treatment plants subject to general pre-treatment regulations.			X	X	

		EU	Directive on Environmental Quality Standards of Water — 2008/108/EC	Limits and EQS for priority substances and pollutants, with objectives for the discharge of dangerous substances.		X	X	X
			Directive on Pollution caused by certain Dangerous Substances discharged into the Aquatic Environment — 2006/11/EC	Inventory of discharges of particularly dangerous substances and reporting obligations, controlling persistent, toxic and bio-accumulable substances (many of these substances are found in the flow back of fracking activities).	X	X	X	X
Land	Surface Pollution/ Spillage	USA	Hazardous Waste Management Standards — Title 43 Federal Regulations 58946	Regulation on the identification and management of waste deemed as hazardous, including mining and extractive industry waste.			X	X
		EU	Directive on the landfill of waste and Waste Framework	Classification of wastes from extractive industries into categories with clear directives on their management, with measures	X	X	X	X

(*Continued*)

Table 12.1. (*Continued*)

		Legal document	Legislation elaboration	Stage of Fracking Development				Other factors considered
				Pre-drill	Drilling	Production	Completion/ Abandonment	
		Directive — 1999/31/EC, 2006/12/EC, 2002/1600EC	to include the waste arising from its associated infrastructure.					
Habitat Degradation	USA	Endangered Species Act of 1973	Protects and promotes conservation by restricting actions that are likely to harm, harass, or pursue any endangered species. Supreme court rules that significant habitat modification is a reasonable interpretation of 'harming', affecting all oil and gas operations in areas where habitat are considered to be critical to listed endangered species.	NA	NA	NA	NA	Economic, Social, Cultural
	EU	Directive on the conservation of natural habitats and of wild fauna and flora — 92/43/EC	The protection of threatened habitats and species, with designed conservation areas considered essential for the maintenance or the survival of a priority natural habitat type or priority species, to promote the maintenance of biodiversity, taking account of economic, social, cultural and regional requirements.	NA	NA	NA	NA	Economic, Social, Cultural

			Strategic Environmental Assessment (SEA) Directive — 2001/42/EC	Integration of environmental considerations in planning and programming levels for a more consistent framework for decision making, by assessing the effects of certain plans and programmes in the environment.	NA	NA	NA	NA	Economic, Social, Cultural
			Natura 2000 — 1992/43/EEC, 1979/409/EEC	List of conservation of natural habitats, fauna and flora, wild birds.	NA	NA	NA	NA	Social, Cultural
Air	Well-site Emissions	USA	Title 40, Code of Federal Regulations Part 49, 51, 52 (Proposed Rule: EPA-HQ-OAR-2013-0685)	Source Determination for certain Emission units in the oil and natural gas sector as part of the Climate Action Plan to reduce emissions of GHG and VOC by cutting methane emissions for new plants.	X	X	X		Climate Change, Political
			Title 40, Code of Federal Regulations Parts 60, 63 Vol 77 Federal Register No. 159 (2012)	Formulation of new source performance standards (NSPS) for production and processing plants, improving on old standards, setting national emissions standards for hazardous air pollutants (NESHAP) for production, transmission and storage, which were previously uncontrolled emission sources. Focus on storage and transportation, beyond the well head.		X	X		Health, Social

(*Continued*)

Table 12.1. (*Continued*)

		Legal document	Legislation elaboration	Stage of Fracking Development				Other factors considered
				Pre-drill	Drilling	Production	Completion/ Abandonment	
	EU	Air Framework Directive — 1996/62/EC, 2008/50/EC, 2001/81/EC	Standards for ambient air quality assessment and management with aims to reduce ambient air pollution and set national emission ceilings for certain atmospheric pollutants.	X	X	X	X	Climate Change, Politi-cal, Health, Social
		Daughter Directives on Air Pollutants — 1999/30/EC, 2000/69/EC, 2002/3/EC, 2004/107/EC	Limit values for sulphur dioxide, nitrogen dioxide and oxides of nitrogen, particulate matter, lead, benzene, carbon monoxide, ozone, arsenic, cadmium, mercury, nickel and polycyclic aromatic hydrocarbons in ambient air.	X	X	X	X	Health, Social
GHG Emis-sions/ Flaring	USA	Title 40, Code of Federal Regulations Part 49, 51, 52 (Proposed Rule: EPA-HQ-OAR-2013-0685)	Source Determination for certain Emission units in the oil and natural gas sector, with plans to reduce emissions of GHG and VOC by cutting methane emissions from the oil and gas sector by 40–45% from 2012 levels by 2025.	X	X	X		Climate Change, Health, Social

		Oil wells are no longer allowed to indiscriminately burn and flare their by-product, methane. Methane has to be captured instead.			
EU	Directive on Emissions from large plants — 2001/80/EC	Limits set on emissions of certain pollutants into the air from large plants.	X	X	Health, Social
	Directive on VOC emissions — 1994/63/EC	Regulations controlling VOC emissions resulting from the storage of petrol and its distribution from terminals to service stations.	X	X	Health, Social

Discussion of Results

By focusing solely on the Federal Laws of the USA and the Directives and Regulations from the European Commission, an even comparison can be done. At the studied level and scale, both agencies are tasked with providing broad recommendations and guidelines, in which the individual States would further develop contextualised laws according to the specific operations for the industry within their jurisdiction.

Several observations can be noted from the comparative analysis done above:

A. The European Commission has regulations with a broader coverage compared to the USA.
B. Current legislation available is targeted to the oil and gas industry in general, with limited applicability to unconventional development.
C. Based on proposed regulations forthcoming, the USA has a more dynamic and adaptive framework compared to the EU.

Observation A

In the analysis of the environmental issues identified in the table, both the EU and USA have legislation that fall along the spectrum from general to topic specific, with some form of directive and regulation for each of subtopic of water, land and air pollution. (See Fig. 12.2 below).

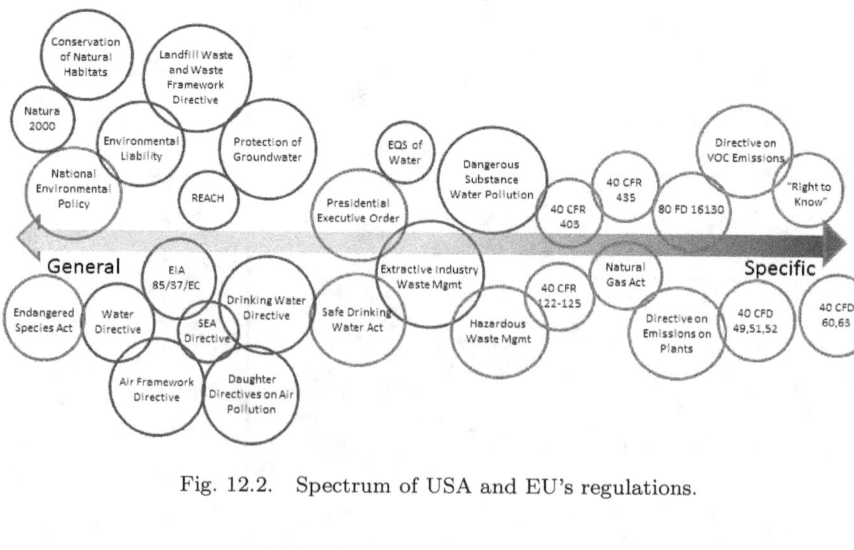

Fig. 12.2. Spectrum of USA and EU's regulations.

In the case for the USA, most of their environmental legislation is focused primarily on water pollution, in particular, groundwater contamination and wastewater management. Vol. 80 Federal Register 16130 calls for public disclosure of the chemicals used in fracking liquids, with pending regulations from the EPA, on subjecting fracking chemicals to toxic substances reporting requirements. A final report on the study between fracking and drinking water, to understand potential contamination risks, is expected in 2016; on which future regulation is to be based on.

However, the USA has no federal regulations to manage water resources, leaving it to individual state governments to manage their watersheds; whereas despite having multiple member states, the EU has a community wide Water Directive (2000/60/EU). In having an umbrella Water Directive that aims to cover all aspects of water use, the EU sees water supply as a shared common resource and considers the sustainable usage of water resources by individuals and businesses an important component in responsible water management and manages to highlight the pertinent water use issue and gives suggestions and recommendations for member states to act on.

In regard to issues of land pollution, the USA is clearly lacking in measures and legislation that are applicable to the extractive industry or to managing indirect impacts that the industry has on surface pollution and habitat degradation. The EU, whilst having more directives, likewise suffers from a lack of regulatory clarity and contextual applicability, as the breadth of those directives does little to help guide the fracking industry.

Air pollution issues have better coverage; with considerations given to climate change and public health. As such, both regions have aims to reduce emissions and manage standards for emitted pollutants. The regulations of the USA, tend to focus on the pre-drill, drilling and production stages of fracking, whereas the EU is focused on the latter stages of production and post-completion. The disparity of breadth in legislation is the most apparent here, where the USA has specifics for source determination and standards regulating new plants and well sites, whereas the EU is dependent on general directives for ambient air quality.

Observation B

Although fracking is basically the extraction of hydrocarbons and is carried out by the same operators of the oil and gas industry, the hydraulic fracking technology that is used and the sheer volume of wells that need to be drilled to exploit the unconventional reserve, sets it apart from its conventional drilling counterpart. Current legislation available and used to regulate the industry is targeted at conventional drilling methods and technology, with limited applicability to unconventional development. This results in environmental loopholes, whereby fracking activities fall between the cracks and may not be appropriately regulated or held liable for environmental degradation.

Gaps in legislation for USA:

Whilst the US EPA is authorised to regulate parts of the fracking practices, the enactment of the Energy Policy Act of 2005 exempted fracking from the Safe Drinking Water Act. Known as the "Halliburton Loophole", two exemptions were introduced. Since then, underground injection of natural gas for the "purpose of storage" and the underground injection of "fluids or propping agents (other than diesel fuels) pursuant to hydraulic fracturing operations" relating to oil, gas, or geothermal production are exempted from EPA's regulation. Only wastewater disposal wells of fluids containing diesel fuels are subjected to the need for an Underground Injection Control permit. As such, although the EPA is able to regulate fracking activities under the Clean Water Act and Clean Air Act, this exemption undermines the regulation of the industry by excluding a significant portion of waste management.

Coal bed methane (CBM) exemptions

The initial study conducted by the EPA in 2004, on the process of fracking and its potential affects for Coal Bed Methane reservoirs, determined that CBM wells poses little to no threat to underground sources of drinking water (Espinosa and Conroy, 2015). In 2010, EPA published a report based on its detailed study of the coalbed methane extraction industry and was in the process of developing controls for pollutant discharges. However, after the collection and analysis of more data, the EPA concluded that the technologies available for pollution control and treatment were not economically achievable for the CBM extraction industry as a whole and made the decision to not develop specific national effluent limitation guidelines for these operations. Dischargers, however, must still comply with National Pollutant Discharge Elimination System requirements and must have a permit before discharging wastewater.

Limited coverage of proposed regulations

Despite acknowledging the need for new regulations, many of the new regulations proposed will only deal with new projects after the law comes into effect, and there is no proposed action to regulate the significant number of existing operations and projects. As oil and gas production tends to have a long life cycle, this means that many of these developments are left to operate with minimal standards.

For example: Title 40, Code of Federal Regulations Part 49, 51, 52 (Proposed Rule: EPA-HQ-OAR-2013-0685) on reducing emissions are only applicable to new plants and operations. Many existing plants and operations are still left regulated.

Gaps in legislation for EU:

Gap in the protection of environment and human health

The Environmental Impact Assessment Directive (2011/92/EU) was previously amended by Directive 4997/11/EC to include natural gas wells into Annex I, in which all natural gas projects would be subjected to a compulsory EIA. However, the defined threshold of 500,000 m^3 daily extraction rate for natural gas wells is for conventional gas wells. For shale gas wells, their lower porosity and permeability means that it is unlikely that shale gas wells will ever reach this threshold and EIAs are not carried out (Trßmer, 2011). A more likely revision of this regulation would be to set the threshold lower, to between 5,000 to 10,000 m^3 of initial extraction volume per day, in order to capture unconventional natural gas wells.

There is also a lack of legislation concerning the wastewater treatment for the extractive industry. Although Directive 91/271/EEC regulates urban wastewater treatment, there is no examination of the capacity of waste water facilities on the more toxic wastewater originating from fracking activities and the technological ability for responsible and effective treatment and disposal of it. Should fracking be considered and adopted in the EU, there is a strong need to look at current capabilities and provide the much needed technology and capacity for the increase in toxic wastewater.

Gap in the regulation of hazardous materials

The hydraulic fracturing chemicals used are often a variety of additives, whose constituent parts are not sufficiently declared due to patents and trade secrets. An assessment of registration dossiers shows no explicit mention of fracking chemicals. REACH (Regulation on the registration, evaluation and authorisation of chemicals) regulates hazardous materials in parts and not additives, and will need to be adjusted and amended in order to work around this limitation. Moreover, much of the water regulation is with regard to water quality that is intended for human consumption with an oversight on pollution to other forms of aquatic bodies outside of natural water sources for drinking.

The Mining Waste Directive also fails to effectively govern the waste from fracking operations as there is no clear indication if fracking is classified under the Industrial Emissions Directive, in which technology available has to be reviewed and a best technology for management of waste be chosen for waste management. Currently, there are no Best Available Techniques Reference Documents (BREFs) for the treatment of fracking wastes or for its drilling equipment (Hawkins, 2015).

Gap in dynamism and legal effectiveness

The set period between reviews of legislation for the EU hinders the adaptability of the legislation to meet unforeseen changes or the introduction of new technology and industry practices. For example, the review of priority substances of the Water Framework Directive is reviewed every four years. Should, at the time of review, hydraulic fracking be seen as a low priority, its related risks and substances would have to be delayed to the next review, even if market forces raise the practice to prominence in the interim.

The regular scheduling of reviews can be beneficial for ensuring that legislations are kept current. The need to keep to the schedule, however, shows an inability to adopt and amend according to the needs of the industry between reviews, and casts doubts on the effectiveness of the legal framework and subsequent enforcement of the associated legislation.

Gaps in legislation for both:

Both study regions failed to consider the mixture of hazardous materials and chemicals that are left in the ground post fracking. Current knowledge with regard to the distribution and impact of these materials is severely limited. As the chemicals will be distributed over time and space in an uncontrollable and unpredictable fashion, it is foolish to leave this aspect unregulated.

When fracking in radioactive rock strata, the issue of radionuclides in the flow back waste water is problematic, as wastewater treatment facilities that are able to treat water for radionuclides in both the EU and USA are rare and almost non-existent in areas where fracking is conducted. This has potentially great health problems, but there are currently no specialised regulations governing the issue in both study areas (Rahm *et al.*, 2013; Brown, 2014).

Observation C

Both study regions have current studies on fracking ongoing and have proposed regulations and/or management strategies for regulating fracking. Relating the current legislation framework back to the suggested framework above, in which regulation adaptation has to be pragmatic and dynamic, the USA appears to be further ahead in adapting dynamically to the need for regulatory amendment and evolution.

USA's current legislation is comparatively more specific than that of the EU's, as seen in Fig. 12.1, with proposed rule 40 CFR 435 looking to enhance wastewater regulations by setting stringent requirements on the wastewater discharged to publicly owned treatment works, for onshore oil and gas activities.

Recognising the significant potential impacts the rise of fracking has, Congress had urged US EPA to conduct a study on the relationship between fracking and drinking water to understand potential contamination risks in 2009, with the final reported slated for completion in 2016. There is also The FRAC Act that is pending, aimed at specifically regulating hydraulic fracturing, offering a comprehensive study and management of all aspects.

The EU has commissioned several studies and reports on the exploration and production of hydrocarbons by fracking (EU Working Document SWD (2014) 22 Final) and has identified several gaps in their legislation. However, there is yet to be concrete suggestions, recommendations and directives dealing with the specificities of the technology. The approach of deliberate consideration of all aspects of fracking before the formulation of guidelines is both laudable from an environmental viewpoint and frustrating from the industry's viewpoint. Without explicit guidelines, there is an uncertainty shrouding the industry, making it risky for industry to investment in the technology and operations within the EU.

The dynamic governance strategies adopted by the USA, appears to be more conducive for practical application. This ability to recognise a rising trend, identify problems and potential problems, and develop preventive regulations and effective mitigation solutions quickly, allows for the industry to flourish, with clear guidelines. Although the industry is allowed to operate initially with limited guidelines, dynamic and pragmatic governance quickly adapts and works with the problems that appear along the way, to formulate better regulations to help guide operators. This approach does not stall industry growth and innovation, but allows for the industry to grow organically.

Limitations of Study

In light of the rapid expansion of unconventional development and the constant evolution of legislation that is struggling to keep up, this study of the current legislation at present is but a snap shot of the legislative landscape.

Although measures have been taken to ensure the best possible and equal comparison, in reality there are no completely identical scenarios for an A to A comparison. Our comparative analysis focused on legislation in the US and EU, but fails to account for the existence of certain factors which favour the creation of the fracking industry in one over the other. The USA has several factors that has helped generate its unconventional boom, which could inhibit replication in the EU. In this section, a brief elaboration of these limitations will be discussed.

Geology and Data Availability

Whilst this study assumes that both study areas have suitable geology, technology and expertise for unconventional development to occur, in reality, the difference in type of geology directly affects the risk of investment for operators.

In the USA, the geology of large shallow material plays imply large technically recoverable resources, with higher return on investment. Moreover, the industry in the USA is also structured with information being publicly available. As such there are plenty of drill core data available for exploration geologists to consult to find the 'sweet spot' of plays. The EU tends to have shale plays that are smaller and deeper, with less material; making fracking technically more difficult. The industry in the EU also lacks wide and easily available data, has limited core data, and lacks a central depository for drill core data to be stored (Lechtenböhmer *et al.*, 2011).

The EU, with its different member states and differing geology, has differing prospectivity of unconventional resources in each country. Some countries such as Poland and Hungary, who have abundant shale basins, are more motivated to develop regulations and promote activity compared to countries such as the United Kingdom, who has limited shale plays.

Economic Regulations

Economic regulations can either support or hinder the growth of the industry. The USA has several economic regulations that encourage unconventional development, which is lacking in the EU.

The 1980 Energy Act, giving tax credits and the introduction of the Intangible Drilling Cost Expensing Rule in the USA, meant that a large portion of the well development cost were subsidised, allowing for small firms with limited cash flow to be able to participate in exploration and production projects. The lack of tax incentives in the EU, with little to no financial dispensations for unconventional operations, sets a high barrier of entry for smaller companies into the industry.

Moreover, the legacy of legislation in the USA, that allows landowners to own property rights of hydrocarbons, creates a financial incentive for private landowners to allow exploration and production to occur on their property. The public has been exposed to oil and gas operations and is used to the close proximity of drilling activities. Land within the EU resides with the state, and landowners receive no compensation or reward for drilling operations in their vicinity that can be disruptive. As such, the public is more resistant to drilling operations in close proximity.

Infrastructure Incentives

With an extensive network of shared pipelines, producers in the USA are able to tap into existing infrastructure to transport their product to readily available domestic markets. In the EU, pipelines are based on "third party access" and producers have to build their own pipelines to transport their products. This increases the cost of production; and as there is yet to be a common EU energy market, the domestic markets are not as readily available.

Structure of Industry

The type of players in the industry and system of licensing acreages are large factors in determining ease of operations and its dynamism. Dominated by small entrepreneurial companies and supported by a dynamic highly-competitive service industry, the fracking industry in the USA is highly competitive and efficient. The EU has traditionally been dominated by large industry players (such as Shell, BP) and its service industry is American-dominated; driving up the costs of drilling an unconventional well. It is estimated that the drilling costs of drilling a Shale well costs between US$6–14 million, compared to US$4 million in the Marcellus play in the USA (Deutsche Bank, 2011). Large companies in the EU tend to be more risk adverse, with strict in-house environmental standards; whereas the smaller companies in the USA tend to be more adaptive and willing to take on risk, as they need not comply to any internal environmental standards.

Recommendations

The success of unconventional development, with its subsequent benefits of energy sufficiency and economic growth in the USA, remains a strong beacon and inspiration for other countries to consider replicating USA's success themselves.

In showing two contrasting approaches to regulating the industry, this study shows that both approaches are imperfect. Although the approach adopted by the USA has had practical success, there are significant gaps in its legislation. On the other hand, while the overly cautious approach of the EU may have stunt the growth of the fracking industry; in reality, the approach is not without merit, especially in light of the uncertain and incomplete knowledge of fracking and its potential impacts.

Moving forward, countries should aim for a more moderate stance, taking the commendable aspects of both approaches and merging them, keeping the need for dynamic governance at the forefront. Countries looking to develop a framework of governance for hydraulic fracturing should first broadly consider the multiple interdisciplinary factors with direct and indirect impacts and influence.

A cost-benefit analysis evaluating the risk of fracking and the rewards gained from exploiting their unconventional reserves has to be conducted for the unique context of the country in question, before setting up a skeleton legislation covering pertinent issues, to allow the industry to operate, whilst actively working on new, comprehensive regulations. These regulations would be a response to the outcomes and challenges of current operations, changing adaptively and dynamically, taking into account the situational context of each country and industry.

We also propose learning from the experiences of other countries who have an active unconventional exploration industry, and encourage open cooperation and consultation between states to share good practices and governance strategies.

Conclusion

Porter and Linde (1995) proposed that regulation serves a variety of purposes — from acting as a signal for resource inefficiencies and potential technological improvements, to reducing uncertainty of investments, motivating innovation and progress, levelling the playing field, or regulating incomplete offsets. For them, regulation is also a means of raising corporate awareness of certain issues. Environmentally, it ensures that the environment is valued.

Despite the rapid expansion and adoption of fracking in unconventional hydrocarbon development, its associated regulations are comparatively still in its infancy. The incomplete knowledge and inconclusive results of studies, further complicates the understanding, regulation and management of the industry. Yet in reforming current regulation to address the concerns and specificity of fracking, it is important to consider the crucial significance of legal, political and administrative structures in its design and implementation (Minogue, 2005). At the crux of the matter, dynamic governance has to shape the approach to ensure continuous relevance and adaptability.

References

Brown, V.J. (2014). Radionuclides in fracking wastewater: managing a toxic blend. *Environmental health perspectives*, **122**(2), A50.

Energy Information Administration (EIA). (2014). Annual Energy Outlook 2014. Retrieved Feburary 15 2015, from http://www.eia.gov/forecasts/aeo/source_natural_gas_all.cfm.

Espinosa, A. and Conroy, D. (2015). Legal Analysis of Hydrologic Fracking Laws and Regulations in the United States. Centro Mexicano de Derecho Ambiental. Retrieved from http://www.cemda.org.mx/wp-content/uploads/2011/12/CEMDA-Fracking-Research.pdf.

European Commission. (2015). Report on the effectiveness of Commission Recommendation 2014/70/EU laying down minimum principles for the exploration and production of hydrocarbons (such as shale gas) using high-volume hydraulic fracturing. Retrieved from http://ec.europa.eu/smart-regulation/roadmaps/docs/2015_env_021_shale_gas_fracking_en.pdf

Haddadian, G. and Shahidehpour, M. (2015). Ripple Effects of the Shale Gas Boom in the US: Shift in the Balance of Energy Resources, Technology Deployment, Climate Policies, Energy Markets, Geopolitics and Policy Development. *The Electricity Journal*, **28**(2), 17–38.

Hanlon, J. (2011). Natural Gas Drilling in the Marcellus Shale under the NPDES Program. Memorandum from Director of EPA's Office of Wastewater Management to the EPA Regions. March 16, 2011.

Hawkins, J. (2015). Fracking Minding the gaps. *Environmental Law Review*, **17**(1), 8–21.

Holloway, M.D. and Rudd, O. (2013). *Fracking: The Operations and Environmental Consequences of Hydraulic Fracturing*. John Wiley & Sons.

Howarth, R.W., Ingraffea, A. and Engelder, T. (2011). Natural gas: Should fracking stop?. *Nature*, **477**(7364), 271–275.

Konschnik, K.E. and Boling, M.K. (2014). Shale gas development: a smart regulation framework. *Environmental science and technology*, **48**(15), 8404–8416.

Kulander, C.S. (2012). Shale oil and gas state regulatory issues and trends. *Case W. Res. L. Rev.*, **63**, 1101.

Lechtenböhmer, S., Altmann, M., Capito, S., Matra, Z., Weindrorf, W. and Zittel, W. (2011). Impacts of shale gas and shal oil extraction on the environment and on human health. Directorate General for Internal Policies. Policy Department A: Economic and Scientific Policy. Retrieved from http://www.europarl.europa.eu/document/activities/cont/201107/20110715ATT24183/20110715ATT24183EN.pdf

Mason, C.F., Muehlenbachs, L. and Olmstead, S.M. (2015). The Economics of Shale Gas Development.

Minogue, M. (2005). Apples and oranges: problems in the analysis of comparative regulatory governance. *The Quarterly Review of Economics and Finance*, **45**(2), 195–214.

Porter, M.E. and Van der Linde, C. (1995). Toward a new conception of the environment-competitiveness relationship. *The journal of economic perspectives*, 97–118.

Rahm, B.G., Bates, J.T., Bertoia, L.R., Galford, A.E., Yoxtheimer, D.A. and Riha, S.J. (2013). Wastewater management and Marcellus Shale gas development: Trends, drivers, and planning implications. *Journal of environmental management*, **120**, 105–113.

Rijke, J., Brown, R., Zevenbergen, C., Ashley, R., Farrelly, M., Morison, P. and van Herk, S. (2012). Fit-for-purpose governance: a framework to make adaptive governance operational. *Environmental Science and Policy*, **22**, 73–84.

Sheng, H.X., Ricci, P.F. and Fang, Q. (2015). Legally binding precautionary and prevention principles: Aspects of epistemic uncertain causation. *Environmental Science and Policy*, **54**, 185–198.

Srebotnjak, T. and Rotkin-Ellman, M. (2014). Fracking Fumes: Air Pollution from Hydraulic Fracturing Threatens Public Health and Communities.

Stevens, P. (2012). The 'shale gas revolution': Developments and changes. *Chatham House briefing paper*, **4**.

Swanson, C.L., Shogren, R.L., Fanta, G.F. and Imam, S.H. (1993). Starch-plastic materials — preparation, physical properties, and biodegradability (a review of recent USDA research). *Journal of environmental polymer degradation*, **1**(2), 155–166.

Teßmer D. (2011). Stellungnahme Landtag NRW 15/621 zum Thema: "Unkonventionelle Erdgasvorkommen: Grundwasser schützen – Sorgen der Bürger ernst nehmen — Bergrecht ändern". Report on legal framework concerning exploitation of shale gas. May 2011.

U.S. Energy Information Administration (EIA) Glossary — Independent Statistics and Analysis. (n.d.). Retrieved November 16, 2015, from http://www.eia.gov/tools/glossary/

U.S. Energy Information Administration (EIA). (2013). *Technically Recoverable Shale Oil and Shale Gas Resources: An Assessment of 137 Shale Formations in 41 Countries Outside the United States.* June 2013. Washington, DC: U.S. Department of Energy.

U.S. Environmental Protection Agency. (2004). EPA 816-R-04-003. Evaluation of Impacts to Underground Sources of Drinking Water by Hydraulic Fracturing of Coalbed Methane Reservoirs.

U.S. Environmental Protection Agency. (2011a). Investigation of Ground Water Contamination near Pavillion, Wyoming. Retrieved from http://www.epa.gov/region8/superfund/wy/pavillion/EPA_ReportOnPavillion_Dec-8-2011.pdf

U.S. Environmental Protection Agency. (2011b). Plan to study the Potential Impacts of Hydraulic Fracturing on Drinking Water Resources. Retrieved from http://water.epa.gov/type/groundwater/uic/class2/hydraulicfracturing/upload/FINAL-STUDY-PLAN-HF_Web_2.pdf

U.S. Environmental Protection Agency. (2015). EPA Proposes New Common sense Measures to Cut Methane Emissions from the Oil and Gas Sector/Proposal Cuts GHG Emissions, Reduces Smog-Forming Air Pollution and Provides Certainty for Industry. Retrieved from http://yosemite.epa.gov/opa/admpress.nsf/bd4379a92ceceeac8525735900400c27/e5f2425e2e668a2b85257ea5005176fa!OpenDocument

U.S. Environmental Protection Agency. (n.d.) Coalbed Methane Extraction Industry. Retrieved from http://www.epa.gov/eg/coalbed-methane-extraction-industry

Uhlmann, D.M. (2011). After the spill is gone: The gulf of Mexico, environmental crime, and the criminal law. *Michigan Law Review*, 1413–1461.

Weiss, E.B. (1992). International Environmental Law: Contemporary Issues and the Emergence of a New World Order. *Geo. LJ*, **81**, 675.

World Bank. (1992). *World Development Report.* New York: Oxford University Press.

World Energy Council (WEC). (2003). *Drivers of the Energy Scene.* A report of the World Energy Council. Retrieved on June 19, 2015 from https://www.worldenergy.org/wp-content/uploads/2013/01/PUB_Drivers_of_the_Energy_Scene_2004_WEC.pdf

World Energy Council (WEC). (2007). Deciding the Future: Energy Policy Scenarios to 2050. *World Energy Council.* London.

Z

Part 5

Urban Studies

13

SUSTAINABLE URBANISM AT NEIGHBOURHOOD LEVEL: POSSIBILITIES AND PRACTICE IN PLANNED RESIDENTIAL AREAS OF DHAKA

NAZIFA ZABEEN SIDDIQUA and OSCAR CARRACEDO

Urbanisation or urban drift is probably the most complex and fastest growing socio-economic phenomenon in today's world. In 2007, for the first time in history, the global urban population exceeded the global rural population, and the world population has remained predominantly urban thereafter.[1] Cities are now practically the drivers of economic growth for a country. As of United Nation's 2014 data, it is projected that, two-third (66%) of the world's population would be living in urban areas by 2050. (UN, 2014) Now, this huge shift of world's population from rural to urban areas does not only mean that the economies of those countries will bloom, it also indicates drastic exploitation of environmental resources to accommodate more "developments".

Generally understood as a shift from a predominantly rural to a predominantly urban society, urbanisation also represents major and irreversible changes in production and consumption, and the way people interact with nature (Allen, 2009). With increasing urbanisation, we have realised how many cities are suffering from similar environmental problems; such as poor air quality, high noise pollution, unhealthy built environments, water scarcity and pollution, enormous waste generation, and failure in waste management. Most importantly, huge amounts of greenhouse gas (GHG) are released, which have started to worsen global climate. Hence, there was an urgent need to re-think, control and guide the urbanisation process.

The "Rio Earth Summit" (1992) first emphasized the importance of achieving sustainability at a local level for greater global benefits. Since then, a good number of neighbourhood sustainability assessment (NSA) tools have been developed worldwide. Because of different culture, climate and urban fabric, the neighbourhood pattern varies widely across the globe. So, ideally the indicators of neighbourhood sustainability should be context-specific. This research begins with an attempt to analyse some of the widely used NSA tools like LEED-ND, BREEAM communities, Green Mark District etc. to compare and find out their sustainability indicators' coverage and limitations. In the next step, a comprehensive study of Dhaka's long historical background, planning initiatives and most crucial sustainability issues was conducted to identify loopholes of conventional planning approach. Based on all these studies, this research proposes an appropriate set of indicators for neighbourhood sustainability assessment, which are designed for Dhaka city's context. The proposed NSA scheme has five broad categories: Urban Metabolism, (Energy efficiency, Water and Waste management) Transportation Planning, Environmental planning, Neighbourhood Design and the last one is Community and Context. Most initiatives by RAJUK, the principle planning authority in Dhaka, concentrate on mitigating the housing demand for the booming population without much focus on adequate infrastructural development and quality of urban environment. The proposed scheme intends to address this loophole by suggesting a wide range of indicators from various aspects of urban sustainability. The scheme will act like a detailed guideline and set standards for the existing or new neighbourhoods.

Dhaka: An Asian Megacity

Although Asia's urbanisation rate (47.5%) is still not as high as Europe and the United States of America (73% and 80% respectively), it accommodates 53% of the world's urban population. The urban population of the world is expected to increase by more than two-thirds by 2050, with nearly 90% of the increase expected to take place in the urban areas of Asia and Africa. (UN, 2014) More than half of the world's total mega-cities (cities with populations over 10 million) are located in Asia and the Pacific region. Of them, the Asian megacities are Tokyo, Delhi, Shanghai, Mumbai, Beijing, Dhaka, Kolkata, Karachi, Manila and Osaka-Kobe.

Although mega-cities contribute to only 11% of Asia's urban population, they are a driving force of regional and global economies. In most of these cities, there is a densification of facilities and infrastructures at the core, which leads to a very imbalanced distribution of population in the city's peripheral and central areas. Also, mega-cities take a detrimental and often irreversible toll on environmental resources. Studies show that cities are currently responsible for producing 80% of total green-house gases.

Bangladesh has witnessed a tremendous rate of urbanisation during the last few decades. The physical growth of its cities is, however, associated with multi-faceted problems. Over time, these cities have grown with huge populations amidst a continuous lack of planned infrastructures. There are many other environmental and socio-economic aspects that require proper attention. Dhaka, the capital of Bangladesh, is one of the fastest growing and most densely-populated megacities in the world. With an area of about 1600 square kilometres, Dhaka is home to a population of about 16 million. But is this population really sustainable?

Most urban-planners and theorists around the world consider compact cities or cities with high population densities as a way to achieve sustainable development. Such compact cities ensure a more efficient distribution of land, water, energy and other environmental resources. By ensuring improved public transport facilities and transit-oriented development, these cities encourage walkability and pedestrian-friendly zones. They restrict urban sprawl and discourage car ownership. The clustering of residential, commercial and various other land uses in these cities, in order to make the best use of limited land, eventually end up in high-density mixed-use zones, which is another requirement for urban sustainability. So, considering all these benefits, the high population density could actually be used to achieve urban sustainability. But unfortunately, the growing pressure of population has led many growing cities towards issues such as unplanned and insufficient infrastructures, poor quality environments, traffic congestion, lack of open spaces, etc. A crucial policy question for Dhaka at this moment, is how to facilitate urban growth without having to pay a devastating environmental cost and risking food security due to the loss of agricultural lands.

However, looking back at the 400-year history of old Dhaka, it is evident that this city used to be a very sustainable one with its abundant environmental resources, geopolitical location, and economic prospects. The answer to this city's sustainability question starts from controlling and planning how it will expand with minimum environmental impacts and maximum economic benefits.

Sustainable Urbanism at Neighbourhood Level

The "Rio Earth Summit" (1992) first emphasised the importance of achieving sustainability at the local level for greater global benefits. Since then, a good number of standards have been developed to promote and assess local-level sustainability, ranging from a building, to a neighbourhood or a city. Neighbourhoods are often termed as "building blocks of the city" (Searfoss, 2011) and they play a major role in shaping urban sustainability.

There is no fixed definition of "neighbourhood" in the literature. Subjectively, the mental borders of a neighbourhood can be defined by local residents (Grant and Guise, 2003). A more rational way to define neighbourhoods is by using administrative boundaries such as existing wards or districts (Barton *et al.*, 2003; Dempsey *et al.*, 2012). Neighbourhood boundaries can also be defined by the catchment areas for local services such as shops, primary schools, and places of worship (Barton *et al.*, 2003). Even within Dhaka city, the size of a neighbourhood may vary according to the location, demography, local culture, residents' living standards, etc. This research identifies "neighbourhoods" as implied by Dover and King (2008): i.e. areas of a walkable size, with a discernible centre, mixed-use, various social groups, street networks, and adequate connectivity. For simplification purpose, the neighbourhoods in the study area were defined as areas within a quarter mile or 5-minute walking radius from the neighbourhood centre.

Sustainable Urbanism, as defined by Douglas Farr (2008) represents a generational shift in how human settlements are designed and developed. Its adaptation as a societal norm requires all of the many participants in the process of planning and developing the built area to work as a single organism towards a shared purpose. Put simply, the pioneers of "Congress for New Urbanism" (CNU) explain sustainable urbanism as "walkable and transit-served urbanism integrated with high-performance buildings and high-performance infrastructure" (Farr*et al.*, 2008). Bio-philia (human access to nature) and compactness are core values of this concept. It comprises three necessary elements: neighbourhoods, districts and corridors (Farr, 2008). This chapter focuses only on the "neighbourhood" scale to identify the "sustainability indicators".

Sustainability assessment of the built environment was initially focused on individual buildings. For more than two decades, the sustainability assessment tools (i.e. BREEAM, LEED, etc.) were used only to evaluate individual building's performance towards various environmental aspects like, energy efficiency, water use and CO_2 emissions. However, since 2005, a good number of neighbourhood sustainability assessment (NSA) tools have been developed worldwide. Because of the difference in culture, climate and urban fabric, the neighbourhood pattern varies widely across the globe. So, ideally the indicators of neighbourhood sustainability should be context-specific. Table 13.1 below contains a brief overview and comparative analysis of five widely used NSA tools.

Table 13.1. Brief overview and comparative analysis of five NSA tools.

Topic	LEED-ND	BREEAM-communities	Green mark district	HQE^2 R	Eco city
Development Date	2009	2009	2009	2001	2002
Country/region(s) where have been applied	USA, Canada, China	United Kingdom	Singapore, China	7 EU countries (Denmark, France, Germany, Italy, Netherlands, Spain, UK)	7 EU countries (Austria, Finland, Germany, Hungary, Italy, Slovak Republic, Spain)
Ratings	**Total Points 110** Certified: 40–49 Silver: 50–59 Gold: 60–79 Platinum: 80–100	**Total points 183** Unclassified <25 Pass ≥ 25 Good ≥ 40 Very Good ≥ 55 Excellent ≥ 70 Outstanding ≥ 85	**Total points 185** 60 to $<$ 75: Green Mark Certified 75 to $<$ 90: Green Mark Gold 90 to $<$ 100: Green Mark Gold Plus 100 and above: Green Mark Platinum	−3 −2 −1 0 1 2 3	A: Innovative B: Best practice C: Advanced D: Normal practice E: poor
Themes	• Smart growth and linkages • Neighbourhood pattern and design • Green Infrastructure and Buildings • Innovation & Design process • Regional priority credit	• Climate & Energy • Resources • Place Shaping • Transport • Community • Ecology & Biodiversity • Business • Buildings • Innovation	• Energy Efficiency • Water Management • Material and Waste Management • Environmental Planning • Green Buildings and Green Transport • Community and Innovation	• Resources & heritage • Local environment • Diversity • Integration • Social life	• Context • Urban structure • Transport • Energy flows • Material flows • Socio-economic issues

Scope of the Study

Since the beginning of the 21st century, architects, planners and theorists have started to design tools for sustainability assessment at the neighbourhood scale. To date, there has been little research to evaluate the efficacy and applicability of the NSA tools in different contexts. This research attempts to analyse some of the widely used NSA tools, such as LEED-ND, BREEAM communities, and Green Mark District, to compare and find out their sustainability indicators' coverage and limitations. An in-depth study of Dhaka's historical background, planning initiatives and most crucial sustainability issues was also conducted to identify the loopholes of the conventional planning approach. Based on these studies, an NSA scheme is proposed to evaluate the possibilities and practice of sustainable urbanism in the planned residential neighbourhoods of Dhaka. The proposed scheme has five broad categories: Urban Metabolism, (Energy efficiency, Water and Waste management) Transportation Planning, Environmental planning, Neighbourhood Design and Community, and Context. The scheme was tested on a planned residential neighbourhood of Dhaka (Sector-13, Uttara, Dhaka). For this, site visits, personal observations, and questionnaire surveys were conducted. Maps and other statistical information were collected from various government agencies and their websites. Scores were given by comparing with the benchmarks for each sustainability indicator. Finally, the research suggests some recommendations which can be beneficial for achieving neighbourhood sustainability in Dhaka.

Contextualizing Dhaka: Urban Form and Spatial Morphology

As one of the oldest large-scale cities of South Asia, Dhaka has undergone many phases of alternation, development and readjustment. Ahmed (1986) indicates that human settlement in Dhaka can be traced back to as early as the 12th century. In its divergent history of more than 400 years, the city has grown from a medieval business hub to a fast-growing metropolis. Dhaka, located in the central region of Bangladesh, is a flat deltaic plain of three large rivers: the Padma (the Ganges), the Bramhaputra (Jamuna) and the Meghna (Kabir and Parolin, 2012). The city is criss-crossed by the tributaries of these rivers. Historically this land had been rich in agriculture because of its fertile soil, sufficient rainfall and suitable climate. With an area of about 1600 square kilometres, the Dhaka Statistical Metropolitan

Area (SMA) is home to 14.54 million people. (BBS, 2015) The elevation of Dhaka is up to 13 metres above mean sea level. (Banglapedia, 2015) Because of its close proximity to the rivers, most parts of Dhaka are prone to annual floods, with water levels reaching 2–4 metres during the rainy season. So, over time, peripheral embankments were constructed in the west and partially in the north and south of the city to protect it from the river floods. (Banglapedia, 2015)

Throughout the years, the physical expansion and urban form of Dhaka have been shaped mostly by the physical landscape in and around the city (Islam, 1996). Economically, geographically and culturally, Dhaka has constantly remained the most advantaged centre of the country. So, there has always been a stress of huge in-migration from different parts of the country (especially from sub-urban and rural regions) of people in search of better jobs and educational opportunities. To meet this increasing population pressure, most parts of the city have grown simultaneously, exhibiting an informal and organic spatial pattern without any comprehensive planning guideline (Islam 1996b; Huq and Alam, 2003; Nilufar, 2010). Nilufar (2009) identifies two distinct spatial patterns in the urban fabric of Dhaka; they are the planned and the unplanned (i.e. organic or informal). In the organically developed areas, land is subdivided through private initiatives and roads are constructed from user demand, instead of following any approved master plan (Nilufar, 2009). The street networks in these areas are winding and complicated; like labyrinths of lanes, by-lanes and *cul-de-sacs* of different scales.

Some parts of the city have been developed as planned residential areas with regular shaped plots and adequate infrastructures to support the basic needs of the residents. Land uses of such areas are determined by the master plans prepared by authorities, and their road networks mostly follow grid-iron pattern. The earlier planned residential areas (Dhanmondi, for example) were designed mostly as residential zone with very basic amenities for everyday life. Over time, these areas have experienced morphological changes in land use pattern to adapt to community needs (Nilufar, 2010); and as a result, the majority of the residential areas have now turned into mixed land use zones (Khan, 2008). From this experience, the more recent planned residential areas (Uttara 2nd and 3rd phase, for example) were designed with reasonably mixed land use patterns. Figure 13.1 gives an overview of the planned residential areas in the overall context of Dhaka city.

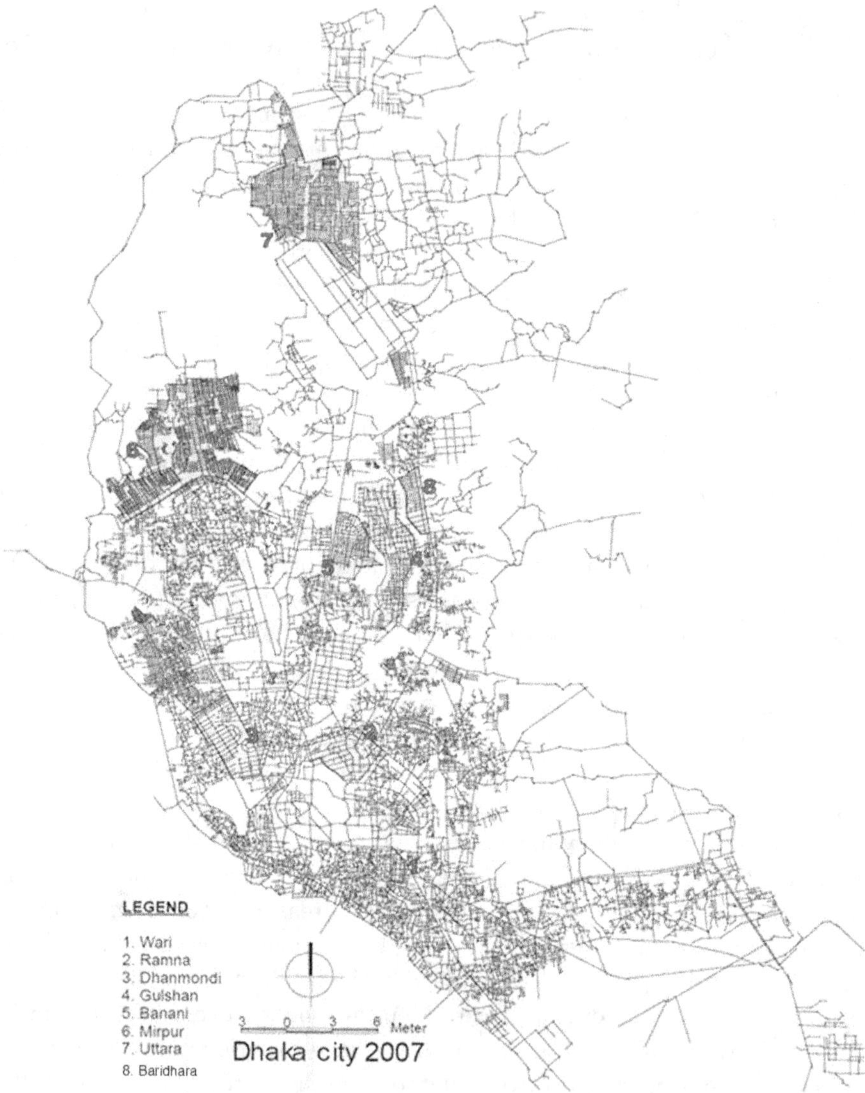

Fig. 13.1. Location of the planned residential areas within the urban grid of Dhaka city (Nilufar, *et al.*, 2009).

At present, there are eight planned residential areas in Dhaka (Fig. 13.1). Of them, Ramna and Wari were planned earlier during British reign. Gulshan, Banani, Dhanmondi, Mirpur, Baridhara and Uttara were planned under the Dhaka Improvement Trust (DIT) during the 1960–1970s.

Why is it Important to Develop a Neighbourhood Sustainability Assessment (NSA) Toolkit for Dhaka?

This research proposes a set of indicators, which will be particularly compatible for the sustainability evaluation of Dhaka's neighbourhoods. There are several reasons why it is necessary to develop an NSA tool for Dhaka. Firstly, there is a gap in literature regarding neighbourhood sustainability issues in Dhaka. Although there is profuse research available on the problems associated with the city's unplanned growth, little attention has been paid to understand Dhaka's sustainability issues at neighbourhood level.

Secondly, each city has very different urban forms and textures, which evolve from their individual history, people and culture. The widely used NSA tools (i.e. LEED-ND, BREEAM community or Green Mark District) have been established in the context of developed countries like USA, UK or Singapore. So, much of these indicators might be inappropriate to use for neighbourhood sustainability assessment in the context of a developing country.

Thirdly, thus far, Detailed Area Plan (DAP) is the only planning guideline for Dhaka which identifies some sustainability issues at the neighbourhood level. But most of these issues have been discussed superficially without any direction on how to mitigate them. This has also been mentioned in the Strategic Environmental Assessment (SEA) on DMDP and the reasons stated were that the DAP consultants themselves did not have enough expertise to provide solutions for environmental issues, and that they were not supplied with baseline data from RAJUK. (DMDP, 2007) As a result, DAP does not include some of the most crucial neighbourhood design guidelines like pedestrian and bicycle networking, mixed land use, storm water management and the demarcation of biodiversity habitats/environmentally sensitive areas.

It is thus important to develop a set of indicators for Dhaka which will cover not only the environmental issues, but also the urban design requirements at the neighbourhood level.

Applicability of the Proposed set of Indicators

The proposed set of indicators is designed for neighbourhood sustainability assessment in the planned residential areas of Dhaka. They can be applied to an existing area to assess its urban sustainability performance and identify the sectors where it is lacking. This can help planning authorities to focus on the particular needs of specific areas. The set of indicators can also be applied to an on-going project or to a master plan for future residential township projects. Dhaka is expanding north- and west-wards. The 3rd phase of Uttara residential model town is under construction. Development of the biggest satellite township of Dhaka, "Purbachal New Town", has also started since January 2013. To keep pace with the rapid urbanisation, the city might expand more in the near future and might initiate more planned residential areas to accommodate its growing population. Also, existing residential areas are undergoing changes and adjustments due to social needs, economic demands or planning initiatives. Hence, it is important to develop a set of indicators to provide detailed guidelines for existing areas as well as projects under-construction and future projects, to ensure that this massive urbanisation does not take a heavy toll on the city's environment. To date, no specific guidelines for neighbourhood planning have been developed by the principal planning authority of Dhaka. If they do take the initiative to develop such guidelines in future, this study can be used as a baseline document towards developing a more suitable neighbourhood sustainability assessment framework for Dhaka.

The proposed set of indicators have been categorised under five broad segments. They are:

1. Urban Metabolism (Water management, Material and Waste Management, Energy efficiency)
2. Transportation Planning
3. Environmental Planning
4. Neighbourhood Design
5. Community and Innovation

The considerations for choosing these five broad categories are briefly discussed in the next section.

Considerations for the Proposed NSA Scheme

1. *Urban Metabolism*

The urban metabolism segment constitutes of the indicators to monitor and regulate the flow of resources and energy within the city's system. This category has three sub-sections:

a. Water Management
b. Material and Waste Management
c. Energy Efficiency

a. *Water management*

One of the biggest problems faced by Dhaka's residents is water logging during heavy rainfall, especially during the monsoon season (May–September), due to poor storm water management and insufficient drainage facilities. Dhaka, being located in the flood plains of three large rivers,[1] experiences about 2000 mm of rain annually, 80% of which falls during the monsoon (Yahya *et al.*, 2010). Dhaka's topography is another reason for this flooding. Dhaka's elevations vary between 2–6 metres above sea level, making the lands vulnerable to inundation during heavy rainfall (Kabir, 2012). Some past seasonal floods were catastrophic; with flood levels up to 4.5m high[2] in certain areas of Dhaka. (BBS, 2015) After 2005, the embankments on the eastern and western parts of the city were constructed. Since 2009, although the city hasn't faced instances of similarly severe flooding, water logging in some parts of Dhaka during the monsoon has become a recurring problem. Sewage is collected by the underground sewer network, which covers around 55% of the city area, and is transported to the various Water Treatment Plants maintained by the Dhaka Water and Sanitation Authority (DWASA). The rest of the sewage, some from industries, is dumped untreated into the city's water sources—although the national policies mandate treatment of sewage before discharge.

On the other hand, severe surface water pollution due to this expulsion of untreated sewage and industrial effluent[3] has resulted in the over-extraction of groundwater in Dhaka; causing the groundwater table to decline at an alarming rate. All these issues have been addressed as major threats on water resource management in Dhaka in recent reports. (DMDP, 2007; DWASA, 2004)

In such circumstances, adequate surface water drainage channels and rainwater harvesting can be excellent solutions for the city's downpour flooding and water shortage problems. Also the treatment of sewage before discharge should be carefully monitored.

[1]The Padma (Ganges), the Brahmaputra (Jamuna) and the Meghna.

[2]During the 1998 flood, about 56% of the city was inundated, including most of the eastern and 23% of the western parts.

[3]Review of several studies (Dasgupta and Marr: 2007) show that industry accounts for 60% of the pollutant load in the Dhaka watershed. A predominant number of industries discharge untreated waste, thus externalizing the cost of cleaning up. As a matter of fact, the garment and apparel industry shifted to Asia as strict environmental regulations in the west were hindering their efforts to minimize the production cost for huge profit making schemes.

b. *Material and waste management*

The Dhaka City Corporation (DCC) — North and South, are the authorised institutions in Dhaka for the collection, transportation and disposal of residential, commercial and institutional Municipal Solid Waste (MSW), which amounts to around 4000 tons/day. According to Dhaka City Corporation's data, 60% of the total generated waste is collected and disposed every day in the two main landfills. (DCC, 2015) The DCC adopted the "Clean Dhaka Master Plan 2005" to ensure an integrated solid waste management system for maximum collection and disposal of wastes. Unfortunately, the master plan does not include any initiative on waste segregation and recycling. At present there are no provisions for Dhaka city residents to separate the various kinds of wastes while disposing of them, and as a result, only a meagre percentage of the total waste is recycled under private initiatives by small-scale paper, glass and plastic industries. The only success story in this sector is from environmental NGO, Waste Concern, which has shown great success in composting biodegradable wastes and generating bio-gas. Furthermore, in the master plan, there is no precautionary direction about the disposal of hazardous and bio-medical wastes. The indicators for "Material and Waste Management" have been developed, keeping in mind all these limitations.

c. *Energy efficiency*

Dhaka's energy infrastructure is mainly dependent on the national electricity grid provided by Dhaka Electric Supply Company (DESCO) in the northern part, and DESA (Dhaka Electric Supply Authority) in the southern part. At present, Dhaka doesn't have any hydro or wind power project for energy generation. However, solar energy generation is considered to be one of the most thriving business sectors of Bangladesh in recent years. Numerous solar power systems (mainly photovoltaic cells on rooftops or building surfaces) have been installed in the city under private initiatives. The Government is focusing more towards installing large scale solar home systems in the remote rural areas of the country where access to the national grid is not available.

2. *Transportation Planning*

Like many overcrowded Asian cities, severe traffic congestion has become a regular nightmare for Dhaka city's dwellers. Every day, especially during peak hours, people spend a significant amount of time on the streets being unproductive and stuck in traffic congestions. This is primarily because of an excessive number of vehicles on roads with generally no separate lanes for slow and fast moving (motorised and non-motorised) vehicles.[4] High car ownership rates and traffic mismanagement along with poor transportation and infrastructure planning are some of the major causes of this problem (Chowdhury, 2013). Another issue that aggravates the situation is the culture of random on-street parking on busy roads, due to the absence of strict law enforcement and sufficient off-street parking spaces.

Although the transportation system in Dhaka is predominantly road based, roads occupy only around 12% of the city's land area; (BBS, 2015) whereas ideally it should be at least 25% (Mahmud and Haque, 2013; Chowdhury, 2013). Besides the road-based transportation system, there are two other modes of transportation operating in the city — a railway[5] and a ferry service[6]. But the potentials of these two modes have been very meagrely explored. Furthermore, the co-ordination among the regulatory authorities for these three modes of transportation is very poor.

The Government of Bangladesh, under the "Strategic Transportation Plan" (STP), has stressed on the development of a multi-modal transport system for the city; with a combination of Bus Rapid Transit (BRT) and a rail-based Mass Rapid Transit (MRT)/Metro. After many years of planning and a feasibility study, the first phase of construction of the country's first metro rail project finally started in 2015. After the Padma Bridge, this is the second largest infrastructure project in the country[7], and is expected to be completed by December 2019. There have been various transportation policies implemented under STP for traffic management and congestion control. But in spite of all these efforts, Dhaka's traffic congestion scenario is not improving, mostly due to the absence of an integrated transport planning approach. Considering all these limitations, the indicators for the "Transportation planning" category have been chosen to ensure a sustainable transportation network within the neighbourhood.

[4]Except for the important avenues and primary roads with 4 lanes or 6 lanes.

[5]A railway line operates in the north–south direction of the city. Bangladesh Railway operates standard gauge passenger trains on this route but the frequency of operation of this mass rapid transport system is relatively low. (BBS, 2015)

[6]The BWTA operates a ferry service to transport passengers from south of Dhaka to Sadar Ghat but this system has no extensions in the north. (BBS, 2015)

[7]According to a study by the Japan International Cooperation Agency (JICA), the metro rail trains will be able to operate every three minutes and carry 60,000 passengers an hour.

The indicators focus on Connectivity, Transit-Oriented Location, Cycling Facilities, Parking, Transportation Demand Management and Walkability.

3. *Environmental Planning*

While developing an area in Dhaka, the preparation of an EIA to identify the potential threats on the environment or to demarcate environmentally-sensitive areas (for example, biodiversity habitats, wetland, marsh land, etc.) has always been neglected. Although the "Bangladesh Open Space and Wetland Protection Act, 2000" and DAP both stress on the protection of natural drainage channels and the conservation of biodiversity habitats, in reality, often the wetlands are filled up to build huge residential or mixed-use projects. This is mainly due to the corruption in law enforcement authorities and the political hold of influential real estate developers. The proposed scheme in this study includes strict guidelines for the protection of wetlands and eco-systems from developers.

Another important indicator under this category is "open space". Public open spaces play an important role both at community and city level, to enhance social interaction and social control. As an adverse consequence of unplanned growth, Dhaka lacks in both the quality and quantity of public open spaces (Nilufar, 2000). About 14.5% of Dhaka city's total land area are open public spaces in the form of parks, playgrounds, etc.; (MoHPW, 2015) whereas megacities should ideally have open spaces making up at least 20–25% of their total area. Furthermore, these open spaces are often occupied by illegal encroachers with political influence to set up slums, tea stalls, community clubs or sometimes even multi-storied high-rise buildings. This happens despite the fact that "Bangladesh Open Spaces and Wetland Protection Act 2000" does not allow parks, open spaces and wetlands to be converted for any other use. (DMDP, 2007)

4. *Neighbourhood Design*

In Bangladesh, neither the planning agencies nor the architects' association have developed any guidelines for "neighbourhood design". Area developments take place based on the master plan provided by the agency in charge of its development. Integration among the various government agencies while developing an area is vital to ensure adequate infrastructural support and amenities for the residents. Unfortunately, RAJUK, the foremost urban planning authority in Dhaka, falters when it comes to coordinating with other agencies.

The inclusion of "Neighbourhood Design" in the proposed scheme is the first step to directing the neighbourhood planning process of Dhaka towards a multi-disciplinary approach. The indicators for this category, which are crucial for a vigorous urban environment, focus on location, density, mixed land-use, affordable and diverse housing, green buildings, recreational spaces, accessibility for people with diverse abilities, educational institutions and work provisions within the neighbourhood.

5. *Community in Context*

Social sustainability indicators remain uncharted in most of the well-known and widely-used NSA tools. One reason for this is that they are not very straightforward to measure. For example, issues such as social interaction or safety are subjective and can be evaluated in many different ways. Nevertheless, some experimental and research-based NSA tools in Europe (such as eco-city and HQE^2R) have included social sustainability indicators like affordable housing, inclusive communities, community health and well-being, social networks, safety, education, and cultural richness. From traditional neighbourhoods of history to successful examples of sustainable neighbourhoods in today's world, it is very clear that, no matter how planned a neighbourhood is, its ultimate sustainability performance greatly depends on social factors. So, this study has suggested the inclusion of some social indicators which are related to safety, social interaction, diverse social groups, local governance, community participation in planning process, public awareness about sustainable lifestyle and context-specific features of the neighbourhood, in the scheme.

Table 13.2 below contains an overview of the proposed NSA tool with all the categories, sub-categories and indicators.

Recommendations

Despite having ample rainfall during the monsoon season, water scarcity and the depletion of the ground water table are major problems in Dhaka. Since 2011, RAJUK has declared the installation of a rainwater harvesting system on the rooftops of all new buildings mandatory. But due to poor legal enforcement, it is not yet strictly followed. This research recommends a more systematic and integrated planning approach of RAJUK and DWASA for large-scale rainwater harvesting and storm water management in Dhaka.

Table 13.2. The proposed Neighbourhood Sustainability Assessment scheme for Dhaka.

PART 1: URBAN METABOLISM		
WATER MANAGEMENT	W1: Rain Water Harvesting System to capture (Pick just one for scoring):	
	25% of total run off	1
	50% of total run off	2
	W2: Surface water drainage channels for 100% area of the neighbourhood.	2
	W3: Treats storm water run-off before discharging into public drains. (Natural/Mechanical treatment)	2
	W4: On-site waste water treatment (Pick just one for scoring):	
	25% of total wastewater reused	1
	50% of total wastewater reused	2
	W5: Water efficient fittings for Public Amenities.	2
	Total Possible Score	10
MATERIAL AND WASTE MANAGEMENT	MW1: Reduction in the quantity of excavated materials transported from/to the neighbourhood by optimising the following (For scoring, award 1 point for each):	
	Reusing at least 50% of the topsoil	1
	Reusing at least 50% of cut and fill material	1
	MW2: Ensures systematic disposal, recycling and overall management of waste within a neighbourhood through the following options (For scoring, award 1 point for each):	1–6
	(a) Provision of at least one waste recycling station within a neighbourhood dedicated to the separation, collection and storage of recyclable materials such as paper, glass, plastics and metals.	
	(b) Provision of at least one drop-off point for potentially hazardous waste.	
	(c) Provision of litter receptacles with integrated recycle containers at public areas	
	(d) Development of a "Community waste strategy" and educating the residents about the importance of waste recycling. (Through forums/posters, etc.)	
	(e) Provision of composting (kitchen and garden wastes) within the neighbourhood and/or at strategic locations.	
	(f) Use of organic waste for energy generation (e.g. production of bio-gas).	
	MW3: Ensures a well-integrated waste conveyance system, taking measures to reduce the odour and other negative impacts on the environment.	1–2
	Total Possible Score	10

(*Continued*)

Table 13.2. (*Continued*)

PART 1: URBAN METABOLISM		
ENERGY EFFICIENCY	E1: Generates renewable energy on-site, providing the following percentage of the project's annual electrical and thermal energy cost (pick just one for scoring):	
	5%	1
	10%	2
	E2: Provides energy-efficient neighbourhood infrastructures and public amenities. (e.g. Motion sensor lights and air conditioning, energy-efficient water pumps, etc.).	2
	E3: Orients the following percentage of total buildings along north-south axes to minimise solar heat gain and maximise natural ventilation.	
	30%	1
	50%	2
	80% or more	3
	Total Possible Score	7
PART 2: TRANSPORTATION PLANNING		
Connectivity	TP1: Is linked to (within 1/2 mile) a well-connected existing street network.	1
	TP2: Includes a street or pathway into the project at least (pick just one of the following for scoring):	
	— Every 800 feet.	1
	— Every 400 feet	2
	TP3: Has high intersections within the project (pick just one of the following for scoring):	
	30 to 50 intersections within 1/4 mile radius	1
	More than 50 intersections within 1/4 mile radius	2
Transit-Oriented Locations	TP4: Is located on a site that is within walking distance (1/2 mile) of high levels of public transit service (e.g. rail or bus rapid transit systems, or other forms of public transport).	2
Cycling Facilities	TP5: Promotes cycling facilities within the neighbourhood (For scoring, award 1 point for each):	
	Has dedicated bicycle lane in all the primary and secondary roads within the neighbourhood.	1
	Provides secure and covered bicycle storage (for at least 10% of non-residential and 30% of residential building occupants), as well as bicycle parking for visitors.	1
Parking Facilities	TP6: Parking Facilities (For scoring, award 1 point for each):	
	Provisions of off-street parking at the side, front or rear of all public buildings (not along the sidewalk).	1

(*Continued*)

Table 13.2. (*Continued*)

	Provides a combination of over/underground multi-storied parking and surface parking (surface parking not more than 25% of development area).	1
Transportation Demand Management	TP7: Transportation Demand Management (For scoring, award 1 point for each).	
	Subsidised transit passes at half of regular price or cheaper. Comfortable and safe vehicle sharing facilities on-site. Employer-sponsored-transit services — such as a shuttle — to off-site employment centres and/or major transit facilities. Separate cost of parking spaces from the price/rent of dwelling units or non-residential spaces (for at least 80% of the buildings in the neighbourhood). A comprehensive transportation demand management (TDM) programme for the neighbourhood, to reduce trips.	1–5
Walkability	TP6: Promotes Walkability (For scoring, award 1 point for each).	
	Standard[8] sidewalks along at least 90% of street length (on both sides of the street). Minimal distance between the sidewalk and most buildings, with mixed-use and non-residential buildings necessarily attached to the sidewalk. Frequent building entries (at least every 75 feet). No blank walls measuring more than 50 feet along sidewalks. Complete (or maximum) street length lined with non-invasive trees. Low design speeds for most tertiary streets (20 mph for residential, 25 mph for non-residential).	1–6
	Total Possible Score	22
PART 3: ENVIRONMENTAL PLANNING		
Erosion and Sedimentation control	EP1: Creates and implements an erosion and sedimentation control plan for construction activities, reducing soil erosion and downstream pollution.	1

(*Continued*)

[8]Standards for Sidewalks in Dhaka is a minimum width of 5 feet, with an even textured surface comfortable for walking and adequate street lights all along the walkway.

Table 13.2. (*Continued*)

Eco Systems and Biodiversity Habitats	EP2: Preserves all heritage trees and at least 25% of other non-invasive trees, especially larger ones.	1
	EP3: Restores (if needed) and conserves pre-existing on-site habitat, wetlands and water bodies in perpetuity.	2
	EP4: Leaves buffers of undeveloped land (at least 50 to 100 feet) around water bodies.	1
	EP5: Implements a long-term (at least 10 years) management plan for onsite wetlands, water bodies, and habitat.	1
Open Space	EP6: Provides at least one Neighbourhood Park or children's playground within walking distance (within 1/4 mile).	2
Urban Heat Island Effect	EP7: Uses one of the following strategies — or a combination of the two — to reflect (instead of absorb) solar heat:	
	Solar-reflective roofs (usually light-coloured) or vegetated roofs.	1
	Shade, open-grid pervious paving, or solar-reflective paving for at least 50% of roads, sidewalks, parking areas, and other "hardscape".	1
	Total Possible Score	10
PART 4: NEIGHBOURHOOD DESIGN		
Location	ND1: Is located on a site that is one of the following (pick just one for scoring):	
	Infill and also a previously developed site	4
	Infill but not a previously developed site	3
	Adjacent to existing development	2
	previously developed site, but not adjacent or infill	1
Density	ND2: Meets the minimum required density (15000 per square kilometre, details in 4.4.3.4).	1
Mixed Use	ND3: Enables walking access (within 1/4 mile) to the following number of land uses, provided within neighbourhood (The possible land uses are named at 4.4.3.4. For scoring, pick only one of the following):	
	4 to 6 uses	1
	7 to 10 uses	2
	11 or more uses	3
Affordable and Diverse Housing	ND4: Provides multiple housing types of different sizes, such as large and small apartments, duplexes, townhouses, and/or single-family homes.	1
Buildings	ND5: All buildings of the neighbourhood are constructed following the RAJUK guidelines (Building construction act, 2008).	2

(*Continued*)

Table 13.2. (*Continued*)

Recreational Spaces	ND6: Enables access (within 1/4 mile walking distance) to indoor or outdoor (or both) recreational facilities open for public use. (Details in 4.4.3.4)	1–2
Universal Design	ND7: Provides Universal Design for either of the following (only one necessary to score a point):	1
	For residential projects, universal accessibility for people of diverse abilities in 50% of dwelling units.	
	For non-residential projects, universal accessibility for people of diverse abilities along 100% of public areas.	
Education	ND8: Is located within walking distance of at least one school (for scoring, award 1 point for each):	
	elementary school within 1/4 mile	1
	high school within 1/2 mile	1
Jobs and Housing Proximity	ND9: Provides local jobs for at least 50% of total employment demand within 1/4 mile walking distance or at a strategic location within 1 mile which can be accessed by green transportation.	2
	Total Possible Score	18
PART 5: COMMUNITY AND CONTEXT		
Innovation in Design and Technology	CC1: Exhibits exemplary sustainability performance in areas not addressed by this rating system. Write in below (for scoring, add up to three):	
	1.	1
	2.	1
	3.	1
Context Specificity	CC2: Context Specific neighbourhood attributes	
	1. Addresses any geographically-specific environmental issue in neighbourhood design.	1
	2. Includes design features to enhance traditional cultural/historical value of the area.	1
Safety	CC3: Ensures safety of residents in "All" public areas of the neighbourhood (details in 4.4.3.5).	2
Community Participation	CC4: Ensures Different Stake holder groups' engagement, feedback and evaluation in different phases of the project (details in 4.4.3.5).	1–2
Local Government	CC5: Has a well-managed community authority to represent community issues to higher administrative level.	2
Public Awareness	CC6: Arranges public awareness and knowledge sharing programmes at regular intervals to inspire residents to pursue sustainable lifestyles.	2
Social groups	CC7: The community accommodates residents from various professional groups (pick just one of the following for scoring):	

(*Continued*)

Table 13.2. (*Continued*)

	4–6 professional groups	1
	7 or more professional groups	2
Social Interaction	CC8: Adequate and spontaneous social interactions among people from various age and social groups, living in the same neighbourhood.	1–2
	Total Possible Score	17

Total Possible Score	94
Certified	48–51
Good	52–62
Very Good	63–73
Excellent	74 or above

This research has suggested some simple waste management interventions at the neighbourhood level in the proposed NSA scheme. Inclusion of litter receptacles with provisions to separate and collect recyclables, composting from bio-degradable wastes, and educating the residents about the benefits of recycling, and examples of simple steps that can contribute significantly to improving the city's overall waste management. Government agencies also need to implement substantial waste recycling programmes. A private–public partnership can expedite this process.

So far Detailed Area Plan (DAP) is the only planning guideline for Dhaka which identifies some environmental issues. But most of these issues have been discussed very superficially without any direction on how to mitigate them. To address this loophole in the conventional planning approach, RAJUK needs to appoint in-house environmental experts to evaluate the environmental impacts of their projects.

Lenient policies and the lack of transparency in legal enforcement have allowed the encroachment of environmentally-sensitive areas in many parts of Dhaka. So, an urgent recommendation is the demarcation of ecologically-sensitive areas in or around the city and publishing it to restrict urban development in those areas.

The proposed NSA scheme provides elaborate guidelines on systematic transportation planning for Dhaka. Walking and cycling have cross-cutting benefits for public health and the urban environment. RAJUK should start facilitating these greener ways of transport to combat Dhaka's traffic congestion and air quality issues. Architects and landscape designers can be appointed to assist the planners of RAJUK to integrate attractive design features for enhanced walkability in the streets of Dhaka.

Another important issue in Dhaka is the lack of co-ordination among the numerous government agencies. Constitutionally, RAJUK has substantial power as the principal planning agency. But the lack of transparency within its own system has often induced misuse of that power. It will therefore, be essential that RAJUK liaise with other sectoral agencies and local governments for systematic urban management.

Bangladesh has a rich culture full of celebrations and festivals. Historically, these festivities have prevailed for a very long time. As a result, people here require quality urban open spaces for social interaction. Unfortunately, these spaces have received little attention from the planners when neighbourhoods are developed. The inclusion of social indicators under this study is an attempt to address this loophole and strengthen the social capital with careful design approaches.

Conclusion

In the pursuit of urban sustainability indicators, time and again, it has been observed that negligence towards natural capitals and the exclusion of environmental issues in city-planning are basically the roots of all other problems. This inspired urban planning concepts like "eco cities" and "zero carbon cities"; with which many developed cities have been more successful in achieving urban sustainability. The challenge facing developing cities is that, issues like over population, financial constraints, and resource scarcity are so colossal that most of the time, environmental issues are not prioritised in planning initiatives. Dhaka is no exception. With a huge population and limited resources, the principle planning authority (RAJUK) often prioritises ensuring housing, roads, amenities, etc. for the citizens, over protecting natural capital.

To date, no specific neighbourhood planning guidelines have been developed by RAJUK or any other agencies in Dhaka. There is also a gap in literature regarding Dhaka's neighbourhood sustainability issues. From that perspective, this study can be considered as the very first attempt to understand Dhaka's sustainability issues at the neighbourhood level. Indeed, there is room for improvement in the proposed scheme. With further exploration and judicious inputs from the relevant authorities, this study can be used to develop a more innovative and contextual neighbourhood sustainability assessment framework for Dhaka.

References

Ahmed, S.U. (1986). *Dacca: A Study in Urban History and Development.* Dhaka: Curzon Press.

Banister, D. (2012). Assessing the Reality- Transport and Land Use Planning to Achieve Sustainability. *The Journal of Transport and Land Use*, **5**(3), 1–14.

Barton, H. (2000). *Sustainable Communities: The Potential for Eco-Neighborhoods.* London: Earthscan.

Barton, H., Grant, M., and Guise, R. (2003). *Shaping Neighborhoods: A Guide for Health, Sustainability and Validity.* London, New York: Spon.

Basiago, A.D. (1996). The Search for the Sustainable City in 20th Century Urban Planning. *The Environventalist*, **16**, 135–155.

Banglapedia. (2015). *The national encyclopedia of Bangladesh.* Dhaka.

BBS. (2015). Dhaka: Bangladesh Bureau of Statistics.

Blum, A. (2007). Research and Demonstration for Assessing Sustainable Neighborhood Development. In M. Deakin, G. Mitchell, P. Nijkamp, and R. Vreeker, *Sustainable Urban Development Volume 2: The Environmental Assessment Methods* (pp. 412–428). New York: Routledge.

BREEAM. (2015, July). *Hompage of BREEAM Communities.* Retrieved from BREEAM Communities Web SIte: http://www.breeam.org/page.jsp?id=372

Charlot-Valdieu, C., and Outrequin, P. (2003). *Brochure HQE2Rn1: Towards a Methodology for Sustainable Neighborhood Regeneration.* The European Commission Community Research.

Charlot-Valdieu, C., Outrequin, P., and Robins, C. (2004). *Brochure HQE2Rn2: The Toolkit for Sustainable Neighborhood Regeneration and European Application Review.* Thr European Commission Community Research.

Choguill, C.L. (2008). Developing Sustainable Neighborhoods. (32, Ed.) *Habitat International*, 41–48.

Chowdhury, A.M., and Faruqui, S. (1989). Physical Growth of Dhaka City. In S.U. Ahmed (Ed.), *Dhaka Past Present Future* (pp. 43–63). Dhaka: The Asiatic Society of Bangladesh.

Cohen, B. (2006). Urbanization in Developing Countries: Current Trends, Future Projections and Key Challanges for Sustainability. *Technology in Society*, **28**, 63–80.

Cole, R.J. (2010). Environmental Assessment: Shifting Scales. In E. Ng, *Designing High Density Cities for Social and Environmental Sustainability* (pp. 273–282). London: Sterling.

DCC. (2015) *Solid Waste Management.* Dhaka: Dhaka city corporation (North) website.

DCC. (2005). *Clean Dhaka Master Plan.* Dhaka: Dhaka City Corporation, The People's Republic of Bangladesh, Japan International Cooperation Agency.

DEGOB. (2005). *Dhaka City State of Environment Report.* Dhaka: Department of Environment, Government of Bangladesh, UNEP, BCAS.

DMDP. (2007). *Dhaka Metropolitan Development Plan strategic environmental assessment (Prepared for World Bank)* Dhaka: Rajdhani Unnayan Katripakahha.

Dover, V., and King, J. (2008). Neighborhood Definition. In D. Farr, *Sustainable Urbanism: Urban Design with Nature* (pp. 127–131). Hoboken, N.J.: Wiley.

Duany, A. (2002). Introduction to the Special Issue: The Transect. *Journal of Urban Design*, **7**(3), 251–260.

Duany, A., Plater- Zyberk, E., and Speck, J. (2000). *Suburban Nation: The Rise of Sprawl and the Decline of the American Dream.* New York: North Point Press.

DWASA. (2004). *Institute of Water Modeling, Concept Paper on Improvment of Sewerage System in Dhaka City, Sewerage Master Plan for Dhaka City.* Dhaka: Dhaka Water Supply and Sewerage Authority.

DWASA. (2007). *Interim Workshop Presentation.* Dhaka: Dhaka Wasa.

Farr, D. (2008) *Sustainable Urbanism: Urban Design with Nature.*

Garde, A. (2009). Sustainable by Design? Insights from US LEED-ND Pilot Projects. *Journal of the American Planning Association*, **75**(4), 424–440.

Geddes, P. (1917). *Report on Dacca City Planning.* Dhaka.

Gibson, R.B. (2006). Sustainability Assessment: Basic Components of a Practical Approach. *Impact Assessment and Project Appraisal*, **24**(3), 170–182.

Gupta, S., and Rayadurgam, I. (2008). Urban Growth and Governance in South Asia. *South Asian Perspective, 2008.*

Hossain, S. (2008). Rapid Urban Growth and Proverty in Dhaka City. *Bangladesh e Journal of Sociology*, **5**(1).

Hunt, D.V., Rogers, C.D., and Jefferson, I. (2008). Application of Sustainability Indicators in Decision-making Processes for Urban Regeneration Projects. *Institution of Civil Engineers-Engineering Sustainability*, **161**, 77–91.

Huq, S., and Alam, M. (2003). Flood Managemet and Vulnerability of Dhaka. (A. Kreimer, M. Arnold, and A. Carlin, Eds.) *Building Safer Cities: The Future of Disaster Risk.*

Hurley, J. (2009). Sustainable or State-quo: Investigating Sustainability Assessment of Residential Estate Development. *State of Australian Cities Conference.* Perth.

Hurley, J., and Horne, R. (2006). Review and Analysis of Tools for the Implementation and Assessment of Sustainable Urban Development. *Paper Presented at the EIANZ.* Adelaide.

Hyder, Z. (1994). *Organic Cities and the Case of Patrick Geddes in Dhaka.* Unpublish Msc Thesis, Massachusetts Institute of Technology.

Islam, N. (1996). Megacity Management in the Asian and Pacific Region. (G. Clarke, and J.R. Stubbs, Eds.) *Megacity Management in the Asian and Pacific Region.*

Islam, N. (1996). Dhaka from City to Mega City: Perspective on People, Places, Planning and Development Issues. *Urban Studies Programme.*

Kabir, A., and Parolin, B. (2012). Planning and Develoment of Dhaka- A Story of 400 Years. *15th International Planning History Society Conference.* Sao Paulo.

Kabir, A., and Parolin, B. (2010). Urban Spatial Structure and Transport Opportunity in Dhaka- a Question of Sustainability. *1st International Conference on Sustainable Urbanization.* Hong Kong.

Kabir, A., and Parolin, B. (2011). Dhaka Mega City: Growth, Congestion, Environment and Planning for a Sustainable Future. *Conference of Asian Planning School Association.* Tokyo.

Khan, N. (2008). *M.Arch Thesis: Study of Morphological Transformation in the Planned Residential Areas of Dhaka City.* Dhaka: Department of Architecture, Bangladesh university of Engineering and Technology.

LEED. (2012, July 15). *Homepage of LEED-ND.* Retrieved from http://www.usgbc.org/DisplayPage.aspx?CMSPageID=148

MoHPW. (2015). Dhaka: Ministry of Housing and Public Works.

MoC. (2005). *Transport Plan.* Dhaka: Ministry of Communications, Government of Bangladesh.

Morshed, A. (2013, August 5). *Archive: The Daily Star.* Retrieved from The Daily Star Web Site: http://archive.thedailystar.net/beta2/news/how-compact-cities-can-save-bangladesh/

Nilufar, F. (2010). Urban Morphology of Dhaka City: Spatial Dynamics of Growing City and the Urban Core. *International eminar on the Celebration of 400 Years of the Capital Dhaka.* Dhaka: The Asiatic Society of Bangladesh.

Nilufer, F. (1997). *PhD Discussion: The Spatial and Social Structuring of Local Areas in Dhaka City — A Morphological Study of the Urban Grid with Reference to Neighborhood Character within Naturally-grown Areas.* London: University of London.

Nilufer, F., and Khan, N. (2009). Spatial Logic of Morphological Transformation: A Paradigm of Planned- Unplanned Areas in Dhaka City. In D. Koch, L. Marcus, and J. Steen (Ed.), *7th International Space Syntax Symposium.* Stockholm: KTH.

Rahman, M.M. (2004). Regionalization of Urbanization and Spatial Development: Planning Regions in Bangladesh. *The Journal of Geo-Environment*, **4**, 31–36.

RAJUK. (2003). *Bid Documentation for Preperation of Detailed Area Plan (DAP) for Dhaka Metropolitan Development Planning (DMDP) Area.* Dhaka: Rajdhani Unnayan Katripakhha.

RAJUK. (2005). *Dhaka Metropolitan Development Plan 1995–2015.* Dhaka: Rajdhani Unnayan Katripakahha.

Rohe, W. (2009). From Local to Global: One Hundred Years of Neighborhood Planning. *Journal of American Planning Association,* **75**(2), 209–230.

Sharifi, A., and Murayama, A. (2013). A Critical Review of Seven Selected Neighborhood Sustainability Assessment Tools. *Environmental Impact Assessment Review,* 73–87.

Smith, M.E. (2010). The Archaeological Study of Neighborhoods and Districts in Ancient Cities. *Journal of Anthropological Archaeology,* **29**(2), 137–154.

Stephen, W. (2004). *Planning for Sustainability: Creating Livable, Equitable, and Ecological Communities.* New York: Routledge.

UN. (2013, November). Retrieved from United Nations Economic and Social Commission for Asia and the Pacific Web Site: http://www.unescapsdd.org/files/documents/SPPS-Factsheet-urbanization-v5.pdf

UN. (2014). *World Urbanization Prospects.* New York: United Nations.

UNDESA. (2001). *Indicators of Sustainable Development: Guidelines and Methodologies.* New York: United Nations.

USGBC. (2011). *LEED 2009 for Neighborhood Development* . Washington, DC: U.S. Green Building Council.

Watson, V. (2009). The Planned City Sweeps the Poor Away. *Urban Planning and 21st Century Urbanization, Progress in Planning,* **72**, 151–193.

WB. (2006). *Bangladesh Country Environmental Assistance, Bangladesh Development Series* . Dhaka: The World Bank.

WB. (2007). *Dhaka Metropolitan Development Plan Strategical Environmental Assessment* . Washinton DC: The World Bank.

WB. (2007). *Dhaka: Improving Living Conditions for the Urban Poor.* Dhaka: The World Bank.

WRPO. (2001). *National Water Management Plan Development Strategy.* Dhaka: Ministry of Water Resources, Government of Bangladesh.

14

HIMALAYAN URBANISATION, DISASTER VULNERABILITY AND THE ROLE OF RISK PERCEPTIONS

PRERNA SHAH and POW CHOON-PIEW

Often revered as the 'Water Towers of Asia', the Himalayas feed seven major rivers of the East; the ecosystem services thus emanating from the Indian Himalayas alone amount to Rs.943 billion (US$15 billion) per year — or one-third of the country's total GDP. Rampant urbanisation is a fairly new phenomenon in this region, which is why the emerging vulnerabilities are unfamiliar and difficult to deal with. Historically, natural resources have acted as a buffer against adverse events but today the carrying capacity of ecosystems has diminished and natural hazards have been transformed into large scale disasters with massive loss to life and property.

Socio-economic, political and environmental factors are the most obvious determinants of this new emerging growth reality. However, this research goes a step further by bringing the psychological dimension into the forefront and investigating the role of perceptions in guiding people's behaviour throughout the disaster process. It adopts a bottom-up approach in trying to understand why people occupy hazard prone areas and how their cognitive limitations, traditions and culture diminish their ability to make rational choices. In the latter half it tries to draw parallels between the response mechanisms of people and aid agencies in a disaster situation.

Introduction

Spawning over seven nations, the Himalayas are home to exotic varieties of flora and fauna, diversified landscapes and extreme weather conditions. Being the source of major perennial rivers, they provide livelihood to 210 million people living in the region and indirect services to 1.4 billion people i.e. one-fifth of the World's population, living in downstream areas (Ning *et al.*, 2013). They also provide a rich base for agriculture, hydropower and tourism. Furthermore, the region includes four global biodiversity hotspots, 488 protected areas, 330 important bird areas, and 60 global eco-regions (Karki and Bhattarai, 2012).

Over recent decades, the Himalayas are experiencing unplanned and unregulated urban growth. Both "pull factors" such as better employment opportunities, infrastructural facilities, access to health and education and "push factors" such as discrimination, poverty, and depletion of essential resources are responsible for this rapid influx. This trend, apart from creating isolated urban agglomerations with a diminishing resource base, has gradually transformed self-sufficient rural communities into competing factions increasingly dependent on the socio-politico and economic volatility of the region.

As one of the youngest mountain ranges in the world, the Himalayas are prone to unstable conditions resulting in earthquakes, landslides and flash floods. Urbanisation has compounded the risks through water pollution, depletion of biodiversity, degradation of forests and disruption of the hydrological system. Today, there is general agreement among scientists that economic growth has diminished the capacity of ecosystems to absorb these stresses, magnifying these instabilities over time and intensifying the vulnerability of the indigenous people to cope with such exigencies.

Nature of Urbanisation in the Indian Central Himalayan Region (IHCR)

The Indian Central Himalayan Region (IHCR) encompasses the 27th state of India — Uttarakhand — it has uneven topography with nearly 93% of its area categorised as 'rugged and mountainous' and about 65% of its area under forests (Kumar and Kalotra, 2011).

Spatial Trends in Urban Population Growth

With an area of approximately 53,483 sq. km, Uttarakhand is home to more than 10 million people (COI, 2011). Over the past decade, the urban

Table 14.1. Development indicators of adjacent Himalayan states.

Primary Indicator	Uttarakhand		Himachal Pradesh		Jammu and Kashmir[1]	
% of urban population	25.67		9.80		24.81	
Secondary Indicators	Rural	Urban	Rural	Urban	Rural	Urban
Poverty	40.8	36.5	10.7	3.4	4.6	7.9
Unemployment	4.1	6.8	5.7	4.9	5.6	6.0

Source: Central Statistical Organisation (CSO), 2006.

[1]It must be mentioned here that Jammu and Kashmir (J&K) states have special autonomy under Article 370 of the Constitution of India. They have a history of civil unrest and insurgency and therefore the figures must be interpreted in that light.

Table 14.2. Urban population by size and class of town in Uttarakhand.

	Population	Cities	Percentage
Class I	1335168	6	44.83
Class II	337540	6	11.33
Class III	535713	19	17.99
Class IV	453903	33	15.24
Class V	281203	37	9.44
Class VI	34787	15	1.17
Total	2978314	116	100.00

Source: Census of India (COI), 2011.

population of the state has phenomenally grown by about 40% which is much larger than the country's percentage decadal growth of 31% (ibid).

Nearly one-third of the population resides in 116 towns and cities (Table 14.2). Hence, the distribution is mostly top-heavy i.e. a few large cities comprise the bulk of the urban population.

Temporal Trends in Urban Population Growth (1901–2011)

The total urban population quadrupled from 7.8% in 1901 to 30.55% in 2011 (COI, 2001–2011). However, the increase was primarily centred in the plain areas/districts of Dehradun, Nainital, Haridwar and Udham Singh Nagar (Table 14.3).

However, these figures should not be interpreted to mean that the externalities arising from unplanned urban growth in the hill areas are in any way inferior to those in the plain areas. In fact, the detriment is much worse because of two reasons. First, many of the hill cities are also popular tourist destinations; hence they cater to a large floating population that is sometimes four to five times more than the total urban population of the city itself.[2] Second, there has been a sharp increase in the percentage decadal growth, rate meaning that the number of people added to the total urban population is substantial in the past couple of decades. Besides, geographic constraints lead to the development of tightly packed urban spaces usually along main arterial routes; this not only places excessive

[2]Nainital, a popular hill station, had a permanent population of about 45000 in 2006 and a floating population of whooping 0.5 million in that year (City Development Plan, May 2007, Jawaharlal Nehru National Urban Renewal Mission, JNNURM).

Table 14.3. Percentage share of urban population and decadal growth rate in districts of Uttarakhand.

Districts	2001 ('000)	2011 ('000)	Decadal growth rate (%)
Uttarkashi	7.77	7.35	−5.67
Chamoli	13.69	15.11	16.54
Rudraprayag	1.20	4.19	263.03
Tehri	9.90	11.37	17.06
Dehradun	52.94	55.90	39.90
Pauri	12.89	16.41	25.37
Pithoragarh	12.94	14.31	16.26
Bageshwar	3.16	3.50	16.51
Almora	8.61	10.02	14.36
Champawat	15.04	14.79	13.52
Nainital	35.27	38.94	38.22
Udham Singh Nagar	32.62	35.58	45.33
Haridwar	30.84	37.77	63.11
Uttarakhand	25.67	30.55	45.27
India	27.81	31.16	31.80

Source: Census of India (COI), 2001–2011.

load on the ecosystem services provided by these areas but also leads to the proliferation of infrastructural arrangements that are often poorly planned, leading to increased vulnerability of the city and its people.

The Occurrence of Disasters

As per the disaster prone map of the country, Uttarakhand has attained a position among the top five states most prone to natural hazards. Prominent among them are:

Earthquakes

Four of the thirteen districts of the state (Pithoragarh, Chamoli, Bageshwar and Rudraprayag) fall completely in Zone V (representing damage risk of ≥IX on MSK scale[3]). The vulnerability of the region to earthquakes is characterised by the fact that during the last century, the region had experienced 11 earthquakes of magnitude greater than 6.0 on the Richter scale (ibid).

[3]Medvedev-Sponheuer-Karnik Scale, a macro-seismic intensity scale with 12 degrees of intensity ranging from 'Not perceptible' to 'Very catastrophic'; Degree IX here corresponds to 'Destructive'.

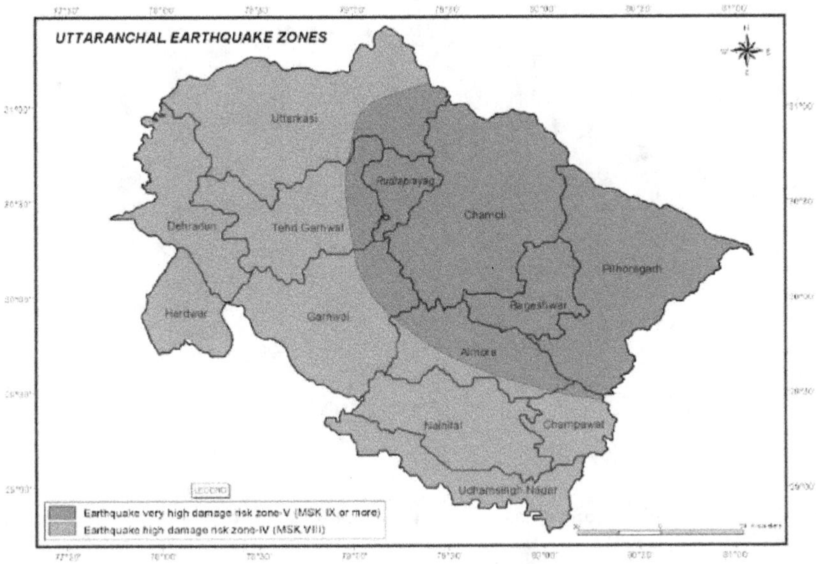

Fig. 14.1. Earthquake prone zones of Uttarakhand (Bureau of Indian Standards, 2002).

Landslides

The geological, topographical and climatic conditions in the State result in frequent landslides especially in the rainy season. Anthropogenic factors such as deforestation, construction of dams, housing schemes, and unscientific construction of roads without proper environmental impact assessments have resulted in huge losses to the State.

The Recent Trend of Excessive Rainfall Events (EREs) or Cloudburst

The IHCR is extremely susceptible to EREs due to its unique topography and altitude-dependent climate. The recent Kedarnath disaster initiated by extreme rainfall[4] was one of the most calamitous disasters in the history of India in terms of human and economic losses. Table 14.4 captures some of the EREs in the past decade; however, this list does not include a lot of events that go unreported due to lack of monitoring or remote and unpopulated regions which hardly pose any threat of human and monetary loss.

---

[4]691.3 mm over a period of three days; mean monsoonal precipitation of 6.97 mm day^{-1} (Indian Meteorological Department (IMD), 2013).

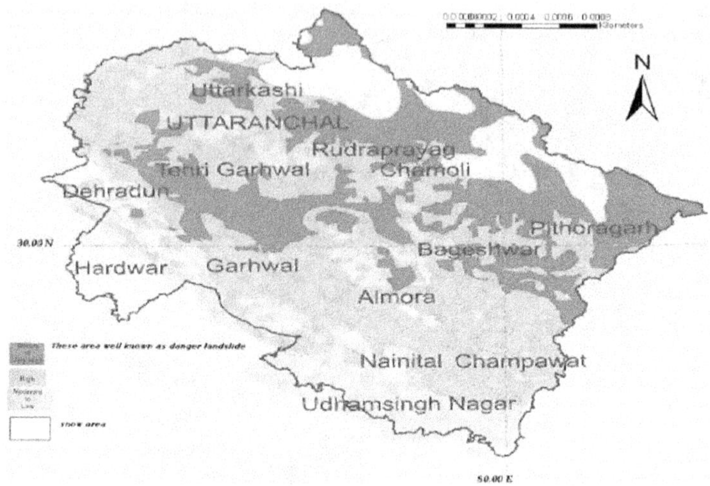

Fig. 14.2. Landslides prone areas of Uttarakhand (Disaster Mitigation and Management Center (DMMC), 2006).

Table 14.4. Excessive rainfall events in Uttarakhand.

Date of ERE	Location
2002	Ketgaon
2004	Ranikhet
July 5–6, 2004	Chamoli
2007	Pithoragarh district
August 6–7, 2009	Pithoragarh district
August 18, 2010	Kapkot, Bageshwar
July 21, 2010	Almora district
Sept 14–15, 2010	Almora district
Aug 2–8, 2010	Bhatwari, Uttarkashi
Sept 13, 2012	Ukhimath, Rudraprayag
June 15–17, 2013	Kedarnath, Pithoragarh
Aug 1, 2013	Kapkot, Bageshwar

The Conceptual Debate on Vulnerability

Birkmann (2006) provides an overview of the different spheres of vulnerability:

a) Vulnerability as a dualistic approach that focuses on the likelihood of harm on the one hand and unusual difficulties in coping and recovering from a hazardous event on the other.

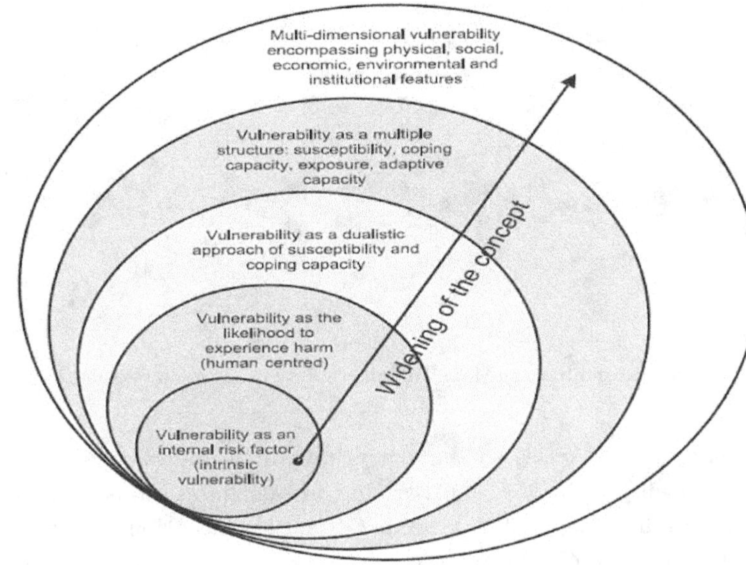

Fig. 14.3. Key sphere of the concept of vulnerability (Birkmann, 2006; National Institute of Disaster Management (NIDM), 2014).

b) Vulnerability as a multiple structure that captures not only susceptibility and coping capacity but also adaptive capacity, exposure and the interaction with perturbations and stresses.
c) Vulnerability as a multi-dimensional phenomenon which accounts for various themes and parameters such as the physical, social, environmental, economic and institutional.

The Pressure and Release Model (PAR)

The PAR model (Wisner *et al.*, 2004) acknowledges that disaster is a result of two opposing forces — the conditions that create vulnerability and the physical occurrence of a hazard. The release is manifest in the reduction of the impact of disaster through reduced vulnerability. In other words, to release the pressure that causes disasters the entire chain needs to be addressed right up to the root causes and not just the proximate causes related to the disaster event.

In the context of the PAR model, *urbanisation* is a dynamic pressure that gives rise to unsafe conditions and increases the vulnerability of a region. It leads to land scarcity as migrants from neighbouring areas move

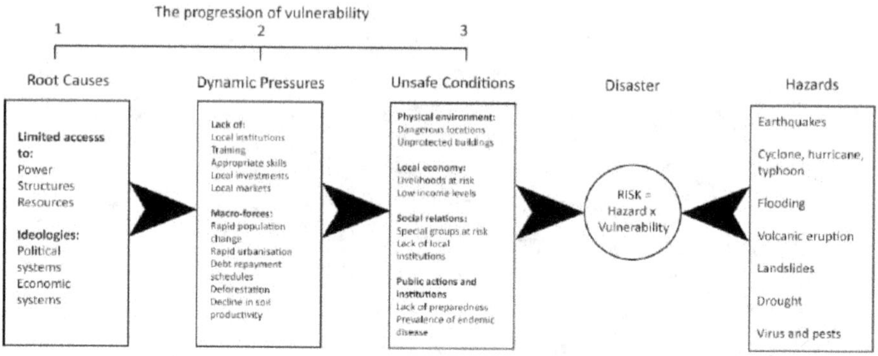

Fig. 14.4. Pressure and Release Model of vulnerability progression (Wisner *et al.*, 2004).

into already packed cities so that they have little alternative apart from living in unsafe conditions, constructing unsafe habitations and working in unsafe conditions (Havlick, 1986). Often this situation is exacerbated in hilly regions where geographic constraints limit the already scarce land area and force communities to live in closely built quarters that disrupt the natural drainage pattern.

Role of Risk Perceptions

The Theory of Bounded Rationality

Simon (1956) introduced the theory of Bounded Rationality which argues that the cognitive limitations of the decision maker forces him to construct a simplified model of the world to deal with it. Because a decision maker is blinded by the limitations of cognition and perception, he is forced to proceed by trial-and-error, modifying actions that do not yield satisfactory outcomes and retaining those that work or fail in future.

Evidence in support of Bounded Rationality (Slovic *et al.*, 1974):

a) *Limited range of alternatives:* the decision makers rarely ever have access to the full range of alternatives due to lack of awareness or local traditions.
b) *Misperception of risk and denial of uncertainty:* people view floods as cyclical events that occurred after fixed intervals (Burton and Kates, 1964), thus shielding them of its ill effects in the near future.
c) *Probabilistic information processing:* the distortions of subjective probabilities are too large, consistent and difficult to eliminate due to which

it is much easier for people to replace the rules of chance by intuitive heuristics.

d) *The law of small numbers:* there is an abnormally high rate of faith in the reliability of small samples (Tversky and Kahneman, 1971) due to which overgeneralisations are made while making inferences of a degree of hazard (Burton and Kates, 1964).
e) *Perception of randomness:* flood plain residents believe that a flood was less likely to occur in year x+1 if it had occurred in year x (Kates, 1962).
f) *Judgments of correlation and causality:* People's prior expectation of a probabilistic relationship leads them to perceive the same relationship between two unknowns, where none may actually exist.
g) *Judgments of probability by availability:* one judges the probability of an event by the ease with which it is imagined or readily comes to mind.
h) *Anchoring and adjustment:* Individuals in the course of making decisions, tend to choose a natural starting point as a first approximation (or anchor) to a judgment and thereafter adjust to accommodate the implications of additional information.

Researching Disaster Vulnerability

Site Selection and Walk-over Surveys

The objectives of this research have been realised in the context of the flash floods in Kedarnath, the severity of which has often been compared to a 'Himalayan Tsunami'. The relevant area (National Highway 109 of over 90 kms) begins from Kedarnath, one of the five popular pilgrimage sites in the State, and stretches up to Rudraprayag, the district headquarters. The three main settlements selected for the study are — Tilwara, Chandrapuri and Kund — all of which lie on either side of the river and were badly affected by the floods.

Walk-over surveys were conducted from one end of the settlement to the other in order to get a good sense of the geomorphology and social environment of the area. A general rapport was made during this time to enable greater ease in future interactions.

Focus Group Discussions (FGDs)

A total of 3 Focus Group Discussions (FGDs) were conducted in each of the settlement areas, consisting of 7–10 persons from different gender, age, and education (Table 14.5). A total of around 70 persons participated in the

Table 14.5. General composition of Focus Group Discussions (FGDs) participants.

	Gender		Age			Education level		
Settlements	Male	Female	18–35	36–50	>50	Literate	Illiterate	Total
Tilwara	14	11	10	9	6	20	5	25
Chandrapuri	17	6	8	10	5	18	5	23
Kund	16	9	9	9	7	22	3	25
Total	47	26	27	28	18	60	13	73

FGDs which usually lasted between 40–50 minutes. It started with a general narrative of the event, gradually followed by a more focused discussion on the research goals.

a) Centred on understanding the rationale behind people's occupation of hazard prone areas, the random sample selected for the first part of the FGDs consisted of a) persons still residing in the floodplain area b) persons who have relocated to safer areas following the disaster, and c) persons residing in safer areas who survived the disaster with minimal loss.
b) The second part of the FGDs attempted to understand the traditions, beliefs and cultures of the locals and how they shape their response to disaster risk and vulnerability. Guided by the understanding that cultures are assimilated over a period of generations, the group consisted of a healthy mix of old, middle aged and young persons.

Semi Structured Interviews

A total of 6 semi structured interviews were conducted across three major stakeholder groups (Table 14.5). The participants were chosen based on their experience in the field and research done after the Kedarnath tragedy. The interviews were based on a set of pre-determined questions but the approach was more or less flexible in order to get a gist of the overall scenario.

The Mandakini River Valley

The Mandakini River Valley is U-shaped valley and leads to the temple of Kedarnath, one of the 12 holy shrines dedicated to Lord Shiva. Gaurikund is the last motor head after which one has to trek a distance of 14 km to reach Kedarnath. The altitude of the river catchment extends from 670 to 6000 m

Table 14.6. List of agencies interviewed across three stakeholder groups.

Stakeholder group	Agencies
Government Representatives	Disaster Mitigation and Management Centre (DMMC), Dehradun
Technical Specialists	Wadia Institute of Himalayan Geology (WIHG), Dehradun Meteorological Centre, Indian Meteorological Department (IMD), Dehradun
Non-Government Organisations (NGOs)	Climate Himalaya, Dehradun *Prakriti* — a mountain environment group, Agastyamuni, Distt. Rudraprayag *Gramin Sudhar Evam Sharmik Sewa Sanstha* (GRASS), Rudraprayag

Fig. 14.5. Location of study area and case study settlements.

Source: Google.

above mean sea level. The climate of this region is governed by monsoons, so much so that the Ukhimath area in the valley is often called the Cherrapunji of Garhwal (Joshi and Kumar, 2006)[5] (one of the two main geographic divisions of Uttarakhand, the other being Kumaon).

The Kedarnath Deluge of 2013

The aforementioned was the worst natural disaster in the history of the Himalayan region. Official sources put the death toll at over 900 persons with 5,748 reported to be missing[6] but many other sources estimate this number to be much higher.

The convergence of the South-East monsoon trough and the westerly disturbance led to the formation of a dense cloud over the Uttarakhand Himalaya. In a period of 5 days between 14 and 18 June, 2013, Uttarakhand received approximately 2000 mm of rainfall (TRMM, 2013) which is more than what it receives in the whole monsoon season. Heavy rainfall also took place over snow clad mountains, leading to the simultaneous melting of ice and snow. This added to the already incessant rainfall and opened the flood gates for disaster.

What Led to the Disaster?

Deforestation

Although the State and Central governments try to deny mismanagement of forest resources by quoting the forest cover in the State which has risen from 23,938 sq. km to 24,496 sq. km in the past 10 years (ibid); this in no way takes into account the effect development projects have had on the fragile ecology of the area. Furthermore, large tracts of broad leaved *Banj* (oak) and *Burans* (rhododendron) have been replaced by *Chir* (pine) plantations, primarily because Pine is a source of industrial timber and resin and hence, a better source of revenue for the State. But it does not provide the same quality of environmental services — the mighty Oak creates a layer of black subsoil that nourishes thick undergrowth and prevents soil erosion even in

[5]Extreme rainfall events and associated natural hazards in Alaknanda valley, Indian Himalayan region.

[6]A month after Uttarahand Disaster (2013, July 15) The Economic Times.From http://economictimes.indiatimes.com/news/politics-and-nation/a-month-after-uttarakhand-disaster-5748-still-missing-not-to-be-declared-dead-yet-cm/articleshow/21085321.cms.

heavy rainfall. In contrast, the Pine grows needles instead of leaves which create a dry flaky bed that does not hold water.

Hydro-electric projects (HEPs)

Soon after the brutal floods in Uttarakhand, the Supreme Court directed the Ministry of Environmental Affairs (MoEA) to set up an expert committee to determine whether hydropower projects influenced the causes and consequences of the flooding. The Report — *Assessment of Environmental Degradation and Impact of Hydroelectric Projects* during the June 2013 Disaster in Uttarakhand — concluded that hydropower development did significantly amplify the damage caused by the disaster. It also recommended the cancellation of 23 hydroelectric projects proposed to be built here due to the irreparable environmental damage they would cause.

Roads and haphazard construction activities

Roads are routinely widened to accommodate the heavy tourist traffic and allow a safe passage for the heavy machinery required to build dams. The unscientific approach destabilises the mountains and makes them prone to landslides. As per the statistics with Uttarakhand State Transport Department (2011), there were 83,000 vehicles registered in the state in 2005–2006; this figure rose to 180,000 in 2012–2013 with cars, jeeps and taxis witnessing the highest increment as they are often the most preferred means of transport for pilgrims. In 2013, the State High Court ordered the state government to demolish all illegal structures built within 200 m of the riverbank but there was little action taken in this regard.

Peoples' Rationale Behind Occupation of Disaster Prone Areas

Generally, the primary reason for people to settle in flood-prone areas is the range of opportunities that are available to them along the road side. They usually set up small businesses (like tea shops, road side *dhabas*-restaurants, tourist service outlets) to earn better incomes during the short (6 months) *yatra*(tourist) season when maximum number of tourists visit Kedarnath. Some young respondents treat this as an interim arrangement to later migrate into bigger urban areas like Dehradun and New Delhi. They yearn for better, modern lives and want to shift base at the first opportunity.

Table 14.7. Factors accounting for occupation for hazard prone (floodplain) of Mandakini river.

Categories	"Push" Factors	"Pull" Factors
Economic	• Stagnating agricultural incomes • Limited livelihood opportunities • Rising debts/costs	• Access to better employment opportunities • Easier to build homes • Better water, sanitation, transport and communication facilities
Social	• Lack of support systems for women, children and elderly • Difficult life	• Better support systems — both community and institutional • Better ties, kinship • Better opportunities to learn and develop livelihood skills • Closer access to education and health institutions
Environmental	• Declining access to common resources (natural springs running dry, limited fodder availability) • Unpredictable weather conditions • Loss of land due to landslides • Human — animal conflict	• Easier access to river water • Fertile land, better crop yields
Political	• Improper rehabilitation due to building of dams/tunnels	• Government policies that focus on development of area around road/ river

The "push-pull" scenarios provide useful insight into the question of why people occupy hazard-prone areas. However, if the goal of policy makers is to predict the behaviour of 'a rational man' then an understanding of his psychological properties and the fallibility of this simplified model designed to deal with real-life situations is imperative. The study divides respondents into three categories:

i. People who continue to occupy flood-prone areas by need or choice (High Risk Takers or HRT)

ii. People who occupied flood-prone areas but have now migrated to (supposedly) safer areas (Experienced Low Risk Takers or LRT^e)
iii. People who always occupied safe areas and therefore, escaped the disaster with minimal loss (Pragmatic Low Risk Takers or LRT^p)

The following section discusses four case studies in an attempt to unravel some of the complexities surrounding the decision-making process.

Case 1: The overly-optimistic mind

The residents generally had the belief that the worst was over, at least in terms of the physical devastation. They perceived no immediate threats and believed they could settle down near the road, although almost all of them foresaw migrating to bigger urban spaces in the near future. Settling near the river was, therefore, an interim arrangement for them.

> "*It is only a matter of time before everyone moves out to urban areas. There is nothing much left here. Until then it is best to settle in areas with maximum economic activity so we can recover from our losses. Besides, the river seems okay now.*"
>
> – Male respondent, 35, Literate

As regulation in the area is still lax, some residents even treat this as an opportunity to grab land areas near the road. It is interesting to note that even though the disaster was of unimaginable magnitude, most people do not see it as a reoccurring phenomenon. As Kates (1962) rightly observes, 'Floods need to be experienced, not only in magnitude but also frequency. Without such repeated experiences, emergency measures to cope with floods hardly take place'.

Case 2: The appeasing mind

There were a few residents who had no choice but to continue living in the floodplain area because they were tied to their fixed assets and did not have enough resources to move out to a new area. Somehow, these people have reconciled to their situation and attribute such devastation to the forces of nature and God.

> "*A lot of sin has been committed in this area. That is why God has punished us in this way. What is the point of leaving now? If we have to die, no one and nothing can save us.*"
>
> – Male respondent, 52, Illiterate

Peoples' denial of uncertainty is evident; they do not see an event of such magnitude happening in the near future or even if they do, they do not feel they have a role in it. Hence, their approach is that of rebuilding and very little of preparedness. In a way they have created a mental schema which has led them to believe that they are safe because in any case they cannot afford to migrate to a different area. This is probably a reflection of Simon's simplified model of the real situation in order to deal with it satisfactorily.

Case 3: The cautious mind

Residents who have borne losses in the aftermath of the disaster perceive a greater risk of hazard. They seem to have learnt important lessons that have enabled them to settle in areas comparatively at a larger distance from the river. The recency effect seems to be at play here; the 2013 disaster is fresh in their memory and so they have tried to take mitigation measures.

> "*The floods were unprecedented. It was like the floodgates of hell were opened upon us. Such events used to happen before as well but something of this scale was unimaginable. Although living near the road is the best option for us we cannot take such a risk again. Had the authorities informed us of the dangers, the magnitude of loss would not have been so huge*".
>
> – Male respondent, 37, Literate

The safety measures undertaken by the locals seem to be limited to spatial distance which in itself is also an 'intuitive guess'; 'how far safe is really safe?' the answer to this question is determined by the limited knowledge and experience villagers have. In short, it can be said that the residents do realise the danger of floods but their response to flood disaster reduction has been limited by lack of awareness. Furthermore, they have little faith in the government aid as many of them have still not received the compensation promised to them. They either bear the loss or wait for non-government agencies to help them reduce their vulnerabilities through activities such as skills-generation and microfinance investments.

Case 4: The rational mind

The study attempts to go a step further by analysing the thought process of people who adopt a comparatively rational approach even at the cost of choosing an unpopular decision within the community. In understanding the flip-side, there is a probability of affecting change in the mindset of people who continue to occupy hazard prone areas. In this case, however, the decision to stay in a safe area is guided by two rival scenarios: a) most of these settlements are either too poor to migrate to new areas or b) they are generally well off since one or two members of these households are earning well in urban areas and adequately providing for their families back home.

Socio-cultural Dimensions of People's Adjustment to Natural Hazards

A study into the cultures, beliefs and traditions of people living in the three settlement areas shows that in some instances culture has helped them manage their risks better but in others it has also increased their vulnerability.

Pre-disaster

The rules for prevention and preparedness seem to have become obsolete in the present-day urban milieu. The indigenous culture fine-tuned to the local ecological conditions has rapidly been eroded by a homogenous westernised culture that takes little account of the local vulnerabilities. This was evident in a few responses of the residents:

> "*Our ancestors always settled in upslope areas because they knew the dangers of flash floods. Sometimes the agricultural fields were located near the river or streams to enhance productivity of land, but the habitations were always in upslope areas. But now agriculture has no income and tourism is a lucrative option. So we don't mind settling here.*"
>
> – Male respondent, 44, Literate

Clearly, the local communities had a sense of where they should settle. The decision of safe, self-sufficient and often times prosperous communities to settle in hazard-prone areas is an indication of a failure in rejuvenating the rural economy and not a collapse of indigenous capabilities of people. Their response is guided by experimentation and experience in an attempt to bring in some stability and security in their turbulent lives.

Another important theme noticed in the people's responses was their indictment of current development activities in the region. Their concern seems plausible if we understand the cultural significance they attach to the local ecology:

> "*The forests belong to Dhari Devi* (local goddess) *We need to protect them so that such tragedies do not occur.*" (In the context of forests destroyed due to road construction and dams)
>
> – Female respondent, 30, Literate

In order to impede the trend of rapidly declining fodder base, a patch of forest used to be dedicated to a local deity. The norms were strictly held due to fear of adverse consequences. This conservation practice not only maintained the natural resource base for the people but also cushioned them from frequent hazards such as landslides and floods. However, since the people now have no control over the forests, they are frequently mismanaged by the authorities and the locals perceive this as a dominant cause of the 2013 disaster. The villagers' close connect with the natural environment is evident in another response:

> "*When the mynah* (local bird) *makes its nest in the lower part of the tree, we know that heavy rain will come. Now, it's hard to predict rain. Disasters happened since the beginning but such disasters were not witnessed even when the rivers were flowing at full mark*"
>
> – Male respondent, 50, Illiterate

The villagers believe that certain animals can sense a mis-happening. This belief may not have a scientific basis but it was a way of reducing uncertainties in people's lives so they can be prepared for future events. By incorporating such beliefs in their culture, people reduce the stress and feel that the impending threats are manageable.

The practice of sustainability embedded in people's culture can have important future policy implications. It can reduce the effort that would otherwise be required to instil the culture of sustainability because the locals already realise its importance.

During/Post disaster

It seems that cultures, beliefs and traditions do not always help people in managing their vulnerabilities. Often, they magnify their risks and expose them to scenarios that rarely ever find mention in disaster mitigation literature.

Faith in a superior force

A number of residents revealed that even when the water had reached critical levels, people refused to abandon their homes. There were religious ceremonies happening at various places as people had immense faith in the local deity. One of the resident whose house was inches away from being swept away quotes:

> "*I vowed to visit Kedarnath barefoot if God spared my family. And he did. I did not leave my house because I knew I could do nothing. My life was in God's hand.*"
>
> – Male respondent, 39, Literate

As miraculous as this escape was, there were many others who lost their lives as a result of such blind faith. The writer's personal experience mentioned in the earlier chapter is also a reflection of a similar faith, not in a superior power but on a higher authority. Ironically though, the communities had little or no hope in getting assistance from official agencies. Such mistrust and doubt among the people speaks volumes of the ineffective disaster mitigation measures undertaken by government authorities in previous hazard scenarios. Perhaps, it also leaves the people with no choice but to turn to God or a higher power/authority to alleviate their ills.

Locals also attribute the disaster to a prevalence of immoral and blasphemous activities:

> "*A lot of sin has been committed in this area. Alcohol and business have killed everybody. This was a punishment from God.*"
>
> – Male respondent, 61, Illiterate

Culture of attachment

Some residents who climbed to higher areas reported watching scores of people being washed away because they dared to go back to their houses to retrieve certain belongings. The willingness of people to undertake such grave risks driven by an unflinching trust in their own immunity (or an optimistic bias?) speaks of a material culture prevalent in modern day societies. There appears to be an unhealthy attachment with one's belongings that lead to such dangerous decisions.

Cohesiveness

Villages are viewed as close knit communities; their existence depends on mutual trust and support in times of adversity. Prior to the disaster, there was no warning sounded by the Indian Meteorological Department (IMD); also the disaster response and mitigation measures were severely lacking. In one of the interactions with an NGO, it was revealed that help did not arrive until the media started capturing images of the destruction; the officials treated it as 'one of the many usual events'. During the disaster, the communities broke the barriers of race, caste, class and religion and came together as one. They improvised on food, shelter, medicine, water and other essential amenities.

> "*There were tourists from all over the country. I put up a tent outside my broken house and fed people for almost a week with the little resources I had left.*"
>
> – Male respondent, 28, Literate

> "*I put up a white board outside my house so passers-by could write their names and I could inform their near and dear ones of their safety.*"
>
> – Male respondent, 34, Literate

> "*The tourists had no experience of such rugged terrain. Apart from helping our own communities, it was our duty to guide them to safer areas. All paths were broken; we had to walk over loose rocks and cross the river with a rope to get to the other side.*"
>
> – Male respondent, 38, Literate

It seems as though the people experience a sense of solace in facing the same challenges together which drives them to such altruistic behaviours. Interestingly though, as soon as the imminent threat is over, a lot of people resort to selfish means. Discussions with residents revealed how many persons started selling food and water at exorbitant prices soon after the disaster.

In conclusion, the concept of ethnocentrism is the biggest threat to risk mitigation measures. We need to realise that those living under risk are best suited to find solutions to mitigate that risk; the experts should not impose their views but should instead provide a choice so that an informed decision can be made based on the local conditions.

Bringing in Birkmann's spheres of vulnerability into picture, if we associate vulnerability to be an "internal side of risk" as viewed in some earlier debates, then the *condition* of say, the ecosystem becomes central. A possible response would be to improve the health of the natural environment but if we incorporate this new element by understanding the importance people assign to various environmental functions then policies can be more contextual and representative of people's needs. Similarly, if we associate vulnerability with unusual difficulties in coping and recovering from a hazardous event, the question of 'why a community is unable to cope' should understand their underlying belief system or the fact that trust-building is also important in the application of policies.

The Incongruence Between Local and Institutional Coping Mechanisms

Government agencies

Interviews were conducted with officials from the state Disaster Mitigation and Management Centre (DMMC), the apex body responsible for implementation of disaster management plans and building capacity of people and personnel. The interviewees understand that they are now faced with issues and challenges that are much different from what they used to be a couple of decades back. Their response to disaster risk mitigation needs to be anticipatory rather than reactive; however, they feel that their capabilities have been limited due to numerous resource constraints and lack of coordination among different agencies. Here are some of the responses from these officials (in italics).

> "*Development activities in the hills have certainly increased the vulnerability of people. But this is not to say that people were not vulnerable before; people in remote areas were as susceptible to harm, only the incidents were not reported due to an underdeveloped media. The dimensions of vulnerability have changed today and there is a 'new kind of vulnerability' we have to deal with.*"

They also realise that the local people have a plethora of knowledge and experience with regard to their environments but their changing geographies have made those solutions redundant and the lack of their involvement in public policies has exposed them to vulnerabilities previously unheard of.

> "*Development should not trigger or aggravate a disaster. To reap maximum benefits, we need to focus on indigenous solutions; traditional rather than alien technology.*"

In many instances though, the interviewees are dismissive of the superstitions and fatalistic attitude of the villagers.

> *"The villagers have rigid faith in God. They refuse to settle in safer areas even when we convey the risks involved."*

> *"People depend on the government for almost everything. Naturally the blame is also entirely put on us when anything goes wrong. Zoning policies are there but nobody follows them. Just like the Indian Penal Code (IPC) which has been there for a century but still rapes and murders happen."*

Technical experts

The experts in the field are involved both in research and consultation. Academic research focuses on geodynamic processes, climate variability, assessment and mitigation of natural hazards while consultancy offers scientific solutions to specific problems in the Himalayan region such as the location of dams or determining the slope angle in the construction of roads.

In the aftermath of the flash floods, the specialists indicate the abundance of research papers and academic journals that warned the authorities of an imminent disaster.

> *"We had published a report warning of the dangers emanating from the outburst of Chorabari glacial lake to downstream areas but it was not heeded and no action was taken."*
> *"The area above 2000m is composed of quaternary deposits which are generally unstable and susceptible to breakage. Rampant construction in such areas and improper drainage practices led to this self-created disaster."*

There is a lack of incentive and a general sense of despair among researchers because of a lack of support from the government in research activities. One of the interviewees elaborated:

> *"There is no encouragement given to research here. Our findings and recommendations are hardly ever integrated into government policies. Whenever there a specific need, expert panel groups are formed and that is about all. Research is never a part of policy."*

The interviewees lament the lack of integration in not just different stakeholder agencies but also within the research and training cohort. They believe this to be a reason for a lack of hard data which is there but nobody knows where to look for it. With regard to the future disaster management scenario, they offer these solutions:

> *"First, there needs to be proper mapping according to the type of hazard. There is lack of this hard data mostly because it is inaccessible. Second, there is a need for instrumentation so data can be collected and validated. Third, a comprehensive plan is required that is adapted to local conditions, implemented on ground and integrated with people's needs."*

NGO groups

The Non-Government Organisations (NGOs) have had the maximum positive impact on the lives of local communities, not just in the post disaster rehabilitation process but also in the more crucial aspects related to livelihood security. This was clearly evident in the writer's interactions with the people.

The views of various NGO groups have been more or less critical of the government role. An inability to provide sufficient livelihood opportunities, destruction of self-sustaining mountain ecology in the guise of development and a lax disaster management system catering to the distribution of relief aid are seen as the biggest barriers to the sustainable development of this region.

> *"The greatest issue faced by hill communities today is the lack of sustainable livelihoods. The creation of assets and the revamp of the local economy are absolutely essential for managing their vulnerabilities. The neighbouring state of Himachal Pradesh has made admirable progress in making horticulture a state brand and synonymous to its name. This has made the locals self-sufficient and resilient towards most mis-happenings. The Uttarakhand state too has abundant resources — it can be marketed as an 'organic state', the locally available Mandua (millet) is in great demand in countries like Japan but there is no formal mechanism to operationalise such systems."*
>
> *"The government approach to disaster management is focused on immediate relief. There is no model to involve the local communities to build disaster resilient societies."*

The NGOs understand the plight of the local communities and believe that their decision to settle in hazard prone areas is guided by a complex interplay of economic, social, natural and political concerns. However, they reveal that in some areas where villagers have been provided with alternative incomes the people have resisted migrating to urban centres. In the opinion of many their decisions if looked upon closely, are to a large extent logical and coherent:

> "*If you ask me, the people's decision to settle in hazard prone areas is not irrational at all. Between a hazard that occurs once in 100 years and an urban space that only provides you with the basic necessities of everyday life — livelihood security, education institutions and medical centres — which one would you choose?*"

They also have a good understanding of the local ecology and their concerns regarding the unsustainable development of the area are synonymous to that of the local people. Apart from the unscientific and unchecked growth, the interviewees blame a lack of coordination among government agencies as a reason for state policies not having achieved its objectives. One example is worth mentioning here:

> "*First, the Transport Department blasts the hillside to carve out roads, metals it, and prepares it for the upcoming season. Then, the Public Works Department realises that new drainage pipes need to be laid so they break the entire stretch of newly constructed road. In the end, the road is broken, the funds granted to build the road are over and the vulnerability of the people to subsequent hazards is still there.*"

This example demonstrates the government's approach as akin to a service provider limited to providing only the service. The issue of maintenance and upkeep is totally ignored. The interviewees maintain that the approach to disaster management follows a similar pattern.

Synthesizing the findings from this study into the Pressure and Release (PAR) model mentioned earlier, the chain of explanation of vulnerability is easily cognizant. People's beliefs and traditions or culture in general can be associated with the *root causes*; in the words of Wisner they are often 'invisible' and 'taken for granted'. Some of the "push" and "pull" factors that are responsible for people's settlement in the Mandakini river floodplain can also be seen as contributing to these root causes — for instance the declining access to natural resources. The dysfunction of the State administration in providing adequate livelihood opportunities to the people and enforcing the zoning policies as set forth by law can also be incorporated into this category. Going a step further, and in the context of diminished livelihoods of people after the 2013 disaster, the continuation of such a scenario can act as a future root cause leading to an increased vulnerability of people to future disasters.

These root causes can be seen as giving rise to the *dynamic pressures* of diminished skills, urbanisation, building of dams and deforestation. People's access to natural resources and declining agricultural incomes can be seen as the root causes that force people to migrate into urban areas. The rapid out-migration due to people's economic and environmentally marginalised condition can be seen as a factor contributing to people's lack of confidence in their local knowledge and a failure to retain essential skills required to deal with a new, unaccustomed environment.

The dynamic pressures give rise to *unsafe conditions*, in this case the need (or choice) to settle in unsafe areas and the inability of the state to provide adequate compensation. Influenced by individuals' limited cognitive abilities, these conditions are intensified by limited access to intangible resources such as support systems (like health and educational institutions) and lack of knowledge (like the ability to prepare or respond in a disaster situation). In the context of the objectives of the study, the economic, socio-cultural, environmental, political and psychological causes (although not clearly a part of this model) constitute one side of the pressure model. The other — the Kedarnath floods of 2013 — constitutes the trigger event whose impact on the vulnerable people led to the disaster.

Key findings of study

Why do people occupy disaster prone areas

People suffer from a misguided belief that the occurrence of the devastating floods just two years back somehow immunises them from future risk for at least a considerable amount of time into the future. In some instances, they also deny or simply accept their susceptibility to a natural hazard in future.

How do people perceive and respond to their vulnerabilities based on their cultures and traditions?

The native culture of conservation has been superseded by a growth process that believes in maximum rather than optimum utilisation of resources. But since the people already have such a close connect with the local ecology, it would not be wrong to say that the widely-advocated principles of sustainable development need only a comprehensive integration into government policy to bear fruit. In this sense, culture seems to offer a supporting role in disaster mitigation measures.

The locals believe in a higher supernatural power that guides their fate so they often do nothing about their adverse circumstance. They also attribute disasters to be a form of punishment for their ill deeds; their control over a disaster is, therefore, practically nil. Arising out of a need for survival or engulfed by a culture of material possessions — the local communities refuse to let go of their belongings even in the face of glaring hazard risks.

How do experts perceive vulnerabilities and respond to disasters?

The government agencies are mindful of their changing roles but find their capabilities limited by resource constraints, insufficient integration, and to some extent the lack of public support. Whether this is a flouting of rules, complacency in action, or a convenient blame-game is doubtful.

The experts in the field criticise the government agencies for failing to integrate research with policy objectives. They also emphasise the need for hazard mapping and advanced technologies for better prediction of future hazardous events.

Lastly, the NGO groups do not seem to categorise the people's decisions as "irrational" in the face of wide scale uncertainties they have to face. Most importantly, they stress the isolated response scenario by different agencies that not only hamper the reconstruction activities but also advance a sense of mistrust among people.

The rising vulnerabilities of the Himalayan people in the face of massive 'developmental aggression' cannot be overemphasised. Additionally, the constraints placed by our cognitive abilities and cultural norms on our decision making abilities prevent us from arriving at optimised solutions. The need of the hour is to develop a Pan-Himalayan development strategy that addresses mountain specificities and also incorporates psycho-social dimensions that are often ignored in the disaster mitigation process.

References

Adger, W. N. (2006). Vulnerability. *Global Environmental Change*, **16**(3), 268–281. http://doi.org/10.1016/j.gloenvcha.2006.02.006.

Akter, S., and Mallick, B. (2013). *The poverty-vulnerability-resilience nexus: Evidence from Bangladesh. Ecological Economics*, **96**, 114–124. http://doi.org/10.1016/j.ecolecon.2013.10.008.

Asch, T. V., Corominas, J., Greiving, S., Malet, J.-P., and Sterlacchini, S. (2014). *Mountain Risks: From Prediction to Management and Governance.* Netherlands: Springer.

Bajpai, A., Wadhwa, N. C., and Mahna, V. K. (2014). Environmental Impact Assessment: Uttarakhand Disaster (June–July 2013), **2**(2), 8–15.

Bharti, V. (2015). Investigation of Extreme Rainfall Events Over the Northwest Himalaya Region Using Satellite Data.

Bharti, V., and Singh, C. (2015). Characteristics of the spatiotemporal pattern of Extreme Rainfall event over the state of Uttarakhand, India. *ICUC9 — 9th International Conference on Urban Climate Jointly with 12th Symposium on the Urban Environment Characteristics.*

Birkmann, J. (2006). Measuring vulnerability to promote disaster-resilient societies: Conceptual frameworks and definitions. *Measuring Vulnerability to Natural Hazards; Towards Disaster Resilient Societies*, **01**, 9–54.

Brenkert A. L., and Malone, E. L. (2003). Vulnerability and Resilience of India and Indian States to Climate Change: A First Order Approximation. *Joint Global Change Research Institute*, (April), 105.

Burton, I., and Kates, R. W.(1964). The perception of natural hazards in resource management. *Natural Resources Journal*, **3**, 412–4.

Busch-Luty, C. (1995). Sustainable Development as a guiding model of ecological economy, *Sustainability: in the natural sciences and social science perspective*, Stuttgart, S. 115–126.

Cardona, O. D. (2003). The Need for Rethinking the Concepts of Vulnerability and Risk from a Holistic Perspective: A Necessary Review and Criticism for Effective, in Bankoff, G., and Frerks, G. (eds.), *Mapping Vulnerability: Disasters, Development and People*, pp. 37–51. London: Taylor & Francis.

Centre for Education and Documentation. (2013). Uttarakhand: Disaster as a Tragedy.

Chopra, R. (2014). Uttarakhand: Development and Ecological Sustainability. India: Oxfam India Ltd.

Ciurean, R., Schröter, D., and Glade, T. (2013). Conceptual Frameworks of Vulnerability Assessments for Natural Disasters Reduction, in Tiefenbacher, J. (ed.), *Approaches to Disaster Management — Examining the Implications of Hazards, Emergencies and Disasters*, pp. 3–32. Croatia: InTechOpen.

Dekens, J. (2007). Local Knowledge for Disaster Preparedness. A Literature Review. Kathmandu: Hill Side Press. Retrieved from http://www.preventionweb.net/files/2693_icimod8fc84ee621cad6e77e083486ba6f9cdb.pdf.

Development, M. (2011). International Conference on Sustainable Mountain Development in the Context of the Green Economy and Rio + 20 A Discussion Brief.

Directorate of Census Operations. (2011). District Census Handbook Rudraprayag. Retrieved from http://www.censusindia.gov.in/2011census/dchb/0966_PART_B_DCHB_0966_RUDRAPRAYAG.pdf.

DMMC. (2012). State Disaster Management Action Plan for the State of Uttarakhand Disaster Mitigation & Management Centre Uttarakhand Secretariat. Retrieved from http://dmmc.uk.gov.in/files/pdf/complete_sdmap.pdf.

Dobhal, D. P., Gupta, A. K., Mehta, M., and Khandelwal, D. D. (2013). Kedarnath disaster: Facts and plausible causes. *Current Science*, **105**(2), 171–174.

Dobhal, D. P., Mehta, M., Kesarwani, K., and Gupta, A. K. (2013). Consequence of Instability Processes Triggered by Heavy Rainfall in Mandakini Valley, Central Himalaya, India, (Retrieved on 16 July 2015).

Douglass, M., Miller, M., Hickey, M., and Bunnell, T. (2013). The Urban Transition of Environmental Disaster Governance in Asia, October 2013, (ARI Working Paper Series No. 210).

GB Pant Institute of Himalayan Environment and Geology(2009). Governance for Sustaining the Himalayan Ecosystem.

Gencer, E. A. (2013). *The Interplay between Urban Development, Vulnerability, and Risk Management: A Case Study of the Istanbul Metropolitan Area.* Netherlands: Springer.

Government of Uttrakhand (2007). Government of Uttarakhand City Development Plan: Nainital Revised, 2007.Uttrakhand: Urban Development Department.

Gupta, A. K., Dobhal, D. P., Mehta, M., Verma, A., Pratap, B., and Kesarwani, K. (2013). Kedarnath Devastation, 1–66. Retrieved from http://www.wihg.res.in/news/report on kedarnath devastation_june 2013.pdf.

Guth, W., and Kliemt, H. (2015). How to cope with (new) uncertainties: A bounded rationality approach. *Economics Discussion Papers*, **2015–46**. Retrieved from http://www.economicsejournal.org/economics/discussionpapers/2015–46.

Anderson, M. B., Woodrow, P. J.(2000). A Framework for Analyzing Capacities and Vulnerabilities. Graduate School of Education, Paper on International Relief, **3704**(617). Cambridge, MA: Harvard University.

Heijmans, A. (2001), Vulnerability: A matter of Perception, Disaster Management. Working Paper 4/2001.

Himalayan Environmental Studies and Conservation Organization (HESCO) Report (2014). Disaster: Nature's Fury: A Preliminary Report on Uttarakhand Disaster.

Hewitt, K. (2009). Culture and risk: Understanding the sociocultural settings that influence risk from natural hazards. *International Centre for Integrated Mountain Development (ICIMOD)*, 1–16. Retrieved from http://scholar.google.com/scholar?hl=en&btnG=Search&q=intitle:Culture+and+Risk+:+Understanding+the+Sociocultural+Settings+that+Influence+Risk+from+Natural+Hazards#1.

Hewitt, K., and Mehta, M. (2014). Rethinking Risk and Disasters in Mountain Areas, *Revue de géographie alpine*, **1**(2012), 1–3.

ICIMOD (2009). The role of local institutions and culture in influencing local adaptation strategies to climate related stresses and hazards, 1–3.

ICIMOD, & Mountain Forum. (2008). Culture and Risk: Socio-Cultural Settings that Influence Risk from Natural Hazards. Participants' Contributions — A Compilation. *Global E-Conference*, (October), 195.

Jayaram, D. (2014). Fallouts of the Uttarakhand Disaster: A critical appraisal., Kansas: Foreign Military Studies Office (FMSO).

Jodha, N. S. (1992). Mountain perspective and sustainability: A Framework for Development Strategies, in Jodha, N.S., Banskota, M., and Partap, T.

(eds.),*Sustainable Mountain Agriculture*, pp. 43–82. India: Oxford and IBH Publishing Co.

Jogia, J., Kulatunga, U., Yates, G. P., and Wedawatta, G. (2014). Culture and the psychological impacts of natural disasters: Implications for disaster management and disaster mental health, *The Built and Human Environment Review*, **7**, 1–10.

Joshi, P. C. (2008). Role of Culture in Disasters. *Culture and Risk: Understanding the Sociocultural Settings That Influence Risk from Natural Hazards.* Retrieved from http://www.mtnforum.org/sites/default/files/publication/files/role_of_culture_in_disasters.pdf.

Joshi, V., and Kumar, K. (2006). Extreme Rainfall Events and Associated Natural Hazards in Alaknanda Valley, Indian Himalayan Region. *Journal of Mountain Sciences*, **3**(3), 228–236.

Kaneria, M., Negi, G. C. S., and Dhyani, P. P. (2015). Comparative Study of Ecosystem Services of Oak and Pine Forests of Western Himalaya in particular relation to Organic Carbon Stock in Soil, in Jain, D. K., Arya, R. C., and Singh, N. P. (eds.), *Climate Change: Socio-Economic and Environmental Issues — Problems and Challenges*, pp. 1–9. New Delhi: Mohit Publications.

Kar, S. (2007). Inclusive Growth in Hilly Regions: Priorities for the Uttarakhand Economy. *Institute of Economic Growth, Uttaranchal Approach Paper for 11th Plan.*

Kates, R.W.(1962). Hazard and Choice Perception in Flood Plain Management, University of Chicago, Department of Geography Research, Paper No. 78. (1962). *Geography*, (78).

Kaul, V., and Thornton, T. F. (2013). Resilience and adaptation to extremes in a changing Himalayan environment. *Regional Environmental Change*, **14**(2), 683–698. http://doi.org/10.1007/s10113-013-0526-3.

Kulatunga, U. (2010). Impact of culture towards disaster risk reduction. *International Journal of Strategic Property Management*, **4**(June 2015), 304–313. http://doi.org/10.3846/ijspm.2010.23.

Kumar, A. (2013). Demystifying a Himalayan Tragedy: Study of 2013 Uttarakhand Disaster. *Journal of Indian Research*, **1**(3), 106–116.

Kumar, J. (2015). Spatial Patterns of Urbanization in Uttarakhand, 2011, *Geography*, (2277), 201–202.

Lei, Y., Wang, J., Yue, Y., Zhou, H., and Yin, W. (2014). Rethinking the relationships of vulnerability, resilience, and adaptation from a disaster risk perspective. *Natural Hazards*, **70**(1), 609–627. http://doi.org/10.1007/s11069-013-0831-7.

Lewis, J., Kelman, L., and Lewis, S. a V. (2011). Is "Fear Itself" the only thing we have to fear? Explorations of psychology in perceptions of the vulnerability of others. *Australasian Journal of Disaster and Trauma Studies*, **2011**(3), 89–104.

Lewis, J. (1976). A review essay on Gilbert F. White: Natural Hazards: Local, National and Global. UK: Oxford University Press.

Lindblom, C. E.(1964). The science of muddling through, in Gore, W. T., and Dyson, J. W. (eds.), *The Making of Decisions*. New York: Free Press.

Macchi, M. (2011). *Framework for Community-Based Climate Vulnerability and Capacity Assessment in Mountain Areas*. Nepal: ICIMOD. Retrieved from http://climatechange-asiapac.com/system/files/resource/icimod-framework_for_community-based_climate_vulnerability_and_capacity_assessment_in_mountain_areas.pdf.

Mamgain, R. (2008). Growth, Poverty and Employment in Uttarakhand. *Journal of Labour and Development*, (39). Retrieved from http://www. ihdindia.org/%5C/Working Ppaers/ 2010-2005/pdf files/39-RPMamgain.pdf.

Mehta, P. S., Sharma, A. K., and Negi, K. S. (2010). Indigenous Knowledge System and sustainable development with particular reference to folklores of Kumaon Himalaya, Uttarakhand. *Knowledge Creation Diffusion Utilization*, **9**(July), 547–550.

Millstein, S. G., and Halpern-Felsher, B. L. (2002). Perceptions of risk and vulnerability. *The Journal of Adolescent Health: Official Publication of the Society for Adolescent Medicine*, **31**(1 Suppl), 10–27. http://doi.org/10.1016/S1054-139X(02)00412–3.

Ministry, T., and Government, F. (2014). Assessment of Environmental Degradation and Impact of Hydroelectric projects during the June 2013 Disaster in Uttarakhand Part I-Main Report, (April).

MSME Development Institute (2014). Brief Industrial Profile of District Rudraprayag. *Igarss 2014*, (1), 1–5. http://doi.org/10.1007/s13398-014-0173-7.2.

MST, G. (2010). The Himalayan Eco-System National Action Plan on Climate Change Mission Document, Government of India, 51.

Naithani, A. K. (1997). Morph tectonics and landslide problems in area among Kedarnath and Okhimath, Garhwal Himalaya., Ph.D. thesis, Science Faculty, Hemwati Nandan Bahuguna (HNB) Garhwal University, India.

Nautiyal, S., Rajan, K. S., and Shibasaki, R. (2005). Interaction of Biodiversity and Economic Welfare — A Case Study from the Himalayas of India. *Journal of Environmental Informatics*, **6**(2), 111–119. http://doi.org/Doi 10.3808/Jei.200500061.

Negi, C. S. (2010). Traditional Culture and Biodiversity Conservation: Examples from Uttarakhand, Central Himalaya. *Mountain Research and Development*, **30**(3), 259–265. http://doi.org/10.1659/MRD-JOURNAL-D-09-00040.1.

NIDM. (2014). Uttarakhand, 1–20. Retrieved from http://nidm.gov.in/pdf/dp/Uttara.pdf.

Ning, W., Rawat, G. S., Joshi, S., Ismail, M., and Sharma, E. (2013). *High-altitude rangelands and their interfaces in the Hindu Kush Himalayas*. Kathmandu: ICIMOD.

Nirupama, N. (2012). Risk and vulnerability assessment: a comprehensive approach. *International Journal of Disaster Resilience in the Built Environment*, **3**(2), 103–114. http://doi.org/10.1108/17595901211245189.

Nirupama, N., Sharma, R., Verma, K., and Panigrahi, S. (2014). A multi-tier hazard-part I: description of the event. *Natural Hazards*, **76**(1), 259–269. http://doi.org/10.1007/s11069-014-1487-7.

Oven, K. J. (2009). Landscape, Livelihoods and Risks: Community Vulnerability to Landslides in Nepal. Thesis, Durham University, UK.

Pant, S., Singh, M., Lal, M., Singh, A., Sharma, A., and Bhandari, S. (2007). Medicinal plants in Himachal Pradesh, north western Himalaya, India. *International Journal of Biodiversity Science, Ecosystems Services & Management*, **3**(4), 234–251. http://doi.org/10.1080/17451590709618177.

Paton, D., and Johnston, D. (2001). Disasters and communities: vulnerability, resilience and preparedness. *Disaster Prevention and Management*, **10**(4), 270–277. http://doi.org/10.1108/EUM0000000005930.

Pennings, J. M., and Grossman, D. B. (2008). Responding to Crises and Disasters: The Role of Risk Attitudes and Risk Perceptions. *Disasters*, **32**(3), 434–448. http://doi.org/10.1111/j.0361-3666.2008.01048.x.

Rautela, P. (2005). Disaster Prevention and Management: Indigenous Technological Inputs for Effective Disaster Management in the fragile Himalayan Ecosystem, *Disaster Prevention and Management*, **14**(2), 233–241.

Pranuthi, G., Dubey, S. K., Tripathi, S. K., and Chandniha, S. K. (2014). Trend and Change Point Detection of Precipitation in Urbanizing Districts of Uttarakhand in India. *Indian Journal of Science and Technology*, **7**(October), 1573–1582.

Price, M. F. (2004). *Conservation and Sustainable Development in mountain Areas*. Switzerland: IUCN.

Proag, V. (2014). The Concept of Vulnerability and Resilience. *Procedia Economics and Finance*, **18**(September), 369–376. http://doi.org/10.1016/S2212-5671(14)00952-6.

Rautela, P. (2005). Increasing vulnerability of the Himalayan urban centers. *Disaster Prevention and Management*, **14**(2), 242. Retrieved from http://proquest.umi.com/pqdlink?did=862992721&Fmt=7&clientId=4574&RQT=309&VName=PQD.

Rautela, P, Sanjwan, K. S., Khanduri, S., Ghildiyal, S., Chanderkala, and Rawat, A. (2014). Geological investigations in Rudraprayag district with special reference to mass instability. Uttarakhand: Disaster Mitigation and Management Centre (DMMC).

Rawat, M. S., Uniyal, D. P., Dobhal, R., Joshi, V., Rawat, B. S., Bartwal, A., and Aswal, A. (2015). Study of landslide hazard zonation in Mandakini Valley, Rudraprayag district, Uttarakhand using remote sensing and GIS. *Current Science*, **109**(1), 1–13.

Rigg, J., and Oven, K. (2015). Building liberal resilience? A critical review from developing rural Asia. *Global Environmental Change*, **32**, 175–186. http://doi.org/10.1016/j.gloenvcha.2015.03.007.

Ross, B. M., and Levy, N. (1958). Patterned predictions of chance events by children and adults, *Psychological Reports*, **4**, 87–124.

Sati, V. P. (2013). Trends of Urbanization and its Implications on Environment and Economy in the Uttarakahnd Himalaya: A Case Study of Dehradun Municipal Corporation. *ENVIS Bulletin on Himalayan Ecology*, **21**, 9–14.

Sati, V. P. (2014). *Towards Sustainable Livelihoods and Ecosystems in Mountain Regions*. Netherlands: Springer.

Serageldin, I. (1995) 'Promoting Sustainable Development: Toward a New Paradigm, in Serageldin, I., and Steer, A.(eds.), *Valuing the Environment: Proceedings of the first Annual International Conference on Environmentally Sustainable Development*, pp. 13–21. Washington D. C.: World Bank.

Simon, H. A. (1956). Rational choice and the structure of environment, *Psychological Review*, **63**, 129–38.

Singh, S.P. (2007). *Himalayan Forests Ecosystem Services: Incorporating in national accounting*. Central Himalayan Environment Association (CHEA). Retrieved from http://lib.icimod.org/record/13207/files/3731.pdf.

Slovic, P. (1987). Perception of risk. *Science*, **236**(4799), 280–285.

Tversky, A., and Kahneman, D.(1971). Belief in the law of small numbers.*Psychological Bulletin*, **76**, 105–110.

UNDP (2013). Issue Brief: Disaster Risk Governance, (January 2010).

von Neumann, J., and Morgenstern, O.(1947).*Theory of Games and Economic Behavior* (3rd ed.). Princeton, N. J: Princeton University Press.

Wachinger, G., Renn, O., Begg, C., and Kuhlicke, C. (2013). The Risk Perception Paradox-Implications for Governance and Communication of Natural Hazards. *Risk Analysis*, **33**(6), 1049–1065. http://doi.org/10.1111/j.1539-6924.2012.01942.x.

Weichselgartner, J. (2001). Disaster mitigation: the concept of vulnerability revisited. *Disaster Prevention and Management*, **10**(2), 85–95. http://doi.org/10.1108/09653560110388609.

White, G.F.(1974). *Natural Hazards: Local, National, Global.* UK: Oxford University Press.

Wisner, B., Blaikie, P., Cannon, T., and Davis, I. (2003). *At Risk: natural hazards, people's vulnerability and disasters* (2nd ed.). UK: Routledge. Retrieved fromhttp://www.preventionweb.net/files/670_72351.pdf.

Ziegler, A. D., Wasson, R. J., Bhardwaj, A., Sundriyal, Y. P., Sati, S. P., Juyal, N., and Saklani, U. (2014). Pilgrims, progress, and the political economy of disaster preparedness — the example of the 2013 Uttarakhand flood and Kedarnath disaster. *Hydrological Processes*, **28**(24), 5985–5990. http://doi.org/10.1002/hyp.10349.

15

THE CHALLENGES OF URBAN FOREST DEVELOPMENT: A CASE STUDY OF URBAN FOREST IN JAKARTA, INDONESIA

SUPRIYANTO and TAN PUAY-YOK

The numerous benefits provided by urban forests for the built environment have prompted Indonesia to stipulate its provision in their national law. However, the implementation of urban forest conservation still faces several challenges. This study aims to provide the scientific basis for the promotion of urban forests in Indonesia. We have selected an example of a successful urban forest at the local level, using Jakarta as a case study, with literature review and an interview approach to collect the data. The results show that conflicting self-interest among related institutions, flexibility of existing laws, inconsistency in the implementation, high development cost, informal settlements and vandalism within the forest, and the severe resistance of local people are serious issues during the process of urban forest development. To increase the awareness among policy makers, the Travel Cost Method (TCM) and Contingent Valuation Method (CVM) have been selectively used to assess the monetary value of the urban forest in Pondok Labu, Jakarta. From the economic calculations, it can be predicted that the total value of Pondok Labu Urban Forest (PLUF) is approximately IDR58.89 billion (US$1 = IDR13,000) in 2015, with an annual visitorship of 7,896 people. This indicates that while urban forests undoubtedly offer economic benefits, there are still several challenges. To overcome these problems, we propose some recommendations including reviewing current regulations, giving more incentives, evaluating the management of urban forest, strengthening the function of related-institutions, and educating society.

Abbreviations

APBD	:	Local Budget
APBN	:	National Budget
AV	:	Average Actual Individual Visit Rate
Bappeda	:	Local Development Planning Agency
Bappenas	:	Ministry of National Development Planning
BPN	:	National Land Board
CP	:	Choke Price
CS	:	Consumer Surplus
CVM	:	Contingent Valuation Method

DBHDR	:	Special Budget for Reforestation
DC	:	Dichotomous Choice
DPPKP	:	Agency for the Fishery, Agriculture, and Food Security Agency in Jakarta
DPU	:	Public Work Agency
FM	:	Family Members
G	:	Gender
IDR		Indonesia Rupiah (currency)
IN	:	Respondent's Income Per Hour
ITCM	:	Individual Travel Cost Method
Kemendagri	:	Ministry of Internal Affairs
Kemen LHK	:	Ministry of Environment and Forestry
Kemen PU	:	Ministry of Public Works
L	:	Amount of Fuel Used by Vehicle
MI	:	Monthly Income
MTE	:	Minimum Travel Expenditure
MVR	:	Maximum Visit Rate
OSE	:	Onsite Expenditure
OTE	:	Onsite Time Expenditure
P	:	Current Price of Fuel per Litre
PLUF	:	Pondok Labu Urban Forest
RTH	:	Green Open Space
RTRW	:	Spatial Planning
S	:	Distance from Home to the Site (km).
S_1	:	Average of Fuel Used in km (53,8 km/L)
Sudin	:	Sub Agency in Jakarta
SUF	:	Srengseng Urban Forest
TCM	:	Travel Cost Method
TP	:	Time Preferred
TS	:	Time Spent
TT	:	Travel Time
TTC	:	Total Travel Cost
TV	:	Type of Visit
V	:	Average of Velocity of Vehicle
WTA	:	Willingness to Accept
WTP	:	Willingness to Pay
Y	:	Visit Rates
YV	:	Total Visit for one-year
ZTCM	:	Zonal Travel Cost Method

Introduction: The Significance of Urban Forest

Urbanisation in Indonesia has grown dramatically from 17.1% of the total population in 1960 to more than 50% in 2009 (Lewis, 2014). In the context of development, urbanisation has two different implications. On the one hand, urbanisation drives multi-dimensional structural transformation that leads to improvements in the quality of life for the low-income population as they move out of rural areas into the city (Annez & Buckley, 2009). On the other hand, people migrating from the countryside to urban areas result in increasing demand for infrastucture, including housing, roads, transportation, energy, water supply, and other human necessities that have widely contributed to problems, such as flooding, pollution, and decreasing quality of water and air (Firman, Kombaitan, & Pradono, 2007). Hence, achieving sustainability in urban development presents a new challenge for policy makers. Consequently, shaping the city with consideration for the ideal ratio between natural landscape and man-made cover is extremely important to achieve urban sustainability. Urban forests may perhaps become one of the most appropriate approaches in overcoming urban issues. Many scientists believe that existing urban forests have many benefits, not only for improving the quality of life but also for contributing significantly to global sustainable development, particularly to maintaining the earth's temperature (Devuyst, Hens, & Lannoy, 2011). Due to the significance of urban forests, the Indonesia government implicitly features urban forests in their national policy by stipulating Act Number 26/2007 regarding Spatial Planning (UU 26), that is elaborated in detail in Government Regulation Number 63/2002 regarding Urban Forests (PP 63).

The Ministry of Forestry legally regulates technical guidance for the implementation of urban forests (Regulation Number P.71/2009, or Permenhut 71). This regulation specifically states that every city around Indonesia should have a forest that covers at least 10% of the urban area and that this should be legalised by local government authorities.

In reality, the number of existing urban forests in most cities around Indonesia is small, and does not fulfill the national standard. However, the central government has been allocating a budget for local governments to accelerate urban forest development. Despite this, progress has been relatively slow. To promote urban sustainability and livability, it is important to understand why the realisation of urban forests has yet to reach the national target and to identify key strategies needed to increase awareness among decision makers about the significance of urban forests.

Within Indonesia, Jakarta is an example of a city with high population growth facing serious environmental issues. Jakarta is also unique due to its status as the city, province, and capital of Indonesia. There are several major environmental problems in the largest city in Indonesia, such as floods, land subsidence, and air pollution. The lack of green open spaces contributes significantly to the environmental issues in Jakarta. Therefore, convincing policy makers of the importance of urban forests is very important to closing the gap between the national mandate and its actual implementation.

Research Approaches

For this study, two different approaches have been selected. The first is a descriptive approach that is used to comprehensively analyse how existing regulations on urban forests are implemented. The second approach is an economic assessment which is selected to estimate the environmental value of urban forests. In this study, the Travel Cost Method (TCM) and Contingent Valuation Method (CVM) have been selected, because theese forests also provide recreational activities, protect the water catchment, and conserve biodiversity.

DKI Jakarta Province (henceforth referred to as "Jakarta") is the location of the research where information about local urban forest development is collected. This study not only reviews the implementation of urban forest development at the national level but also examines efforts by the local government to implement the national mandate. For the monetary valuation of an existing urban forest, Pondok Labu Urban Forest (PLUF) in South Jakarta is selected as the study site due to several considerations. Firstly, PLUF does not have direct economic value, in that its presence does not generate any income. Secondly, based on preliminary assessments, academic research of this forest has yet to be conducted, either in an economic or environmental analysis. Thirdly, the logistics of conducting this study are relatively uncomplicated, both in terms of site accessibility and communicating with its officers. Lastly, due to budget and time constraints, PLUF is a good choice to represent monetary value of an urban forest in Jakarta and Indonesia without compromising the credibility and reliability of the environmental valuation study.

Three techniques were used to collect the data. First, a review was done on how the programme has been working. Second, interviews were carried out with a wide range of stakeholders directly or indirectly contributing to

Fig. 15.1. Location of Pondok Labu Urban Forest, South Jakarta.

Fig. 15.2. Pondok Labu Urban Forest, South Jakarta.

the urban forest development. Last, questionnaire surveys were conducted on the environmental valuation method, especially for TCM and CVM.

Definition of Urban Forest in Indonesia

The general definition of 'urban forest' is a canopy composed of trees that appears as a forest in a metropolitan area (Loeb, 2011). While greenery is commonly found in many parts of a city, the establishment of one single forest is relatively uncommon. An ambiguity can arise when a forest located at the border of urban and rural areas is sometimes considered an urban forest. Indeed, Careless (1954) believed that such urban forests developed over time by expanding towards surrounding countryside and other cities. Indonesia officially defines an urban forest in PP 63 (2002) as a stretch of land with dense trees situated in urban areas, either on state land or private land that has been officially established as urban forest by the authorised government, with a total minimum area of 0.25 hectare, and is part of a green open space (RTH). The authorising government involves the mayor, regent, and governor. Those on private land should have a legal certificate from the managing authority (e.g. national land board or related local environment or forestry government/agency) which declares that the land cannot be converted into other land uses for at least 15 years, and that its canopy of trees should cover a density of 70% in order to provide ecological services, such as water catchment and pollutant absorbtion.

Implementation of Urban Forests

According to data from the Ministry of Internal Affairs, the total land area in all cities in Indonesia is roughly 193.56 million ha (Permendagri 6, 2008). Based on the national mandate in which every city should provide at least 10% of its land for forests, more than 2.15 million ha of urban forests should have been developed in Indonesia (Dirjen BPDASPS, 2011). In reality, the area covered by existing urban forests around Indonesia, which have been successfully established as of 2009, is only 20,261.89 ha (Dirjen RLPS, 2009). Hence, there is a significant gap between the existing area of urban forests and the ideal coverage. To reduce that gap, the Ministry of Forestry increased urban forests around Indonesia by 4,034.6 ha since 2010 (up till 2014). From 2015 to 2019, the government plans to develop urban forests by at least 200 ha per annum (Dirjen BPDASPS, 2015).

In Jakarta, there are only 14 sites of urban forests, covering a total of 149.18 ha, which are officially legalised by the Governor. However, the existing green open space (RTH) for forests is 646.44 ha (Bappeda Jakarta, 2012; DKP Jakarta, 2011). This means there is a gap of 487.26 ha of RTH, which are not legally recognised as urban forests. Some of these forests are owned by other parties, such as ministries, enterprises, and society (Kusuma, 2015).

To increase the number of urban forests, Jakarta has successfully bought 22.18 ha of land for urban forests at 16 sites from 2004 to 2013, and 21.6 ha at 15 locations since 2014 (Jakarta Government, 2015). In 2015, there were some 159 spots of land around Jakarta that would be converted into RTH, and the estimation of the budget spent on this project is about IDR2.3 billion (Khairany, 2015). Moreover, this big city promises to reach 10% of the minimum land for urban forests by stipulating it in its Spatial Planning (RTRW) 2030. This commitment is not only to provide new urban forests in some parts of the city, but also to increase the quantity and quality of existing urban forests. For instance, urban forests in Kembangan Utara will be extended to 2.8 ha from the previous land of 1.9 ha, with continuous land acquisition (Kusuma, 2015; Yossy, 2015).

Challenges of Urban Forest Development

Institutional Challenges

Indonesia applies political decentralisation by giving more power to the public for decision-making. Consequently, in terms of urban forest development, the approach selected to implement this programme is the delegation

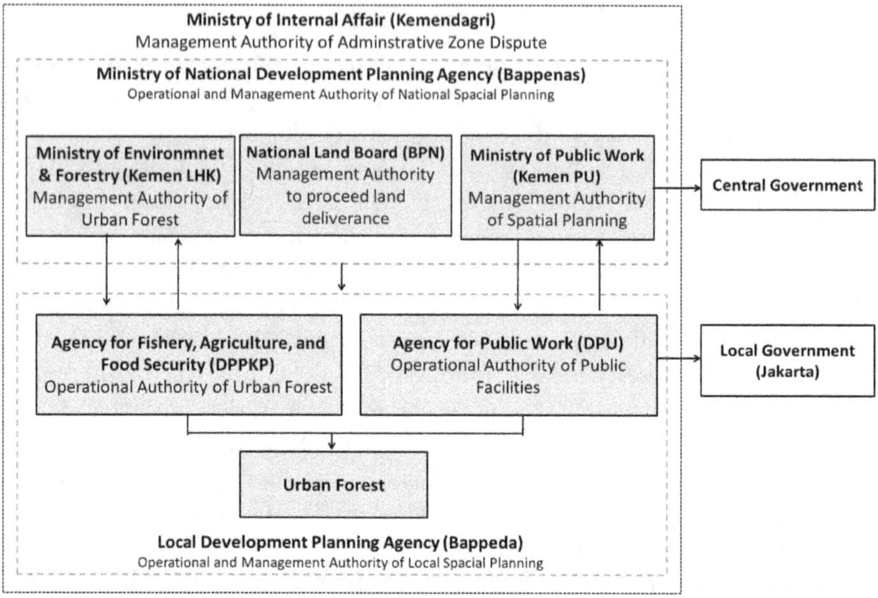

Fig. 15.3. Illustration of the relationship between the central and local governments in terms of urban forest development.

from central government to the local level (PP 38, 2007). The illustration of the relationship between the central and local governments in terms of urban forest development is shown in Fig. 15.3.

The central government involves the Ministry of Environment and Forestry (Kemen LHK), Ministry of Public Work (Kemen PU), Ministry of National Development Planning Agency (Bappenas), Ministry of Internal Affairs (Kemendagri), and the National Land Board (BPN). Kemen LHK is the primary management authority of urban forests, with the right to decide the standards and procedures of urban forest development and to stipulate the technical guidance of urban forest development on how the local level governments implement the national mandate (Permenhut 71, 2009).

As the management authority of spatial planning, Kemen PU regulates the standards and principles for urban planners, on how to give a clear design of space integrated with socioeconomic and environmental attributes. The Ministry of Public Works Regulation Number 5/2008 states clearly that forests, as a part of public spaces, should be provided in urban areas in order to increase the quality of the environment.

Bappenas has the authority to initiate the framework of the national agenda that drives national development planning, which includes the areas of human resource, economy, infrastructure, natural resource, environment, education, finance, and law (Keppres 4, 2002). In terms of environmental management, this ministry has the main responsibiity to provide guidance on how to utilise, conserve, and manage nature, in the form of environmental programmes. Urban forests are one of the examples of this national environmental programme.

The presence of Kemendagri is to mediate possible disputes among administrative regions, particularly when urban forests are located in more than one city/regency/province causing conflicts among the management systems. This has been reiterated in article 198 of the Act (Number 32 of 2004) that *"In case of any dispute between the province, inter-province and regency/city, either in or out of its region, the Ministry of Internal Affairs should take over this case"*. This situation might happen when the urban forest of interest has a high economic value and can likely generate money to support the operational government at the local level. An example of conflict owing to an administrative boundary issue is the plan for Cilacap Regency to build a fishery harbour to support economic activities in the west part of Bodho River, which is the line border between Cilacap Regency and Kebumen Regency. This sparked a dispute with Kebumen Regency, who felt it had the right to build the seaport (Ag, 2003).

BPN is the authorised institution to process land deliverance for urban forests. This institution has a right to issue legal documents on the status of land, particularly to avoid any claim from another party who feels it has a right to manage that land. However, the National Land Board is a central institution and any conflict relating to land at the local level will be resolved by this institution through its office representative in every city/regency.

The local levels for implementing urban forests are the city/regency, and province. Jakarta, however, is administered at the provincial level because of its status as the capital of Indonesia (PP 63, 2002). These parties have a main duty as operational authority to be the executant of urban forests, including funding, planning, design, planting, monitoring, and management. More importantly, the local level has a right to legalise the status of land as forest or otherwise.

The operational and management authority to implement urban forests in Jakarta lies in the Fishery, Agriculture, and Food Security Agency (DPPKP), which is assisted by Sub Agency (Sudin) in every district. The main difference in responsibility between DPPKP and Sudin is that

DPPKP has the authority to build and manage urban forests of land areas greater than 5 ha, whereas Sudin deals with urban forests smaller than 5 ha (Kusuma, 2015). In practice, the operation of Sudin is under supervision from DPPKP, and DPPKP is under the guidance of Kemen LHK.

At the local level, mainly in Jakarta, there are also other agencies that directly or indirectly contribute significantly to the process of urban forest development. These include the Local Development Planning Agency (Bappeda), and Public Works Agency (DPU).

The major function of Bappeda is to formulate how urban forests can be established together with the physical development of commercial areas, public services, and housing. To guide how to carry out urban forest development, the Jakarta government released Local Regulation Number 1 of 2012 regarding Spatial Planning (RTRW) 2030. Article 6 of subarticle (5) states assertively that "*green open space (RTH) development is to meet the ecological balance through the conservation of biodiversity, emission abatement, and water catchment*". That is to say, urban forests as a part of RTH, has strategic ecological functions for a sustainable city.

DPU serves as an important agency of urban forest development at local levels by providing public facilities, such as roads for access to the urban forest. In Jakarta, the jogging tracks and gates within the forest are provided by DPPKP, while the lighting is the responsibility of DPU (Kusuma, 2015).

Is this management system really effective for urban forest development? If we take a look back at the main purpose of decentralisation, which is to deliver good governance at the local level, the implementation of urban forests should be more successful, especially to attain urban sustainability. Why has this realisation not met the national target yet? First, there are many management authorities, either at the central or local level, that are driven by self-interest when deciding which one of them has the legal right to build urban forests. For instance, a conflict like that between Kemen LHK and Kemen PU may happen when the geographical location of the urban forest borders a lake, or toll road, roadside, and so on. However, these ministries have distinctly different jobs. It is sometimes baffling which party should take responsibility. This is because public facilities are in the control of Kemen PU, while tree planting is the domain of Kemen LHK. Another example is the dispute between Kemen LHK and the local agency. Although the law states clearly that the operational level of urban forests lies at the local level, several regencies/cities sometimes still compete for funding to develop urban forests, especially those with poor budgets. To tackle these issues, cooperation by all parties (the local level, non-governmental organisation, and the enterprise) involved is immediately needed to increase the quantity of urban forests. An example of the implementation of an urban forest involving many parties is PT JIEP Urban Forest in Pulogadung, Jakarta. This development was based on a proposal from a private company (PT JIEP) that was willing to unconditionally allocate its land in an industrial area to be a green open space for a forest (Didit, 2011). The design and selection of tree species in this forest was carried out by the local government.

Second, because of the execution of urban forests at the local level, including land deliverance and funding, several local governments face difficulty in terms of budget. That is the major reason why they are reluctant to build urban forests. The solution to overcome this issue is the allocation of more funds to the poor city or regency.

Spatial Planning Challenges

Development of urban forests is one of the national environmental agendas, and should be in accordance with spatial planning. The definition of spatial planning in RTRW, based on UU 26 (2007), is a process to decide the spatial structure, including economic, social, and environmental considerations. This means that the function of RTRW is to synchronise all national interests of development to utilise space wisely without compromising the environment.

In terms of the scope of authority, RTRW can be differentiated into two levels. Nationally, RTRW involves the management of space based on the national constitution, to achieve the country's goals of development with sustainable principles, and to minimise possible negative impacts of the development. This scope includes all national aspects with consideration for society, economy, and environment. Next the local level, which has two different levels of management systems: province, and city/regency. Each has its own right to select the best approach in terms of RTRW based on their existing strengths and weaknesses in order to achieve the national goals of providing better public service locally.

One of the concrete forms of RTRW in the urban area is green open space (RTH), which should be at least 30% of an urban area. RTH consists of forest, yard, park, square, and greenbelt. As a part of RTH, the minimum land for urban forests should be 10% of the urban area. However, under PP 63 and Permenhut 71, this percentage of land for urban forests could be adapted to the geographic features of the city. In other words, there is flexibility in the minumum land allocated for urban forests. This might cause urban planners to place urban forests as second priority in urban developmental programmes. Indeed, the absence of a penalty in RTH for not meeting the minimum quantity of urban forest affects the implementation of urban forest development.

The second issue is that some of the cities and regencies have not had the fixed-urban RTRW yet. The recent data from Kemen PU per 26 May 2015, showed that 25 out of 34 provinces and 413 out of 493 regencies and cities have completed urban spatial planning. The high economic value of land, limited open space, and political situation in their urban areas might cause some of them not to legalise RTRW. The absence of RTRW makes the conversion of critical land to other uses easier. For example, land that play an important role in water catchment could be easily developed into a business area. For instance, the 1990s saw the conversion of massive plots of land into commercial areas because of the speculative land-business in Indonesia (Firman, 2000). In 2001, the Mayor of Bandung unveiled plans to build an integrated-tourism area, including an apartment building, commercial centre, art centre, and restaurant in Babakan Siliwangi Urban Forest in order to increase the economic value of the land and increase local income. The government appointed PT Esa Gemilang Indah as the private enterprise to carry out this scheme (Ema, 2008).

Third, the issue of the slow expansion of urban forests is due to inconsistencies on the part of local governments — as also stated by Nugraha (2012). He believed that the root of all environmental problems originate from policies made by local governments that violate spatial planning, and that local regulation (Perda) legalised by policy makers is only a formality, because local governments are often inconsistent. Moreover, there was also a lack of coordination among agencies. Evidence of this was also visible in East Java, where the province already has spatial planning, but its implementation is not in accordance with regulation. For example, though a mall development in Malang did not meet the urban spatial planning requirements, it was still built without any penalty from the authorised government (DPD RI, 2010). Another empirical example is that in Kutai Barat, Kalimantan Timur discovered that 91.02% of its project did not fit RTRW, which was only 4.49% of the appropriate project, while 4.45% of the project was conducted before RTRW (Dilang, 2008).

Funding Challenges

The development of urban forests may be the most expensive project compared to that of other forests, such as private forests or industrial forests. There are two main reasons behind this. First, the provision of land has made it difficult for urban planners to implement urban forest development. There is usually strong competition for the available land for residential or industrial uses. Based on economics, market price is determined by demand and supply, and higher demand leads to higher cost. Consequently, governments will have to spend more money for land acquisition. The primary challenges of land release can be classified according to the type of city. The first type is the "Planned City", which can be subdivided into (a) cities that grow naturally without any intervention or planning initially, but with RTRW to manage the uncontrolled development subsequently; and (b) cities built from scratch with the guidance of city planning. For a planned city, both before and after, the city built may not meet any difficulty in delivering land in the case of funding availability. The second is the "Unplanned City", which does not have RTRW. The main problem here is that it may face challenges for land acquisition, mainly because the land may be used for other purposes, such as housing. For example, in 2014, the government faced opposition from illegal squatters and failed to free the state land around Ria Rio Lake in Jakarta for urban forest despite offers of financial compensation (Yusuf, 2014).

Second, the selection of tree species is another important part that leads to the high cost of urban forest development. The main reason is that for the purpose of aesthetic value, urban forests need to have a minimum height of trees, specific species, and resilience towards marginal land. The guidance to select trees for urban forests can be found in both PP 63 and Permenhut 71, in which every plan for urban forest development should be with appropriate trees. This is to guarantee the success of planting and to minimise any possible additional costs of replanting trees.

Funding for urban forest development could come from either public or private sources. There are three different types of public funding: (1) National budget (APBN) refers to yearly funding planning that is approved by national parliament for a one-year period, from 1 January to 31 December, with taxes and grants as the main sources. APBN for urban forests is officially managed by Kemen LHK, which is then distributed to the local government. (2) Local Budget (APBD) is the budget planning that is legalised by local parliament for a one-year period, from 1 January to 31

December, at the level of province and regency/city, and is operated by the local agency related to the forestry/environment. (3) Special budget for reforestation (DBHDR) constitutes funding specifically to bearing the cost of land (purchase) and forest rehabilitation. This funding originates from the shared profit of the wood company that is distributed to the central government. The aim of this funding is to stimulate reforestation as well as to empower people surrounding the forest. Another funding for urban forests relies on the private sector. This comes from the corporate social responsibility of companies and society. These sources have a significant role in increasing urban forests, particularly for local governments that lack the budget to build urban forests—for instance, an urban forest in Banda Aceh was entirely built by funding from a banking company.

Social Challenges

Most locations of urban forests in densely populated communities will likely attract many people to use them for their own interests. This might potentially threaten the function of the urban forest itself. There are several social challenges facing urban forests. The first of which are informal settlements within urban forests, which can damage vegetation and infrastructure — as seen in the houses illegally built (and occupied) in the urban forest in Balikpapan, East Kalimantan (Dep, 2014). A second possible challenge is a land-owner dispute between the government and people who claim to have the right to that forest. This mostly occurs when the government has abandoned and or ignored the land for years — during which people have settled for several generations; they thus feel that they have a legal right to the land (Kusuma, 2015). A recent case of such a conflict is where scores of people claimed some parts of Srengseng Urban Forest in Jakarta as their own land (Setiawan, 2015). The third is resistance from local people who have not agreed to sell their land for public space. The fourth social challenge is vandalism and immoral activities, which can damage the existing vegetation and infrastructure. Tony (2015) describes how some urban forest visitors, mainly teenagers, have been caught vandalizing trees, writing on gates, and leaving garbage within the forest.

What is the Actual Monetary Value of PLUF?

The results of the calculations obtained from two different monetary valuation approaches selected were very different. This is because the technique and strategy used to measure the economic value of urban forests reflect on how the researcher collects the data from respondents. The Travel Cost Method (TCM) is one of the revealed-preference models that looks into consumer behaviour and their monetary expenditure, and that information is used as an indirect proxy of the value of the environment they enjoy. The Contingent Valuation Method (CVM) is a stated-preference model in which people are asked directly of their willingness to pay for or to accept proposed money for improving the quality of the environment. From those definitions, it is assumed that TCM is the actual economic value of PLUF in a time period without any entrance fees, and CVM is the additional economic value of PLUF in a one-year period if the government implements ticket fees for every visitor. The total monetary value of PLUF can be computed by combining these values.

Based on the results of the monetary valuation of PLUF using TCM and CVM, the total estimation could be numerically acquired. With a total area of only 2,02 ha, PLUF is predicted to have as many as 7,896 visitors per year. From that annual average visitorship, the total monetary estimation of PLUF is IDR23.29 billion. This is numerically acquired from the sum of customer surplus of TCM and the willingness to pay for CVM. The summary of the result is presented in Table 15.1.

Table 15.1. Estimated monetary value of PLUF.

No	Method	PLUF (2.02 ha)
1.	TCM	
	Model	Individual TCM
	Variable CS	MTE (the actual minimum amount of money spent for reaching the forest and going back home); OTE (the time allocation to enjoy the forest); and OSE (the cost to buy food and drink during enjoyment in the forest)
	Respondent	200 visitors
	Mean CS	IDR20,079.47
	Visitors/year	7,896 people
	Total CS	IDR23.24 billion
2.	CVM	
	Model	Dichotomy choice
	Respondents	200 visitors
	Mean WTP	IDR6,380.10
	Visitors/year	7,896 people
	Total WTP	IDR50.37 million
3.	Total monetary value (TTM)	IDR23.29 billion
4.	TTM/ha	IDR11.53 billion
5.	Date of study	2015

Evaluation of Land Value of PLUF

To understand the land value of PLUF, the definition of land value, based on the dictionary of investopedia (2015), is "*the value of a piece of property, including both the value of the land itself as well as any improvements that have been made to it. Land values would increase when demand for land exceeds the supply of available land, or if a particular piece of land has an intrinsic value greater than neighboring areas (e.g., oil can be found on the land).*" From that definition, it is clear that land value in PLUF, including everything available in there, is both tangible and intangible. Tangible value refers to items that can be directly seen with the naked eye, such as trees, gazebos, and jogging tracks, while intangible items include the opportunity to admire nature, to relax and to improve the quality of the environment. Due to limited time and budget, this study has yet to calculate the value of the trees in PLUF. Alternatively, we have the estimated existence value (non-use value), option value, and use value. Those values can be elaborated as follows.

1. Existence value

   According to data from the Agency for Fishery, Agriculture, and Food Security of Jakarta, the local government bought the land for PLUF for IDR26.55 billion in 2010. I deduced from statistics that the inflation rate per year is relatively 6.01% (BPS, 2015). In 2015, the price of land in PLUF was predicted to be about IDR35.55 billion. This value is generally known as the "existence value" with market price, because its value is decided by the equilibrium of demand and supply. This is also known as the "non-use value". The major reason behind this argument is that people take care of this site, not only because of its presence, but also because they wish to own it (Markandya *et al.*, 2002).

2. Option value

   In 2012, infrastructures costing as much as IDR78.5 million were built in PLUF (DKP Jakarta, 2012). Because of the continuous depreciation of those infrastructures every year, we proposed an annual depreciation rate of about 10%. This predicted the value of the infrastructure in PLUF in 2015 to be IDR57.22 million. Practically, the presence of those infrastructures cause this forest to be recognised as an option value that is obtained from market price. The uncertainty and irreversibility of nature are two major components that construct it as an option value (Markandya *et al.*, 2002, p.293).

3. Use value

   The result of the estimation of environmental monetary valuation in PLUF is approximately IDR23.29 billion. This non-market price, which is acquired by the monetary valuation method using TCM and CVM, is usually called a use value. What it says is that those who want to enjoy this forest could directly gain a benefit — such as landscape, the quality of air, or the available facilities — from its existence. Therefore, the environmenal services provided by this forest could attract many people to visit this site. This phenomenon could be identified as opportunity value, in which policy makers use this value to consider improving either the quality or the quantity of urban forests. Interestingly, this opportunity value is significantly higher than market price value; however, it has combined the value of land and the value of infrastructures. As a result, environmental value in PLUF is a prospective business that could contribute to economic growth.

If we add the price of land and its infrastructures, as well the environmental value, the total existing value of PLUF is IDR58.89 billion (1US$=IDR13,000) in 2015. This is not a small amount of money, especially when the area of PLUF is only 2.02 ha. If there is a continuous inflation every year, the monetary value of this forest would be higher. Moreover, in this study we have yet to estimate the value of trees, so in future studies, this should be calculated to gain the complete value of PLUF.

Conclusion

With the literature analysis and unstructured interview with selected informants from different backgrounds, we have identified several significant challenges in terms of urban forest development. First, its implementation involved many parties, both from the central level and local level. At the central level, this development required participation from the Ministry of Environment and Forestry (Kemen LHK), Ministry of Public Works (Kemen PU), Ministry of National Development Planning Agency (Bappenas), Ministry of Internal Affairs (Kemendagri), and the National Land Board (BPN). The local level included at least three agencies, in which the names of those agencies were based on the characteristics of the province/regency/city. For example, in Jakarta, urban forest development involves the Fishery, Agriculture, Food Security Agency (DPPKP),

Public Works Agency (DPU), and Local Development Planning Agency (Bappeda).

Second, the decentralised system in Indonesia relegates the operation of urban forest development to the local government (city/regency/province). This meant they have the authority to build urban forests entirely, including funding, delivering land, planting trees, and constructing infrastructures, in which those activities are under the supervision of Kemen LHK. In practice, this political system creates self-interest among stakeholders, and cities/regencies with limited budgets may be reluctant to build urban forests. The third challenge is that while current laws demand that every city should have urban forests covering at least 10% of its total urban area, its flexibility permits adjustments to the geographical features. As such, stakeholders can establish an area less than the stipulated 10% because of the limitation of land in their city. Furthermore, there is no law enforcement in the existing regulations if the city does not provide the minimum land for an urban forest. Fourth, inconsistency in the implementation of Spatial Planning (RTRW) in the city by policy makers is a serious challenge. RTRW seems to only be a formality that is largely ignored, rather than as a major guidance to develop the urban environment. Fifth is the high cost of urban forest development in comparison to building other forests, such as industrial or private forests. One of the most expensive parts of developing urban forests is the provision of land, because land in the metropolitan area has a high economic value. Additionally, landowners may not be willing to surrender their land to the government, despite high compensation. Last, the social challenges from people who stay illegally and commit vandalism within the forests, and disputes between the government and society present critical issues for urban planners.

In terms of environmental monetary valuation, we estimated that the average annual visitorship per year in PLUF was about 7,896 people, and the economic value in that forest was predicted to be IDR23.29 billion. This number was derived from CS of IDR23.24 billion and WTP of IDR50.37 million. CS could be assumed as the actual economic value in PLUF, in a period of time without any entrance fee scenario, while WTP could be illustrated as the additional value for a one-year period if the government plans to implement a ticket charge for every visitor. Indirectly, this value indicated that this forest contributed significantly to Jakarta's economic growth, because more visitors meant more money spent in PLUF. Surprisingly, the total environmental value in PLUF was considerably lower than the value of investment from the government to buy the land and to build infrastructure. If we consider the annual rate of inflation and depreciation of the initial capitals, its value in 2015 is roughly IDR35.60 billion. Finally, the whole monetary value of PLUF in 2015 was approximately IDR58.89 billion (1US$=IDR13,000).

In accordance with the findings presented above, quantitative analysis shows that urban forests have an undoubtedly promising opportunity for economic growth. However, in practice, there are several serious challenges in their development that lead to slow progress in terms of the national target. Therefore, policy makers should consider some of the following recommendations:

1. Reviewing the current regulation

   Both PP 63 and Permenhut 71 should be carefully revised; mainly, the article stating that "the minimum urban forest in every city could be adjusted with geographical characteristics". This might allow city planners to neglect urban forest development. Another important point is that a penalty clause for omitting the lowest standard of urban forest should be included. The type of penalty that could be applied may be a discontinuation of funding assistance. This could motivate local governments to incorporate urban forests into city planning.

2. Giving Incentives

   Incentives can influence a person or organisation to act. While the current regulation proposes incentives for positive contributions to the forest, the details are determined by the local government. For example, Jakarta did not particularly have any law about incentives towards urban forests. To tackle this issue, the Ministry of Environment and Forestry should give an incentive to local governments that have successfully implemented urban forests. The form of incentive could either be recognition or financial support. The society or enterprise that is willing to relinquish their land for urban forests should also be given an incentive, such as tax rebates or business opportunites.

3. Evaluating management of urban forests

   The finding of quantitative data shows that the development of urban forests is a really new prospective business, especially in locating the strategic area, having complete facilities and growing visitors. By charging for some special facilities, such as gazebos and bicycle paths, and by leasing some locations to be used as food courts and for parking space, revenue which can be used for improvement of the quantity and quality of urban forest, can be generated. This could be applied to improve the additional value of the urban forests themselves.

4. Strengthening of institution

   The decentralised system's involvement of many parties can be good for the implementation of urban forests. If this system is carried out correctly, in accordance with the constitution, it will bring many benefits. Local governments with limited income could initiate collaboration with third parties, such as societies or enterprises. Empirical evidence has shown that many urban forests have been successfully sustained by involving other groups. To minimise the ineffectiveness of the system within institutions due to challenges such as corruption, conflict among stakeholders, and duplication of jobs and budget, enforcement and monitoring are important to ensure the success of urban forest development.

5. Educating people

   Education has an important contribution in shaping people's behaviour. How they interact with environment, whether they protect or destroy it, depends on the values inculcated in them. The discovery of illegal settlements in the forests and the vandalism of trees and facilities show us that education is urgently needed to teach the importance of conserving urban forests. The best form of education is to include the subject of the environment in the curriculum in schools and by campaigning through social media.

References

Abidin, H. Z., Andreas, H., Gumilar, I., Fukuda, Y., Pohan, Y. E., and Deguchi, T. (2011). Land subsidence of Jakarta (Indonesia) and its relation with urban development. *Natural Hazards*, **59**(3), 1753–1771.

Afandi, S. H., Samdin, Z., and Shuib, A. B. (2013). Review of valuation from a non-market perspective: travel cost method for rural tourism. *Worldwide Hospitality and Tourism Themes*, **5**(4), 329–341.

Ag. (2003, September 29). Cilacap-Kebumen Minta Bantuan ke Topdam IV. Semarang, Central Java, Indonesia: Suara Merdeka. Retrieved September 5, 2015, from http://www.suaramerdeka.com/harian/0309/29/dar27.htm.

Akhir, D. J. (2015, August 28). Harga BBM 1 September Tidak Naik. Jakarta. Retrieved August 29, 2015, from http://economy.okezone.com/read/2015/08/28/19/1203988/harga-bbm-1-september-tidak-naik.

Annez, P. C., and Buckley, R. M. (2009). (M. Spence, P. C. Annez, & R. M. Buckley, Eds.) Retrieved December 20, 2014, from www.worldbank.org:

http://siteresources.worldbank.org/EXTPREMNET/Resources/489960-1338997241035/Growth_Commission_Vol1_Urbanization_Growth.pdf.

Anning, D. (2012). *Estimation of the economic importance of beaches in Sydney, Australia, and implications for management.* Sydney, Australia: University of New South Wales.

Anonymous. (1992). Pembangunan Hutan Kota di Indonesia. *Media Konservasi,* IV(1).

Anonymous. (2014). Menghirup Segarnya Udara Sejuk di Babakan Siliwangi. Bandung. Retrieved July 2, 2015, from http://sebandung.com/2014/02/babakan-siliwangi/.

Ardyanto, A. (2015, April 1). *Sejarah Kebun Raya Bogor — kerjasama antar bangsa?* Retrieved July 6, 2015, from www.lovelybogor.com: http://lovelybogor.com/sejarah-kebun-raya-bogor-kerjasama-antar-bangsa/.

Bappeda Jakarta. (2012). *Presentation of Spatial Planning: Urban Forest Development in DKI Jakarta Province.* Jakarta: Bappeda Jakarta.

Blackwell, B. D. (2003). *The economics of coastal foreshore and beach management: Use, safe bathing facilities, erosion and conservation.* Brisbane: The University of Queensland.

Blackwell, B. D. (2007). The value of a recreational beach visit: an application to Mooloolaba Beach and comparisons with other outdoor recreation sites. *Economic Analysis and Policy,* 37(1), 77–98.

BNPB (National Disaster Mitigation Agency). (2015). Retrieved January 4, 2015, from http://dibi.bnpb.go.id/DesInventar/dashboard.jsp?countrycode=id&continue=y&lang=ID.

BPS (Statistics Indonesia). (2015). *Monthly Indonesia Consumers Price Indices and Inflations.* Retrieved October 1, 2015, from Statistics Indonesia: http://www.bps.go.id/linkTabelStatis/view/id/907.

Brown, W. G., and Nawas, F. (1973, May). Impact of Aggregation on the Estimation of Outdoor Recreation Demand Functions. *American Journal of Agricultural Economics,* **55**(2), 246–249.

Brown, W. G., and Nawas, F. (1973). Impact of Aggregation on the Estimation of Outdoor Recreation Demand Functions. *American Journal of Agricultural Economics,* **55**(2), 246–24.

Budiharjo. (2002). Paradigma Pengelolaan Sumberdaya Air dalam Era Otonomi Daerah. In R. J. Kodoatie, and H. A. Indonesia, *Pengeloaan Sumberdaya Air Dalam Otonomi Daerah.* Yogyakarta: Andi Press.

Bulkin, I. (1995). Infrastucture Requirements under Conditions of Development: The Dynamics of Urbanization in Indonesia, 1990–2010. New York, USA: Cornell University.

Careless, J. M. S. (1954). Frontierism, Metropolitanism, and Canadian History. *Canadian Historical Review,* **35**(1), 1–21.

Carson, R. T., Flores, N. E., and Meade, N. F. (2001). Contingent Valuation: Controversies and Evidence. *Environmental and Resource Economics,* **19**(2), 173–210.

Clawson, M., and Knetsch, J. L. (1966). *Economics of outdoor recreation.* Baltimore: Johns Hopkins Press.

Davis, D. F., Golicic, S. L., and Boerstler, C. N. (2011). Benefits and challenges of conducting multiple methods research in marketing. *Journal of the Academy of Marketing Science*, **39**(3), 467–47.

Dep. (2014, October 7). Hutan Kota Dijadikan Tempat Tinggal. Balikpapan: Balikpapan Post. Retrieved December 1, 2015, from http://www.balikpapan pos.co.id/berita/detail/137353-hutan-kota-dijadikan-tempat-tinggal.html.

Devuyst, D., Hens, L., and Lannoy, W. D. (2011). *How Green is the City? Sustainability Assessment and the Management of Urban Environment.* USA: Columbia University Press.

Dictionary Investopedia. (2015). *Land Value.* Retrieved August 23, 2015, from http://www.investopedia.com/terms/l/landvalue.asp.

Didit. (2011). Kawasan Industri Pulo Gadung Miliki Hutan Kota. Jakarta: Berita Kota. Retrieved July 1, 2015, from http://www.jakarta.go.id/v2/news/2011 /03/Kawasan-Industri-Pulogadung-Miliki-Hutan-Kota#.VoNadBV97IV.

Dilang, M. (2008). *Implementasi rencana tata ruang wilayah (RTRW): Studi kasus indikasi program pembangunan Kutai Barat Provinsi Kalimantan Timur.* Yogyakarta: Magister Perencanaan Kota dan Daerah UGM. Retrieved September 20, 2015, from http://etd.repository. ugm.ac.id/index. php?mod=penelitian_detail&sub=PenelitianDetail&act=view&typ=html& buku_id=38643.

Dirjen BPDASPS. (2011). *Statistics of Forestry.* Jakarta: Dirjen BPDASPS.

Dirjen BPDASPS. (2015). *Presentation of Urban Forest.* Jakarta: Dirjen BPDASPS.

Dirjen RLPS. (2009). *Statistics of Forestry.* Jakarta: Dirjen RLPS.

Djaja, R., Rais, J., Abidin, H. Z., and Wedyanto, K. (2004). Land Subsidence of Jakarta Metropolita Area. *3rd FIG Regional Conference.* Jakarta.

DKP Jakarta. (2011). *Informasi Kehutanan Dinas Kelautan dan Pertanian Provinsi DKI Jakarta.* Jakarta: DKP Jakarta.

DKP Jakarta. (2012). *Pengawasan Pembangunan Sarana Hutan Kota Pondok Labu.* Retrieved September 6, 2015, from http://www.pengadaan.net/eproc/ 94602/6/pengawasan-pembangunan-sarana-hutan-kota-pondok-labu.html.

DKP Jakarta. (2015). *Pengawasan Pembangunan Sarana Hutan Kota Pondok Labu.* Jakarta: DKP Jakarta.

DPD RI. (2010, July 27). *Masalah Perkotaan Disebabkan Inkonsistensi Pemerintah Dalam Rencana Tata Ruang.* Retrieved December 15, 2015, from Dewan Perwakilan Daerah Republik Indonesia: http://www.dpd.go.id/ artikel-masalah-perkotaan-disebabkan-inkonsistensi-pemerintah-dalam-ren cana-tata-ruang.

Ema. (2008). Sepenggal Cerita Babakan Siliwangi. Jakarta: Detik. Retrieved July 2, 2015, from http://news.detik.com/wisata/993175/sepenggal-cerita-babakan-siliwangi/1.

Fielding, N. G. (2009, May). Going out on a Limb Postmodernism and Multiple Method Research. *Current Sociology*, **57**(3), 427–447.

Firman, T. (2000). Rural to Urban Land Conversion in Indonesia during Boom and Bust Period. *Land Use Policy*, 17(1), 13–20.

Firman, T., Kombaitan, B., and Pradono, P. (2007, December 3). The Dynamics of Indonesia's Urbanisation, 1980–2006. *Urban Policy and Research*, **25**(4), 433–454.

Galih, H. (1945). Warna. *Geography Journal*, 40.

Gault, R. H. (1907). A History of the Questionnaire Method of Research in Psychology. *The Pedagogical Seminary*, **14**(3), 366–383.

Gum, R. L., and Martin, W. E. (1975). Problems and Solutions in Estimating the Demand for and Value of Rural Outdoor Recreation. *American Journal of Agricultural Economics*, **57**(4), 558–566.

Gum, R. L., and Martin, W. E. (1975). Problems and Solutions in Estimating the Demand for and Value of Rural Outdoor Recreation. *American Journal of Agricultural Economics*, **57**(4), 558–566.

Hanemann, W. M. (1994). Valuing the Environment Through Contingent Valuation. *The Journal of Economic Perspectives*, **8**(4), 19–43.

Hanley, N., Shogren, J. F., and White, B. (2013). *Introduction to environmental economics* (2nd ed.). Oxford: Oxford University Press.

Hearne, R. R., and Santos, C. A. (2005). Tourists' and Locals' Preferences Toward Ecotourism Development in the Maya Biosphere Reserve, Guatemala. *Environment, Development and Sustainability*, **6**(3), 302–318.

Hesseln, H., Loomis, J. B., and Gonzalez, A. (2004). Comparing the economic effects of fire on hiking demand in Montana and Colorado. *Journal of Forest Economics*, **10**(1), 21–35.

Holzer, T. L., and Johnson, A. I. (1985). Land Subsidence Caused by Ground Water Withdrawal in Urban Areas. *GeoJournal*, **11**(3), 245–255.

Honda. (2015). Daftar Konsumsi BBM Sepeda Motor Honda. Jakarta. Retrieved August 29, 2015, from http://www.hondacengkareng.com/daftar-konsumsi-bbm-sepeda-motor-honda/.

Investopedia. (2015). *Land Value*. Retrieved October 15, 2015, from Investopedia: http://www.investopedia.com/terms/l/landvalue.asp.

Jakarta Government. (2015). *Luas Hutan Kota Jakarta per Lokasi beserta Keterangan Lainnya*. Retrieved September 1, 2015, from Pemerintah Provinsi DKI Jakarta: http://data.go.id/dataset/luas-hutan-kota-jakarta-per-lokasi-beserta-keterangan-lainnya.

Kemen PU (Ministry of Public Work). (2015, May 26). *Status of Local Regulation regarding Local Spatial Planning*. Retrieved May 26, 2015, from http://www.pu.go.id/site/view/3.

Keppres 4. (2002, January 7). Keppres (Indonesia Presidential Decree) Number 4 of 2002 regarding the Amendment of Presidential Decree Number 108 of 2001 for Organizzation and the Duty of Echelon I of Indonesia Ministry. Retrieved September 12, 2015, from http://kepustakaan-presiden.perpusnas.go.id/uploaded_files/pdf/government_regulation/normal/keppres_4_2002_mega.pdf.

Khairany, C. (2015, April 2015). DKI targetkan pembebasan 159 titik lahan untuk RTH. (D. Saputra, Ed.) Jakarta, Indonesia. Retrieved July 2, 2015, from http://www.antaranews.com/berita/493058/dki-targetkan-pembebasan-159-titik-lahan-untuk-rth.

Kunu, P. J., and Lelolterry, H. (2010). Penggunaan Lahan dan Evolusi Pembangunan Lahan di Provinsi DKI Jakarta. *Journal Agroforestry*, V, 3.

Kusuma, D. (2015, August 4). Urban Forest in Jakarta. (Supriyanto, Interviewer)

Lee, C. -K., and Han, S. -Y. (2002). Estimating the use and preservation values of national parks' tourism resources using a contingent valuation method. *Tourism Management*, **25**(5), 531–540.

Lewis, B. D. (2014). Urbanization and Economic Growth in Indonesia: Good News, Bad News and (Possible Local Government Mitigation. *Regional Studies*, **48**, 184.

Lindsey, J. K. (1999). A review of some extensions to generalized linear models. *Statistics in Medicine*, **18**(17–18), 2223–2236.

Lockwood, M., and Tracy, K. (1995). Nonmarket economic valuation of an urban recreation park. *Journal of Leisure Research*, **27**(2), 155.

Loeb, R. E. (2011). Old growth urban forests. New York, USA: Springer.

Markandya, A., Harou, P., Bellu, L. G., and Cistulli, V. (2002). *Environmental Economics for Sustainable Growth.* Northampton: Edward Elgar Publishing Limited.

Mursidin, Priyo, T., Achlil, M. R., Yuliarsana, I. N., Soewondo, R. A., Wartam, *et al.* (1997). *35 Tahun Penghijauan di Indonesia.* Jakarta, DKI Jakarta, Indonesia: Dirjen Reboisasi dan Rehabilitasi Lahan, Departemen Kehutanan.

Mustafa, S., Riaz, K., and Perveen, Q. (2015). Stability of Linear Regression Models. *Science Internationa*, **27**(1), 73–76.

Nas, P. J., and Prawito. (2002). Java and De Groote Postweg, La Grande Route, the Great Mail Road, Jalan Raya Pos. *Bijdragen tot de Taal-, Land- en Volkenkunde*, **158**(4), 707–725.

Nassaji, H. (2015, March). Qualitative and descriptive research: Data type versus data analysis. *Language Teaching Research*, **19**(2), 129–132.

Ng, N. (2014). The Endless Wait for Hong Kong's Rooftop Dwellers. Hongkong. Retrieved December 1, 2015, from http://hkstories.jmsc.hku.hk/2014/11/27/the-endless-wait-for-hong-kongs-rooftop-dwellers/.

Nowak, D. J., and Greenfield, E. J. (2012). Tree and impervious cover change in U.S. cities. *Urban Forestry & Urban Greening*, **11**(1), 21–30.

Nugraha, A. (2012). Kebijakan Pelembagaan Penataan Ruang Daerah. *Stiami Journal*, 1–16.

Peluso, N. L., Afiff, S., and Rachman, N. F. (2008). Claiming the Grounds for Reform: Agrarian and Environmental Movements in Indonesia. *Journal of Agrarian Change,* **8**(2–3), 377–407.

Perda 1. (2012, January 12). Perda (Local Regulation of Jakarta) Number 1 of 2012 regarding Spatial Planning (RTRW) 2030.

Permen PU 5. (2008, May 28). Permen PU (Minister of Public Work Regulation) Number 5 of 2008 regarding the guidance for Green Open Space in Urban Area. Retrieved September 10, 2014, from http://dtrb.pekanbaru.go.id/?wpfb_dl=383.

Permendagri 6. (2008). Permendagri (Minister of Internal Affair Regulation) Number 6 of 2008 Regarding Code and Data of Admistrative Area. Jakarta, Jakarta: Ministry of Internal Affairs.

Permenhut 71. (2009, December 9). Permenhut (Minister of Forestry Regulation) Number 71 of 2009 regarding Technical Guidance for Urban Forest. Retrieved September 5, 2014, from http://www.dephut.go.id/uploads/files/P71_09.pdf.

PP 38. (2007, July 9). PP (Government Regulation) Number 38 of 2007 regarding Distribution of Power between Central, Provincial, and District Government. Retrieved September 10, 2015, from http://prokum.esdm.go.id/pp/2007/pp_38_2007.pdf.

PP 63. (2002, November 12). PP (Government Regulation) Number 63 of 2002 regarding Urban Forests.

Pushumas Kemenhut. (2012, November 28). *Presiden S B Y Tanam Pohon di Puncak Acara Hari Menanam Pohon Indonesia dan Bulan Menanam Nasional.* Retrieved June 5, 2015, from http://www.ppid.dephut.go.id/berita_foto/browse/53.

Ranis, G., and Stewart, F. (2006). Decentralisation in Indonesia. *Bulletin of Indonesian Economic Studies*, **30**(3), 41–72.

Risnandar, C. (2015, June 18). *Hutan kota.* Retrieved July 1, 2015, from https://jurnalbumi.com/hutan-kota/.

Rudi, A. (2015, June 4). Kecepatan Berkendara di Jakarta Tinggal 5 Km Per Jam. (Kistyarini, Ed.) Jakarta: Kompas. Retrieved August 29, 2015, from http://megapolitan.kompas.com/read/2015/06/04/11193151/Kecepatan.Berkendara.di.Jakarta.Tinggal.5.Km.Per.Jam.

Sainani, K. L. (2013). Understanding linear regression. *PM & R: the journal of injury, function, and rehabilitation,* **5**(12), 1063–1068.

Santoso, M., Lestiani, D. D., and Markwitz, A. (2012). Characterization of airborne particulate matter collected at Jakarta roadside of an arterial road. *Journal of Radioanalytical and Nuclear Chemistry*, **297**(2), 165–169.

Santoso, S. (2015). *SPSS 20: Pengolah Data Statistik di Era Informasi.* Jakarta: Kompas Gramedia.

Setiawan, S. (2015, November 27). Banyak Warga Klaim Lahan Hutan Kota Srengseng. Jakarta. Retrieved December 1, 2015, from http://www.beritajakarta.com/read/21250/Banyak_Warga_Klaim_Lahan_Hutan_Kota_Srengseng#.VoOOURV97IU.

Tate, J. W. (1998). The Hermeneutic Circle vs. the Enlightenment. *Telos*, **1998**(110), 9–38.

Tipple, G., and Speak, S. (2006). Who is homeless in developing countries? Differentiating between inadequately housed and homeless people. *International Development Planning Review*, **28**(1), 57–84.

Tony. (2015, August 12). Urban Forest in Pondok Labu, Jakarta. (Supriyanto, Interviewer)

UU 13. (2003, March 25). UU (Act) number 13 of 2003 regarding Man Power.

UU 26. (2007, April 26). UU (Act) Number 26 of 2007 regarding Spatial Planning. Jakarta.

UU 32. (2004, October 15). UU (Act) number 32 of 2004 regarding Local Government. Retrieved July 5, 2015, from http://www.kpu.go.id/dmdocuments/UU_32_2004_Pemerintahan%20Daerah.pdf.

Walsh, R. G., Ward, F. A., and Olienyk, J. P. (1989). Recreational demand for trees in national forests. *Journal of Environmental Management*, **28**(3), 255–268.

Ward, F. A., and Beal, D. (2000). *Valuing Nature with Travel Cost Models: A Manual.* Cheltenham, UK: Edward Elgar.

Whittington, D. (1998). Administering contingent valuation surveys in developing countries. *World Development*, **26**(1), 21–30.

Willis, K. G., and Garrod, G. D. (1991). An Individual Travel-Cost Method of Evaluating Forest Recreation. *Journal of Agricultural Economics*, **42**(1), 33–42.

Xue, D., Cook, A., and Tisdell, C. (2000). *Biodiversity and a Travel Cost Evaluation of Tourism: The Case of Changbaishan Mountain Biosphere Reserve, China.* Brisbane: Department of Economics, The University of Queensland.

Yin, Y., Zhang, K., and Li, X. (2006). Urbanization and land subsidence in China. *China Geology Survey*, p. 31.

Yossy, A. (2015, January 6). Hutan Kota Kembangan Utara Bakal Diperluas. Jakarta, Indonesia: BeritaJakarta. Retrieved July 5, 2015, from http://beritajakarta.com/video/play/11103/Hutan_Kota_Kembangan_Utara_ Bakal_ Diperluas.

Yulief, E. M. (2008). *Penilaian Ekonomi di Hutan Kota Srengseng.* Jakarta: University of Indonesia.

Yusuf, M. (2014, November 2). Warga Waduk Ria Rio Tolak Relokasi. Jakarta: Tribun Jakarta. Retrieved October 1, 2015, from http://www.tribunnews.com/metropolitan/2014/11/02/warga-waduk-ria-rio-tolak-relokasi.

Zhang, F., Wang, X. H., Nunes, P. A., and Ma, C. (2015). The recreational value of gold coast beaches, Australia: An application of the travel cost method. *Ecosystem Services*, **11**(2), 106–114.

www.ingramcontent.com/pod-product-compliance
Lightning Source LLC
LaVergne TN
LVHW010538300926
849162LV00002B/169